The Unattainable Attempt to Avoid the Casus Irreducibilis for Cubic Equations

Sara Confalonieri

The Unattainable Attempt to Avoid the Casus Irreducibilis for Cubic Equations

Gerolamo Cardano's De Regula Aliza

Springer Spektrum

Sara Confalonieri
Bergische Universität Wuppertal
Germany

Dissertation Université Paris 7 Diderot, 2013

ISBN 978-3-658-09274-0 ISBN 978-3-658-09275-7 (eBook)
DOI 10.1007/978-3-658-09275-7

Library of Congress Control Number: 2015933836

Springer Spektrum

Printed on acid-free paper

Springer Spektrum is a brand of Springer Fachmedien Wiesbaden
Springer Fachmedien Wiesbaden is part of Springer Science+Business Media
(www.springer.com)

Foreword

The present work comes from my dissertation [55], supervised by Marco Panza (*directeur*) and Massimo Galuzzi (*co-directeur*), defended at the Université Diderot Paris 7 on October 12, 2013. It has been improved according to the remarks and observations that I have received from the viva examiners Sabine Rommevaux (*président*), Veronica Gavagna (*rapporteur*), Klaus Volkert (*rapporteur*), Paolo d'Alessandro, Jeffrey Oaks, and my two supervisors. The dissertation was originally composed of two parts. Only the first one is contained in the present work. The second part is a compared Latin transcription of the 1570 and 1663 editions of the *De Regula Aliza* and a partial English translation; this has been omitted since it will hopefully give rise to a critical edition of Cardano's treatise.

For their support and help, I want to express my gratitude to Marco Panza and Klaus Volkert. I am indebted to Massimo Galuzzi, Veronica Gavagna, and Sabine Rommevaux for their outstanding advice. A special thank goes to Paolo D'Alessandro, Jeffrey Oaks, David Rabouin, Jackie Stedall, Pier Daniele Napolitani, the SPHERE group in Paris, Jens Høyrup, and Antoni Malet.

Notations and more

For a modern reader it is almost impossible – or at least it is really hard work – to account for Cardano's time mathematics without reverting to some sort of formalism. This is why I choose to rewrite it using a certain amount of symbols that he did not employ. On the one hand, these symbols betray – as each translation – Cardano's text since they go more or less far away from the codified language of that time. On the other hand, I believe that the advantages for the reader in terms of precisely grasping its contents are to be preferred over the ultimate fidelity to printed words.

Choosing to use the appearance of our mathematics to talk about Cardano's one does not mean that even the least part of our understanding belongs to Cardano. I only employ our notations as a good stenography, or as a "convenient linguistic trick",[1] to avoid cumbersome and sometimes ambiguous expressions. As such, they are not essential. In other words, I do not believe in those "sartorics of mathematics"[2] that consist in clothing a unique, timeless mathematics in the fashionable idioms of a historical period. Cardano's mathematics and ours are different – they belong to completely different universes. It is true that our mathematics also let us understand the mathematical content of Cardano's arguments, but still Cardano's mathematics must be considered as a whole and contain subtleties that must be understood in their own context.[3]

[1] See [88, page 207].

[2] See [80, page 3].

[3] As Karine Chemla says, "*resituer l'objet dans le contexte ds procédures qui le mettent en œuvre s'avère essentiel pour en déterminer la nature [...]. Ces éléments sont à considérer dans les ensemble qu'ils forment, et non pas uniquement à l'état de pièces détachées sous le prétexte que seules ces pièces seraient pour nous significatives*", see [53, page 102].

Nevertheless, notations are not only a clothing, but they have a real value. This cannot be prescriptive, but rather practical. We have a mathematical object and then we point at it in more or less effective ways: there are good and bad notations, where 'good notation' means that makes it easier (or, at least, not more difficult) to handle that object. Then, the effectiveness of the following notations in rephrasing Cardano's mathematics (and only its effectiveness!) is the reason for which I choose to use it.

A noteworthy part of Cardano's mathematics that we analyse in the following concerns equations. I call 'equation' what we now call 'equation' – basically a standardised way of writing that involves an equality and that has a certain shape, depending on one or more unknown quantities and on their powers.[4] For instance, a cubic equation is an equality such as '$x^3 = a_1x + a_0$', where x varies and a_1, a_0 are constant. Or one can also say 'a cube equal to some things and a number', as it was the case in Cardano's mathematics. Similar expressions are used for linear, quadratic, quartic, or whatever else degree equations. We remark that, in spite of this terminology, Cardano's equations are seen as problems, rather than mere writings, since they are for the most part originated from there.

Concerning geometrical objects, I use a double notation (except for points, which will be denoted by capital letters). According to the first notation, 'AB' stands for the segment delimited by the points A and B. In the same way, 'AB^2' stands for the square the side of which

[4]Note that varied definitions have been given by historians and philosophers of mathematics. For instance, Jens Høyrup maintains in [69, page 84] that, generally speaking,

> [a]n "equation" is the statement that some complex quantity (for instance, the area A of a square) defined in terms of one or more simple quantities (in the example, the side s), or the measure of this complex quantity, equals a certain number or (the measure of) another quantity.

Concerning instead Arabic-speaking algebra, Albrecht Heeffer argues in [89, page 89] and [66, pages 58-59] that an "equation is *the act of keeping related polynomials equal*. Guglielmo de Lunis and Robert of Chester have a special term for this: *coaequare*", from which the term 'coequal polynomials' stems.

is AB, 'AB^3' for the cube the side of which is AB, '$ABCD$' for the parallelogram the sides of which are AB and CD, and '$ABCDEF$' for the parallelepiped the sides of which are AB, CD, and EF. When needed, I write '(A)' for a segment (and in this case the segment is identified by the diagram), '$(AB)^2$' for the square identified by its diagonal AB, '$(AB)^3$' for the cube identified by the diagonal AB of one of its faces, and '(AB)' for the parallelogram identified by its diagonal AB. Finally, '$(A\,B\,\ldots\,N)$' stands for the gnomon identified by the points $A, B, \ldots, N$. I use this standard notation when, with respect to the purpose of Cardano's text, one essentially needs to characterise the geometrical object at issue by its position compared to the one of other geometrical objects.

Note that it can happen that the relative positions of geometrical objects play a role, still without being essential. This happens, for instance, when a diagram is used to fix the reference[5] of the relevant object, in order to apply to it the results that we already have at disposal. A good way to check if the relative position of geometrical objects covers an essential role (or not) is to reputedly change the position of the involved objects and verify whether the argument still maintains its validity. If the answer is 'yes' – that is, when the position is not essential and only the relative sizes of the objects matter – I will use a different notation. In fact, in this case we better refer to the measure of the considered object, not to the object itself. I write '$\overline{AB}$' for the measure of the segment delimited by the points A and B, '$\overline{AB}^2$' for the measure of the square constructed on the segment that measures $\overline{AB}$, '$\overline{AB}^3$' for the measure of the cube constructed on the segment that measures $\overline{AB}$, '$\overline{AB}\,\overline{CD}$' for the measure of the parallelogram the sides of which respectively measure $\overline{AB}$ and $\overline{CD}$, and '$\overline{AB}\,\overline{CD}\,\overline{EF}$' for the measure of the parallelepiped the sides of which respectively measure $\overline{AB}, \overline{CD}$, and $\overline{EF}$. When needed, I write '$\overline{(A)}$', '$\overline{(AB)}^2$', '$\overline{(AB)}^3$', and '$\overline{(AB)}$ for the measures of the above

[5] "Very often – most often – letters are not completely specified. So how do we know what they stand for? Very simple: we see this in the diagram", see [79, pages 19-26].

objects. This happens, for instance, when a surface is not named after its vertexes, but is denoted only by a letter, or when a square or a parallelepiped is named after its diagonal. In this context, 'A' simply stands for the point A since we do not speak of the measure of a point. Quantities of such a kind can appear as coefficients in equations.

Concerning operations, I use the standard, ambiguous (to a certain extent) notation. I write '$\ldots + \ldots$' for the sum, '$\ldots - \ldots$' for the difference, nothing for the product, '$\frac{\ldots}{\ldots}$' for the division, '$(\ldots)^n$' for the nth power, and '$\sqrt[n]{\ldots}$' for the nth root. This notation is ambiguous since it denotes operations that are different depending on which kind of objects one is dealing with. This means that we are not allowed to sum, subtract, multiply, or divide two objects of different kinds, such as numbers and geometrical objects. So, as long as one does not dispose of a conceptual environment in which he can perform operations on quantities in general,[6] distinguishing different kinds of objects becomes of the utmost importance. In addition, note that Cardano never multiplies or divides a geometrical object by another one. He only multiplies or divides geometrical objects by numbers. Note moreover that, always concerning geometrical objects, he takes the nth-powers and n^{th}-roots only up to $n = 3$. With regard to proportions, I write '$a : b = c : d$' or '$\frac{a}{b} = \frac{c}{d}$', where a, b, c, d are quantities. In particular, I take care to use '$a : b = c : d$' whenever a, b, c, d are geometrical quantities. In this case, '$a : b$' should not be intended as the result of the division of a by b. If a, b, c, d are numbers or non-positional, geometrical objects I like better '$\frac{a}{b} = \frac{c}{d}$' to lighten the outfit of the text.

Using this notation, I account for some of Cardano's proofs. My main concern being not to betray the spirit of the text, I put into square brackets the additions and comments that are mine. To refer to propositions, rules, or proofs in Cardano's texts, I write '(⟨*Book title in short*⟩ ⟨*Chapter*⟩.⟨*Paragraph, when needed*⟩)'. I usually write the book title in capital letters, the chapter in Roman numerals, and the paragraph in Arabic numerals. For example, (A I.1) refers to the

[6] See [86, pages 17-30] and [87].

first proposition of Chapter I of the *De Regula Aliza.* All the examples, except when specified, are by Cardano himself. On page 415, there is a list of all the cubic equations in one unknown considered by Cardano in the selection of his works that we will analyse.

So far concerning my notations in the present work. But what about Cardano's own notations? As it is well known, Cardano has neither a stable nor an overall formalism. He employs some stenographic expressions, which still represent the language in which the statements are expressed. This is why it has been sometimes addressed as 'syncopated algebra'.[7] Cardano's stenography boils down to a small amount of symbols. First of all, there are operations: '*p:*' or simply 'and [*et*]' stand for the sum [*plus*] as well as '*m:*' stands for the difference [*minus*]. Note that, according to a common agreement of his time, Cardano writes, for instance, '$1\frac{1}{2}$' to mean '$1+\frac{1}{2}$'. '*R*' usually stands for the square root [*radix*] and '*Rcu:*' stands for the cubic root [*radix cubica*]. '*RV:*' stands for the universal root [*radix universalis*], which is the (usually, square or cubic) root of the sum or difference of two terms, like RV: cub. 1 + R 2 or $\sqrt[3]{1+\sqrt{2}}$.

Sometimes we find a completely different kind of stenography. At the end of the *Aliza* (in Chapters LIII and LVII), referring for instance to $x^3 = 20x + 32$, Cardano writes '*20 d. p. R p. 32*' to mean '20 divided into the part and the root that produce 32 [*divisum in partem et radicem producentes 32*]', '*32 p. 20 cum p. 32*' to mean 'what produces 20 with what produces 32 [*producentis 20 cum producente 32*]', '*Ag. R p: 20 p: n: 16*' to mean 'the aggregate of the roots of the parts of 20 that reciprocally multiplied produce 16 [*aggregatum radicum partium 20, quæ mutuo ductæ producunt 16*]' (or the half of 32), '*32 p: 20 c. p. 32*' to mean 'what produces 20 with what produces 32 [*producens 20 cum producente 32*]', '*R 20 p: d. 32*' to mean 'the root of 20 plus 32 divided by the same root [*R 20 p: diviso 32 per ipsam radicem*]', and finally '*R 20 f. 32*' to mean 'the root of 20 with a fragment of 32 [*R 20 cum fragmento 32*]' (a 'fragment

[7] See [78] and [51, page 180]. See also Albrecht Heeffer's criticism of this terminology in [67].

[*fragmentum*]' is 'what comes forth from a division [*quod ex divisione prodit*]'). For more details see page 346 below.

Concerning – in particular – equations, Cardano presents them in words, employing the common Latin terminology of the time. The coefficient of the term of degree zero is a 'number [*numerus*]' and the unknown is a 'thing [*res*]' or a 'position [*positio*]', this last being abbreviated by '*pos.*'. Concerning the higher degrees of the unknown, we have the 'squares [*quadratus*]' (written '*quad.*' or '*q.*') or the '*census*' for degree two and the 'cube [*cubus*]' (written '*cub.*' or '*cu.*') for degree three. The prime degrees of the unknown bigger than three that Cardano mentions are the 'first related (*relatus primus*)' for degree five and the 'second related (*relatus secundus*)' for degree seven. Unknowns of higher, non-prime degrees are obtained by multiplication of prime factors, such as for instance the 'square square (*quadrati quadratus*)' for degree four, the 'square cube (*quadrati cubus*)' for degree six, the 'square square square (*quadrati quadrati quadratus*)' for degree eight, the 'cube cube (*cubi cubus*)' for degree nine. When the unknown goes together with a coefficient, Cardano uses the plural form, for instance 'some things [*res*]' or 'some squares [*quadrati*]'.

What is fundamental in order to get an understanding of Cardano's equations is to identify the quantities that are equated. Cardano's coefficients are non zero, positive numbers. In the greatest majority of his examples, they are rational (even natural) numbers, but there are a handful of cases in which he also allows irrational numbers (see the table in Appendix C, on page 415). Whenever we would write a negative coefficient, we find that in Cardano's equations they have been moved to the other side of the equal. Moreover, Cardano only considers monic equations. These are quite standard criteria to write equations at the time. Cardano is interested in finding the real solutions. He has a strong preference for positive solutions (which he calls 'true [*vera*]'), but sometimes, especially when dealing with substitutions, he also needs to consider the negative ones (which he calls 'fictitious [*ficta*]').

With regard especially to cubic equations, Cardano takes into account only cubic equations that do not lower in degree *via* a substitution and that cannot be solved by taking the third root (that is, he does not include the two-term equations like $x^3 = a_0$). We have then the following classification for cubic equations (the terminology is obviously mine):

- equations lacking in the second degree term (also called 'depressed'):

$x^3 + a_1x = a_0$	or	*cubus et res æquales numero,*
$x^3 = a_1x + a_0$	or	*cubus æqualis rebus et numero,*
$x^3 + a_0 = a_1x$	or	*cubus et numerus æquales rebus;*

- equations lacking in the first degree term:

$x^3 = a_2x^2 + a_0$	or	*cubus æqualis quadratis et numero,*
$x^3 + a_2x^2 = a_0$	or	*cubus et quadrati æquales numero,*
$x^3 + a_0 = a_2x^2$	or	*cubus et numerus æquales quadratis;*

- complete equations:

$x^3 + a_2x^2 + a_1x = a_0$	or	*cubus quadrati et res æquales numero,*
$x^3 + a_1x = a_2x^2 + a_0$	or	*cubus et res æquales quadratis et numero,*
$x^3 + a_2x^2 = a_1x + a_0$	or	*cubus et quadrati æquales rebus et numero,*
$x^3 = a_2x^2 + a_1x + a_0$	or	*cubus æqualis quadratis rebus et numero,*
$x^3 + a_0 = a_2x^2 + a_1x$	or	*cubus et numerus æqualis quadratis et rebus,*
$x^3 + a_1x + a_0 = a_2x^2$	or	*cubus res et numerus æquales quadratis,*
$x^3 + a_2x^2 + a_0 = a_1x$	or	*cubus quadrati et numerus æquales rebus.*

Since Cardano only considers equations with positive (real) solutions – or, at least, he has a very strong preference for them[8] – the equations $x^3 + a_1x + a_0 = 0$, $x^3 + a_2x + a_0 = 0$, and $x^3 + a_2x^2 + a_1x + a_0 = 0$ do not appear in the classification. Indeed, we can easily see thanks to Descartes' rule of sign that they only have negative solutions.

[8] To this end, see here, Section 3.1.1 on page 56 and Section 3.1.2 on page 60. It must be said in fact that, even though Cardano sticks to the tradition concerning negative numbers (see for instance "but the subtraction is made only of the smaller from the bigger, in fact it is entirely impossible to subtract a bigger number from a smaller [*detractio autem non fit nisi minoris a maiore: nam maiorem numerum, a minore, detrahere omnino est impossibile*]" in [19, Chapter XI, page 18]).

Sometimes – and this often happens in the *De Regula Aliza* – Cardano writes an equation such as $x^3 = a_2x^2 + a_0$ under the shape $x^2(x - a_2) = a_0$. We remark that this does not mean that he allows negative coefficients, since $a_0 > 0$ implies that $x - a_2 > 0$ (it would have been so if he had written $x^3 - a_2x^2 = a_0$, but it is not the case). In the end, there are thirteen families of cubic equations. Such a classification provides an effective way to distribute cubic equations in families so that it is easier to speak about them. We remark that a classification in the same vein can already be found in al-Khayyām's algebra.

Recall that I use our formalism only as long as it is a more effective stenography. In particular, by writing the equations as in the above table, I do not obviously mean that Cardano is interested in the zeros of the function associated to the appropriated polynomial (intended, in modern terms, as an infinite, eventually zero sequence of coefficients) – it would simply sound ridiculous. We should rather recall that Cardano's equations originated from problems. A polynomial is a mere formal writing, and even if one wanted to allow operations like the product and the sum in it, they would be at large operations between real numbers, which – if one wants to be fully pedantic – are not exactly the same operations that Cardano uses, since he does not have 'the field of the real numbers' available.[9] He only has a kind of a system of numbers, which is by the way not close. More precisely, Cardano takes the freedom to consider that, when he performs some algebraic operations on rational numbers, other numbers that he knows how to deal with are produced.[10] I rather agree with [81] and

[9] Under the same viewpoint and for the sake of brevity, I will sometimes use notations concerning equations, especially the 'discriminant Δ_3', referring to Cardano's equations. In this case, the discriminant is not to be intended as the symmetric function $\alpha_3^3(x_1 - x_2)^2(x_2 - x_3)^2(x_1 - x_3)^2$ of the roots x_1, x_2, x_3 of a polynomial $\alpha_3x^3 + \alpha_2x^2 + \alpha_1x + \alpha_0$. Rather, it is a stenographic writing for the number that is under the square root in the formula (see here, Section 2.5 on page 30).

[10] For instance, the operation of taking the square root applied to 2 gives $\sqrt{2}$, and Cardano knows how to further perform algebraic operations on it.

This remark opens a great scoped issue, which I am not going to study here: how can we characterise Cardano's numbers?

[82] (transposed from Arabic algebra context to Cardano's case), for whom equations are equality between collections of certain objects taken a certain number of times.[11] In this way, a family of equations bears by itself a certain structure and it is not only a formal writing that must be filled in with the coefficients. More precisely, the fact that a certain term is on one side or the other of the equal depends on the inner structure of a family of equations, and not on the sign of the coefficients (since they are always supposed to be positive).

[11]"[T]he two sides of an equation in the medieval Arabic algebra are aggregations of the algebraic 'numbers' (powers) with no operations present. Unlike an expression such as our $3x+4$, the Arabic polynomial 'three things and four *dirhams*' is merely a collection of seven objects of two different types. Ideally, the two sides of an equation were polynomials so the Arabic algebraists preferred to work out all operations of the enunciation to a problem before stating an equation", see [82, page 169], or "Medieval algebraists conceived of polynomials differently than we do today. For us, a polynomial is constructed from the powers of x with the operations of scalar multiplication and addition/subtraction. In other words, it is a linear combination of the powers. By contrast, Arabic polynomials contain no operations at all. In the expression 'four things', the 'four' is not multiplied by 'things'. Instead, it merely indicates how many things are present. Think of 'four things' like 'four bottles'. Further, the phrase 'four things and five *dirhams*' entails no addition, but is a collection of nine items of two different kinds. Think of it like 'four bottles and five cans'. The *wa* ('and') connecting the things and the *dirhams* is the common conjunction. It does not take the meaning of the modern word 'plus' ", see [81, page 548].

In Cardano's case, since he allows irrational numbers to be coefficients, there could be a little inconsistency in taking an irrational number of times a certain object. Anyway, this could be also seen as a first step toward a further development and, as such, inconsistent in a certain respect with the usual way (see [83]).

Contents

1. Introduction

Solving cubic equations by a formula that involves only the elementary operations of sum, product, and exponentiation of the coefficients is one of the greatest results in 16th century mathematics. At that time, the solution of quadratic equations was a well-mastered issue and the attention was driven on searching a generalisation to the cubic case. This was achieved by Girolamo Cardano's *Ars Magna* in 1545. Still, a deep, substantial difference between the quadratic and the cubic formulae exists: while the quadratic formulae only involve imaginary numbers when all the solutions are imaginary too, it may happen that the cubic formulae contain imaginary numbers, even when the three solutions are all real (and different). This means that one could stumble upon numerical cubic equations of which he already knew three (real) solutions, but its cubic formula actually contains some square roots of negative numbers. This will be later called the '*casus irreducibilis*'.

In the present work, I give an insight on the latest of Cardano's algebraic works, the *De Regula Aliza* (1570), in which he tried to overcome the difficulty entailed by the *casus irreducibilis*. I comply with the following plan. In Chapter 2, I contextualise from both a non-mathematical and a mathematical viewpoint Cardano's mathematical works, with closer attention to the ones that deal with cubic equations. In particular, I introduce the *Aliza*, which is a puzzling and controversial treatise under many respects. Then, I give a first estimate in time of the book and report on its readers all along the centuries. Regarding to the mathematical side, I retrace how we solve nowadays quadratic, cubic, and quartic equations. Concerning in particular cubic equations, I also discuss the *casus irreducibilis* using Galois theory.

Due to the exceptionally problematic nature of the *Aliza*, I consider fully worthwhile to relieve at most the interpretative effort while dealing with this book. This is achieved by a prior analysis of the topics related to cubic equations that can be found in Cardano's other mathematical works. This also helps in setting the mathematical context in which the *Aliza* fits. In particular, I firstly analyse the most advanced results in the *Ars Magna* (1545, 1570) and then go and search for their origins in the *Practica Arithmeticæ* (1539) and in the *Ars Magna Arithmeticæ* (probably composed between 1539 and 1542). This is of the utmost importance in decoding afterwards the *Aliza*. In Chapter 3, I discuss in detail all the solving methods in the *Ars Magna* for cubic equations. This gives us a quite accurate picture of their mutual interdependencies, which is fundamental in order to understand the impact of the problem set by the *casus irreducibilis*. Moreover, I discuss a bunch of fringe topics, like transformations of equations and particular solving methods, which help in completing the frame. I also take into account Cardano and Ferrari's treatment of quartic equations. The main annoyance with Cardano's solving methods (or, better, with Cardano's enhancement of Tartaglia's poem) is that they are completely opaque concerning the way in which they had been discovered. They work, but their proofs belong to that varied family of proofs that let glimpse no hints on how one could have possibly come across them. Then, if we want to know how Cardano used to deal with equations, we must consider some other results, like the fringe topics that I have mentioned above. Some of these topics and others (and also a few solving methods for cubic equations) can be found in two of Cardano's previous works. This is why in Chapter 4 I will analyse some parts of the *Practica Arithmeticæ* and the *Ars Magna Arithmeticæ*. As said, all this work is necessary to set the mathematical environment in which the *Aliza* is located. On the one hand, the study of the other mathematical treatises by Cardano is useful to get oriented while reading the *Aliza*, in particular when the guide of a solid, general architecture is missing – and this will frequently happen. On the other hand, we obviously need to know the contents of those treatises, especially of the *Ars Magna*

Arithmeticæ and of the *Ars Magna*, in order to fully grasp the explicit (and implicit) references that are made to them in the *Aliza*. But there is more. The targeted analysis of the other mathematical treatises by Cardano has the valuable effect to show the overall development trend of his inquiries on equations. The *Practica Arithmeticæ* is an early work in the *abaco* old-school tradition, where only a few pages out of the whole book are devoted to equations. These are substantially seen as a side-effect of proportions. In the *Ars Magna Arithmeticæ* the situation begins to change. We still have a consistent part that deals with irrational numbers in the Euclidean style of *Elements*, Book X – the so-called 'arithmetisation' of Book X. Nevertheless, these results on irrational numbers are used to study the shape of the irrational solutions of some cubic equations, and gradually the equations become a subject of inquiry by themselves. In the *Ars Magna Arithmeticæ*, we also find a handful of results on equations that will later appear in the *Ars Magna*, but they are much less refined. At the end of this development trend, the *Ars Magna* is completely devoted to equations. They are the subject of the whole book, from the beginning to the end – and this will also be the case of the *Aliza*. In the *Ars Magna*, complete and coherent methods to deal with quadratic, cubic, and quartic equations are expounded. Cubic equations are, in particular, systematically treated: a geometrical proof, an algorithmic rule (that is possibly translated into a formula), some examples, and sometimes other collateral rules are given. In this treatise, there is a well-developed context (I am referring, for instance, to the transformation rules) in which these equations are immersed. Little by little we are led to conceive the picture of the interdependency of all sorts of equations one on another and, in this way, the main problem of Cardano's construction arises: it is entailed by the *casus irreducibilis*. At this point, we will be ready to plunge into the *Aliza*.

In Chapter 5, I propose my interpretation of this treatise. I have made a choice of the most meaningful chapters of the *Aliza*, where 'meaningful' sometimes barely means 'understandable'. Once abandoned the (vain) hope to coherently interpret the whole book as if it were a well-structured treatise (which it is likely not, in fact), my aim has been, more reasonably, that of finding some common threads capable of guiding the interpretation of these parts of it, at least. These threads are joined together by the shared purpose to overcome the problem entailed by the *casus irreducibilis*.

Some appendices follow. I firstly relate on Cardano's life, and in particular on the controversy with Tartaglia. Afterwards, I provide the list of all the internal and external references in and to the *Aliza* and, finally, the list of the equations that Cardano considers there and in the *Practica Arithmeticæ*, *Ars Magna Arithmeticæ*, *Ars Magna* as numerical examples.

Far from offering an ultimate account of the *De Regula Aliza*, the present work is a first step towards a better understanding of a book of one of the most outstanding scholars of the 16th century.

2. Putting in context Cardano's works and mathematics

This chapter is an introduction to Cardano's works from both a non-mathematical and a mathematical viewpoint. In the following, I give an overview of Cardano's mathematical treatises, including the few manuscripts still available, and discussing in particular the writings in which cubic equations are dealt with. At a second stage, I introduce the *Aliza* and provide a first estimate in time. It is nevertheless worthwhile to address from now a few issues concerning this very controversial book, such as the meaning of the title and its supposed aim. Its detailed analysis will be presented in Chapter 5. Finally, I account in this section for the secondary literature concerning the *Aliza*. Being such an unfamiliar book, it had a handful of readers during the four and a half centuries that have passed. Most of them only take episodically into account a very few topics. Pietro Cossali, instead, made a quite detailed analysis at the end of the 18th century, which I discuss. Finally, in order to fully grasp Cardano's reasoning, I recall what our mathematics can nowadays say about quadratic, quartic, and especially cubic equations.

2.1. Cardano as a mathematical writer

Cardano has been a very prolific writer on a great variety of topics. Although on two occasions he destroyed almost one hundred thirty books, a consistent number of treatises and manuscripts has come to us. In the latest version of his autobiography, which was completed in 1576, a few months before his death, he lists one hundred thirty-one published treatises and one hundred eleven manuscripts. At Cardano's death, they were left[1] to Rodolfo Silvestri, one of his good friends in Rome and a former pupil of him. Subsequently, Cardano's writings were handed to another of his pupils, Fabrizio Cocanaro, and, in 1654, the majority of the manuscripts ended up at a Parisian bookseller. Eventually, they were bequeathed to Marc-Antoine Ravaud and Jean-Antoine Huguetan, who, with the help of the physician Charles Spon, published in 1663 in Lyon Cardano's *Opera Omnia*. It consists of ten volumes and one hundred twenty-seven treatises in all.

In the *Opera Omnia* there are thirteen mathematical treatises, some of which are published there for the first time, while others had previous editions. Chronologically listed according to the date of the first publication, then in the order in which they appear in the *Opera Omnia*, they are:[2]

- *Practica Arithmeticæ* (Milan 1539, Lyon 1663),
- *Ars Magna* (Nuremberg 1545, Basel 1570, Lyon 1663),
- *De Proportionibus* (Basel 1570, Lyon 1663),
- *De Regula Aliza* (Basel 1570, Lyon 1663),
- *De Ludo Aleæ* (Lyon 1663),
- *De Numerorum Proprietatibus* (Lyon 1663),
- *Libellus qui dicitur Computus Minor* (Lyon 1663),
- *Ars Magna Arithmeticæ* (Lyon 1663),
- *Sermo de Plus et Minus* (Lyon 1663),
- *Encomium Geometriæ* (Lyon 1663),
- *Exæreton Mathematicorum* (Lyon 1663),
- *De Artis Arithmeticæ Tractatus de Integris* (Lyon 1663),

[1]See [46].
[2]See [44].

- *De Mathematicis quæSitis* or *Paralipomena* (Lyon 1663).

They are all contained in the fourth volume of the *Opera Omnia*, which is devoted to mathematics, except for the *De Ludo Aleæ* in the first volume, and for the *De Artis Arithmeticæ Tractatus de Integris* together with the *Paralipomena* in the tenth volume. We remark that in the fourth volume of the *Opera Omnia* two spurious writings, the *Operatione della Linea* and the *Della Natura de Principij e Regole Musicali*,[3] have also been counted in.

We have moreover three mathematical manuscripts available. The Manuscript Plimpton 510/1700 at the Columbia University Library in New York is an Italian translation of the *Ars Magna* entitled *L'algebra*. The Manuscript N 187 at the Biblioteca Trivulziana in Milan, which was firstly entitled *Supplementum practicæ Hieronimi Castillionei Cardani medici mediolanensis in arithmetica* and then *Hieronimi Castillionei Cardani medici mediolanensis in libruum suum artis magnæ sive quadraginta capitulorum et quadraginta quæstionum*, partially attests to the *Ars Magna Arithmeticæ*. Finally, there is the Manuscript Lat. 7217, called *Commentaria in Euclidis Elementa*, at the Bibliothèque Nationale de France in Paris. It is a treatise in nine books on geometry, which has many common points with one of Cardano's unfinished projects, the *Nova Geometria*, which should have likely completed and enhanced the *Elements*.[4]

We must also consider another work that Cardano would have liked to write, but in the end he never wrote. It is the *Opus Arithmeticæ Perfectum*, probably conceived between the 1530s and the 1560s. It should have been an encyclopedic, arithmetical work in fourteen books, a sort of counterpart to the *Nova Geometria* (for more details, see below, Section 2.3). Finally, there are some non-mathematical

[3]Contrary to all the others, they are written in Vernacular. The *Operatione della Linea* is actually a work by Galileo Galilei contained in his *Le Operazioni del Compasso Geometrico e Militare* (1606). Concerning the *Della Natura de Principij e Regole Musicali*, see [91, page 97]. Also Ian MacLean includes them in a list of manuscripts written in Vernacular and not acknowledged by Cardano himself, see [48, page 60].

[4]For more details, see [63].

treatises, which anyway contain some mathematics as examples. They are the logical treatise *Dialectica* (Basel 1566, Lyon 1663) and the philosophical writing *De Subtilitate* (Lyon 1550, then fifteen editions more).

Among this huge amount of pages, we will only consider the works that deal with equations and, more precisely, principally with cubic equations. As said, they are the *Practica Arithmeticæ*, the *Ars Magna* (of which we have also available an English translation made by Richard Witmer in 1968), the *Ars Magna Arithmeticæ*, and – obviously – the *De Regula Aliza*. Except for the *Ars Magna Arithmeticæ*, they were all published during Cardano's lifespan. The *Ars Magna* also had a second edition, and all the mathematical treatises were posthumously republished in the *Opera Omnia*.

During the 14th and 15th centuries in the region of northern Italy (where Cardano grew up and subsequently lived for the first part of his life), a significant, even if modest, tradition of *scuole d'abaco* developed.[5] It is plausible[6] that Cardano had some contacts with this environment, since at that time these were the only places where one could learn algebra, or rather what was called so at that time. Anyway, there is no evidence that he had been himself an *abaco* master. It is moreover very likely[7] that Cardano pit himself against the more widespread Italian treatises, namely the inheritors of the *abaco* treatises like the *Nuovo Lume* by Giovanni Sfortunati and the *Libro de Abacho* by Pietro Borghi, and obviously the *Summa de Arithmetica, Geometria, Proportioni et Proportionalita* by Luca Pacioli. It indeed seems that the real link between Cardano and the *abaco* school was Gabriel Arator, one of Pacioli's students.

The general image of practical mathematics conveyed by these treatises is a double-sided one. On the one hand, there are miscellaneous problems solved case by case where the calculation skills are

[5] See [62, page 273].

[6] See [61]. Moreover, Cardano frequently mentions a certain Gabriel Arator, who assigns to himself the name of *magister*, see [2, Chapter LI, paragraph 17] and [62, page 275].

[7] See [62, page 276].

overwhelming. On the other hand, there is an encyclopedic work mainly focused on theory, that is Pacioli's treatise. The 1539 *Practica Arithmeticæ* – the first work in which Cardano deals with cubic equations – can be inserted in this tradition. Even if it is written in Latin and not in Vernacular, it contains all the major features of an *abaco* treatise: topics, choice of problems, and also terminology. It sometimes provides original reworkings of some techniques in the *abaco* tradition. Moreover, Cardano himself made the connection with Pacioli, even if in a conflicting way, which was a common publishing strategy after that the *Summa* appeared (given the turmoil that this work caused in the long term). In the *Practica Arithmeticæ*, cubic equations are only one among various problems and are seen as tightly depending on proportions.

According to the above chronological list, the *Ars Magna* should follow. Actually, thanks to some recent reappraisals of Cardano's mathematical writings,[8] we have come to the knowledge that the *Ars Magna Arithmeticæ* was preexistent to the *Ars Magna*. For a long time, the *Ars Magna Arithmeticæ* has been considered as a minor work and in any case a late work.[9] Cardano himself did not mention this treatise in his autobiographies and never had it printed. The (supposedly chronological) position of the published version in the framework of the 1663 *Opera Omnia* contributed to the misunderstanding, since it was wrongly placed after the *Ars Magna*. Nowadays it is commonly agreed that *Ars Magna Arithmeticæ* had been conceived before – or at least at the same time at – the *Ars Magna*. Massimo Tamborini suggests[10] that the *Ars Magna Arithmeticæ* and the *Ars Magna* were at the beginning planned together, and then separated and revised, maybe in order to publish the *Ars Magna* as soon as possible. In fact, in his 1544 autobiography, Cardano spoke of an "Ars Magna, which contains sixty-seven chapters",[11] whereas both the three editions of

[8]See the edition project of Cardano's works by Maria Luisa Baldi and Guido Canziani [44] and the conference proceedings [71], [45], [47], and [48].

[9]See [50], [74, page 298], and [22, Footnote 20, pages xvii-xviii].

[10]See [93, pages 177-9].

[11]"*Ars magna, continet sexaginta septem capitula*", see [24, page 131].

the *Ars Magna* and the *Ars Magna Arithmeticæ* published in 1663 both consist of only forty chapters. In this context, Cardano was talking about his project of an *Opus Arithmeticæ Perfectum*, crediting this "*Ars Magna*" to be its tenth volume. We take a more detailed look at the *Opus Arithmeticæ Perfectum* in Section 2.3.

Ian MacLean prefaces his edition of Cardano's autobiographies[12] by a commented chronology of Cardano's treatises. While speaking about the description of the *Opus Arithmeticæ Perfectum* given in Cardano's 1557 autobiography,[13] he identifies the tenth volume of the *Opus Arithmeticæ Perfectum* or the "*De regulis magnis, atque ideo ars magna vocatur*" with the *Ars Magna Arithmeticæ* itself. But MacLean's interpretation is confusing. In fact, the full 1557 quotation says: "De regulis magnis, and for that reason it is called ars magna. And only this one among the others is published",[14] whereas the *Ars Magna Arithmeticæ* has never been printed during Cardano's lifespan. Moreover, in the footnote to the full quotation, MacLean contradicts, and (fairly) identifies the same "De regulis magnis" with the *Ars Magna*.

Finally, Veronica Gavagna argues[15] that the *Ars Magna Arithmeticæ* had been composed between 1539 and 1542. She reasons as follows. Firstly, the *Ars Magna Arithmeticæ* is dedicated to "Philippum Archintum, bishop of Borgo Santo Sepolcro", who was in office between 1539 and 1546. Secondly, the Manuscript N 187 (1r-62v) at the Biblioteca Trivulzian in Milan, which partially attests to the *Ars Magna Arithmeticæ*, should have been at the beginning an addition to the *Practica Arithmeticæ*, as its first title '*Supplementum practicæ Hieronimi Castillionei Cardani medici mediolanensis in arithmetica*' suggests. Then, the title was modified into the final heading '*Hieronimi Castillionei Cardani medici mediolanensis in libruum suum artis magnæ sive quadraginta capitulorum et quadraginta quæstionum*'. When collated,

[12] See [24].
[13] See [24, page 65].
[14] "*De regulis magnis, atque ideo ars magna vocatur: atque hic solus ex omnibus editus est*", see [24, page 185].
[15] See [65].

the *Supplementum* and the *Ars Magna Arithmeticæ* show a word by word correspondence in entire chapters, especially the ones related to cubic equations. Then, it is quite likely that the latter originated from a reorganisation of the former. Moreover, in the *Practica Arithmeticæ* we read this:

> in the book titled Supplementum Practicae, where I solved algebraic equations of whatever degree, possible and impossible, general and not; so in that book there is everything people want to know in algebra. I have added new algebraic cases, but I can't publish them in the Practica because is too large [...] furthermore Supplementum is the crowning achievement of the whole of algebra and it is inspired to book X of the Elements.[16]

And this:

> there is a section lacking, that I can't add for the sake of brevity, because this volume is already too large: since this [new] book is a completion of algebra, its title will be Ars magna. Here you find the rules to solve all equations and all my outcomes obtained by the arithmetisation of book X. In particular, I have added some beautiful rules and two new cases.[17]

[16]This is a translation by Gavagna. "*[L]ibello qui dicit supplementum practice in quo ostendi omnia capitula algebræ possibilia et impossibilia usque in infinitum et quæ sint generalia et quæ non itaque non est aliquid desiderabile in tota arte quantacumque difficile quod non habeat radicem dantem cognitionem in illo libro et addidi plura capitula nova in ipso et non potui edere in ipsunt propter nimiam magnitudinem huius libelli eo quod est impressus in forma parva licet liber ille non transiendit tria aut quattuor folia et est consumatio totius artis et est exctractus ex decimo euclidis*", see [2, Chapter LXVIII, Paragraph 18, no page numbering].

[17]This is a translation by Gavagna. "*[D]eest in opusculum quod ob exiguam formulam cum in nimiam liber hic auctus sit magnitudinem adiici non potuit, ad artis totius complementum hoc artis magne titulo dicat: in quo universorumque capitulorum algebræ usque in infinitum inveniendi formula descripta est et quod super euclidis decimum ad normam numero rerum reducti inveneram congessi*", see [2, *Ad lectorem*, no page numbering].

We will see in Section 4.2 that this description – a book that deals with cubic and quartic equations, which is entitled "Ars Magna", and that contains the arithmetisation of *Elements*, Book X – fits the *Ars Magna Arithmeticæ*. Moreover, at the very beginning of the *Ars Magna Arithmeticæ*, we read this:

> [a]fter writing the Practica, it has seemed to me that it was necessary illustrating the cases, considered by the majority impossible, that I have found thanks to the proofs contained in our three books on Euclid, except for two rules,[18]

which is exactly the *incipit* of the Manuscript N 187, except that here the words 'the former has been revealed by *magister* Nicolò Tartaglia, the latter by Ludovico Ferrari' are struck through. Therefore, the *Supplementum* comes truly to be the link between the *Practica Arithmeticæ* and the *Ars Magna Arithmeticæ*, which follow one the other in this order. These two evidences would moreover place the minimal time limit of the composition of the *Ars Magna Arithmeticæ* on 1539. Gavagna's third remark is that Scipione del Ferro is never mentioned in the *Ars Magna Arithmeticæ*, whereas it is in the *Ars Magna*. Since Cardano and Ferrari discovered Scipione's contribution in 1542 (see below, Appendix A), this suggests to fix the maximum time limit on this year. But Gavagna goes even further. Her main thesis is that the *Ars Magna Arithmeticæ*, originally an addition to the *Practica Arithmeticæ*, was the first writing that should have released the cubic and quartic formulae. The time passing by, this work became so considerably filled with results that it conflicted with the project of the *Opus Arithmeticæ Perfectum*. Thus, it should have been reorganised

[18] This is a translation by Gavagna. "*Post compositionem Libri Practicæ visum est mihi necessarium ea ostendere, quæ a pluribus impossibilia iudicata sunt, quæ omnia nos invenimus ex demonstrationibus trium librorum a nobis supra Euclidem, exceptis duabus regulis harum*", see [13, page 303]. The mentioned "three books on Euclid" are the *Nova geometria*, a treatise on geometry started in 1535, which should have been composed at the beginning by three, then seven, nine, and finally fifteen books.

in two parts. The first part devoted to irrational numbers should have formed Book III of the *Opus Arithmeticæ Perfectum*, while the second one on equations became the core of the *Ars Magna*. This is indeed the structure of the *Ars Magna Arithmeticæ* as we know it nowadays. We consider therefore the *Ars Magna Arithmeticæ* as an intermediate step between the *Practica Arithmeticæ* and the *Ars Magna*. In the following, we come across several links between the two treatises that go in this direction, confirming Gavagna's hypothesis. Even though the *Ars Magna Arithmeticæ* is originated from the *Practica Arithmeticæ*, a lot of work had been done in the meanwhile concerning cubic equations. As Gavagna indicates and as we will see, the *Ars Magna Arithmeticæ* marks the rupture with the *abaco* schools and opens the way to equations as a proper subject of inquiry.

Finally, in 1545 came what has been considered Cardano's masterpiece, the *Ars Magna*. There, the equations, and in particular the cubic ones, play the main role. One can only grasp the importance of the *Ars Magna* in opposition to what came before. At last, all the scattered methods, rules, and special cases that had been discovered up to that time gave birth to a systematic treatment of cubic equations, including geometrical proofs of the solving methods.

2.2. A puzzling book

Nevertheless, in the *Ars Magna* the things are revealed to be not so plain as Cardano wished. As we will see, when a cubic equation has three real distinct solutions, it happens that square roots of negative numbers appear in the formula – which is the *casus irreducibilis*, as it has been lately called.

Twenty-five years after the publication of the first edition of the *Ars Magna* came the *De Regula Aliza*, in which Cardano conveyed his hope to overcome the above problem. In the 1570 *Ars Magna*, Chapter XII, Cardano himself provides the reader with this tip (for the full quotation see the next section on page 18). This is a key point, since my whole interpretation of the book starts from the link that is established there between the *Aliza* and the *casus irreducibilis.*

Unfortunately, the *Aliza* is an extremely cryptic work, starting from the title. It immediately strikes any reader who is trying to get acquainted with the book. What is the meaning of the term '*aliza*' in the title? This is neither a common Latin word nor was one of the brand-new mathematical words loan-translated into Latin at that time. Moreover, this term does not seem to belong to Cardano's own mathematical terminology. As far as I know, only three other occurrences can be found in the whole of Cardano's writings, and only one of these refers to the book itself. In chronological order, the first reference is in the *Ars Magna*, Chapter XII,[19] while speaking of the *casus irreducibilis.* There, we read of an "*aliza* problem" (in 1545) or of a "book of *aliza*" (in 1570 and 1663), which are supposed to amend the cubic formula in the case $\Delta_3 < 0$. The second occurrence steps outside the topic of cubic equations and is a mere mention. It is in the 1554 (and following) edition(s) of the *De subtilitate*, while Cardano deals with the "reflexive ratio [*proportio reflexa*]".[20] Nevertheless, I could not retrieve the corresponding passage in the *Aliza.* The last occurrence is in the *Sermo de plus et minus*, which contains a specific mention of *Aliza*, Chapter XXII (see here, Section 5.6 on page 385). Thereafter, except in the title page of the *Aliza*, the term is nowhere else mentioned. We also remark that Cardano does not pay any attention to trying to explain such a rare term, and this is telling about his overall attitude towards his public.

In 1799 Pietro Cossali hinted that the term '*aliza*' means 'unsolvable'[21] In 1892, Moritz Cantor related that Armin Wittstein suggests that this term comes from a wrong transcription of the Arabic word

[19]See [4, Chapter XII, page 251] or see here, page 114.

[20]More precisely, Cardano makes reference to the *Ars Magna* and to the *De Regula Aliza* ("*ex demonstratis in Arte magna et regula Aliza*") concerning the ratios of the side of a regular heptagon to any diagonal subtended to two or three sides, see [5, Book XVI, page 428].

[21]"De Regula Aliza, *cioè* De regula Irresolubili", see [57, volume II, page 441] or [60, Chapter I, paragraph 2, page 26].

'*a'izzâ*' and means 'difficult to do', 'laborious', 'ardous'.[22] In 1929 Gino Loria advised as a common opinion that the term comes from a certain Arabic word that means 'difficult'.[23] None of the above hypotheses were supported by a precise etymology. Very recently, Paolo D'Alessandro has confirmed Cossali's hypothesis.[24] The term '*aliza*', or '*aluza*', is likely a misspelling based on the Byzantine pronunciation 'ɑ́liʤjɑ' of the Greek word 'ἀλυθεῖα', which is composed by the negative particle 'α' and by the feminine singular passive aorist participle of the verb 'λύω', 'to unbind', 'to unfasten', 'to loosen', 'to dissolve', 'to break up', 'to undo', 'to solve'.[25] Thus, '*aluza*' means 'non-solvable'. This etymology soundly agrees with Cardano's words in *Ars Magna*, Chapter XII, where the *Aliza* and the *casus irreducibilis* are linked. Note that, if we understand '*irreducibilis*' as the fact that the cubic formulae (which should convey three real roots) cannot reduce to some real radicals, then in Cardano's mathematics '*irreducibilis*' completely overlaps with 'non-solvable'.

If the enigma of the title can be relatively easily fixed, the same cannot be said about other difficulties. Basically, there is one fundamental justification for the high unfamiliarity of the *Aliza* among modern readers. Spoiling the whole surprise, Cardano did not reach his aim (or at least the aim that concerns the *casus irreducibilis*) in the *Aliza*. In very truth, he could not reach it, but this had only been showed three centuries later.[26] The book then tells of a failed

[22]"*[D]ie dritte führte den nie und nirgend erklärten Titel De regula Aliza, der durch unrichtige Trasscription aus dem arabischen Worte a'izzâ (schwierig anzustellen, mühselig, beschwerlich) entstanden sein kann [Diese Vermuthung rührt von H. Armin Wittstein her.], und alsdann Regel der schwierigen Fälle bedeuten würde*", see [52, page 532].

[23]"*[...] il titolo, sinora inesplicato* De Regula Aliza *(secondo alcuni* aliza *deriverebbe da una parola araba significante* difficile*)*", see [74, p. 298].

[24]By personal communication.

[25]See [73] and [85, page 367]. If it is so, then note that '*aluza*' and 'analysis' have a common origin.

[26]Indeed, it has been proved that imaginary numbers have necessarily to appear in the cubic formula when the equation has three real, different roots by Pierre Laurent Wantzel in 1843, Vincenzo Mollame in 1890, Otto Hölder in 1891, and Adolf Kneser in 1892. Paolo Ruffini also provided an incomplete proof in

attempt – rather, of an unattainable attempt – so that it is not surprising to find a good amount of looseness inside.

There are anyway other extrinsic reasons that contributed to worsen the situation and that could depend on the previous justification. However common the typos and mistakes in calculations were at the time, the book is particularly filled with them. Sometimes the misprints concern the correspondence of the lettering between the diagram and the text, and then we can only hope that the text is sufficiently enough devoid of incongruities to manage to correct the wrong letter(s). Sometimes we find some lettering in the text and no diagram at all. In other cases, there must have been full lines that were skipped at the page break. Moreover, Cardano's peculiar style in writing in Latin is not the plainest one. Speaking the mathematics as Cardano and his contemporaries did without a complete and stable formalism implied a certain amount of ambiguity, and Cardano was not used to simplify the task to his readers. The most favourable case is when a general statement is accompanied by numerical examples, so that we can untangle the text relying on calculations. Apart from that, Cardano's general observations are usually long, convoluted sentences with more than one possible interpretation. Unfortunately, the 1663 edition cannot help the reader: Spon, who revised Cardano's writings, was not a mathematician and the best that he could do was to proofread the grammar and to correct here and there some clear mistakes.

The *Aliza*'s obscurity is displayed by the lack of any global ordering in the flowing of the chapters' topics. At the beginning of Chapter 5, I provide a general overview of the kinds of issues that are dealt with in the *Aliza*. A same topic is usually scattered all along the book, with repetitions and logical gaps, so that in the end the reader feels highly disoriented. Under these conditions, the *Aliza* is more pragmatically provided as a miscellany rather than as a unitary book. To resume, using Pietro Cossali's words:

1799. See [42], [33] and [34], [30], [31], [27], and [36] and [37]. I plan to develop further this topic in a next work.

> [t]o the obscurity of that book caused by the difference of language, another [obscurity] is coupled, [which is the one] produced by an infinite, very pernicious mass of errors both in the numbers in calculations and in the letters in figures and related proofs. And another one, which is absolute and intrinsic, has to be added. It is the lack of order due to which it is very hard and uncomfortable to pull the strings, to see the result, and to evaluate the discovery. We spot at Cardano trying, opening new roads, retracing his steps, turning to one side or another according to his intellect's suggestions. In a word, the book is the action of attempting, not the ordering of the discoveries. And in the end it is one of Cardano's books, that his contemporaneous Bombelli called obscure in the telling.[27]

It is exactly because of these long-range incongruities that one needs to master as good as possible the contents of the other treatises by Cardano that deal with equations. It is an obviously recommended practice to contextualise a work in the frame of the other writings of the same author. But more than that, in this particular case where we miss the guide of a general structure, we need all the short, coherent problems, exercises, and mathematical notes that can be gathered from Cardano's other works (and in particular from the *Practica Arithmeticæ*, the *Ars Magna Arithmeticæ*, and the *Ars Magna*) to get oriented while reading the *Aliza*. This is why we devote two whole chapters to those treatises and other contextual matters.

[27] "*Alla tenebria di quel libro a noi cagionata dalla diversità del linguaggio se ne accoppia un'altra prodotta da un'infinita perversissima folla di errori, e di numeri ne' calcoli e di lettere si' nelle figure, si' nelle relative dimostrazioni; e bisogna eziandio aggiungersene altra assoluta, ed intrinseca proveniente da mancanza di ordine, per la quale riesce cosa faticosissima, e malegevolissima l'unire le fila, vedere il risultato, valutare il discoprimento. Si scorge Cardano, che tenta, che si apre nuove strade, che ritorna sulle battute, che si volge or da un lato, or dall'altro seguendo i suggerimenti varij dell'ingegno; in una parola il libro è l'atto del tentare, non un ordine delle scoperte; e finalmente è un libro di Cardano, che da Bombelli pur suo coetaneo fu denominato nel dire oscuro*", see [57, volume II, page 442] e [60, page 27].

2.3. From the "*aliza* problem" to the *De Regula Aliza*: changes in the editorial plan of the *Opus Arithmeticæ Perfectum* and an attempt to loosely assign a date

Before addressing the mathematical details, it is fully worthwhile to spend a few words to clarify as much as possible the chronological boundaries of the *Aliza*. Indeed, this work has appeared to be a miscellany of mathematical writings, notes, remarks, and observations that shine out neither for consistency nor for cohesion – at least, at first sight. Then, the question that one immediately asks is whether it is possible to date these writings and, if it is so, to when – even if approximately. If one takes a look at the (supposedly chronological) order in which the books are displayed in the fourth volume of the *Opera Omnia*, he will notice that the *Aliza* comes immediately after the *Ars Magna Arithmeticæ*, which in turn follows the *Ars Magna*. But this is completely ineffective, since, as we have seen in the preceding section, the *Ars Magna Arithmeticæ* has been very likely composed before the *Ars Magna*.

It is a very hard issue to give a reasonable estimate in time of the *Aliza* starting from the internal relations, namely from the mutual relationship of the contents of the chapters. At present, I content myself with checking the external references to this book. Let us firstly address the references to the *Aliza* that are present in the other mathematical writings by Cardano. As already said, the term '*aliza*' appears only four times in the whole of Cardano's mathematical writings, once in the title of the book itself, once in the *Ars Magna*, Chapter XII, once in the *De Subtilitate* (1550), and once in the *Sermo de Plus et Minus*. I leave aside the reference in the *De subtilitate* (see above, Footnote 20 on page 14), since it is a very brief mentioning of a certain result from the *Aliza*, which is not better specified and that I could not retrieve. I analyse the reference in the *Sermo de Plus et Minus* in Section 5.6 on page 385. Here follows the quotation from the 1545 *Ars Magna*'s edition:[28]

> [w]hen the cube of one-third the coefficient of x is greater that the square of one-half the constant of the equation, which happens whenever the constant is less than three-fourths of this cube or when two-thirds the coefficient of x multiplied by the square root of one-third the same number is greater than the constant of the equation [that is, for short, when $\Delta_3 < 0$], then the solution of this [$x^3 = a_1x + a_0$] can be found by the aliza problem which is discussed in the book of the geometrical problems[29]

and from the 1570 and 1663 editions:

> [w]hen the cube of one-third the coefficient of x is greater that the square of one-half the constant of the equation, which happens whenever the constant is less than three-fourths of this cube or when two-thirds the coefficient of x multiplied by the square root of one-third the same number is greater than the constant of the equation [that is, for short, when $\Delta_3 < 0$], then consult the Aliza book appended to this work.[30]

These quotations not only allow to fix the *Aliza*'s overall subject topic, but they also couple (in 1545) the name '*aliza*' with a certain "book of

---

[28]See also below, Footnotes 49 and 51, page 114.

[29]"*At ubi cubus tertiæ partis numeri rerum excedat quadratum dimidii numeri æquationis, quod accidit quandocunque numerus æquationis est minor $\frac{3}{4}$ cubi illius, vel ubi ex $\frac{2}{3}$ numeri rerum producitur in R $\frac{1}{3}$ esiusedm numeri maior numerus numero æquationis, tunc hoc dissolvitur per quæstionem Alizam, de qua in libro de quæstionibus geometricis dictum est*", see [4, Chapter XII, page 62] or see [22, page 103].

[30]"*At ubi cubus tertiæ partis numeri rerum excedat quadratum dimidii numeri æquationis, quod accidit quandocunque numerus æquationis est minor $\frac{3}{4}$ cubi illius, vel ubi ex $\frac{2}{3}$ numeri rerum producitur in R $\frac{1}{3}$ esiusedm numeri maior numerus numero æquationis, tunc consules librum Alizæ hic adjectum*", see [10, Chapter XII, page 31v] and [12, Chapter XII, page 251] or see Footnote 5 in [22, page 103].

the geometrical problems". Let us store away for now this information. Note moreover that in 1545 the "*aliza*" is merely a "problem", while in 1570 it has reached the extent of a whole "book" (and it must be so, since in the latter quotation Cardano makes reference to one of the books appended to the new *Ars Magna*'s edition). Therefore, the *Aliza* likely arose as a problem, more precisely as a geometrical problem, and then grew bigger and bigger until it became a book. This could (at least partially) explain the lack of order in its structure. But, much more important is the fact that the *Aliza* is firstly published together with a book – the *Ars Magna* – that contains, even though in a few lines, the key to interpret it. This is fundamental in order to understand how Cardano possibly regarded the *Aliza*, namely as the treatise that (hopefully) contains the solution to the problem entailed by the *casus irreducibilis*.

With regard to the other way round, that is the references to other Cardano's mathematical writings that appear in the *Aliza*, it is implausible that they could help in locating the *Aliza* in time. It is possible – actually, likely – that Cardano subsequently added some references in a text that had already been written a while ago. We should remark, indeed, that he mainly made reference to the *De Proportionibus* and *Ars Magna*, which were published together with the *Aliza* in 1570.

So far, so good, but we did not progress much. Cardano already had in mind a certain "*aliza* problem" around 1545 – which makes sense, since he already most have come across the *casus irreducibilis* developing his cubic formulae in the *Ars Magna*. Afterwards, we can check for the references to the *Aliza* in Cardano's non mathematical treatises, and there we are extremely lucky. We do not only have his autobiography *De Libris Propriis*, but, with Cardano being the true graphomaniac that he was, he left five versions of it, written over a period of more than thirty years. They are [3] (which is also in the *Opera Omnia* in [14]), [23] (which dates back to 1550), [6] (which is also in the *Opera Omnia* in [15]), [7] (which is also in the

Opera Omnia in [16]), and [17] (which dates back to 1576).[31] There, Cardano recalled not only his personal life, but also his career and his writings. Referring to his autobiography, Cardano says that

> [t]he book was often modified [...]. Thus I set forth for myself in it a certain image of everything that I had written, not only as an aide-memoire, and a mean of selecting those books which I would finish and correct first, but also to set down why, when and in what order I wrote what I did [...]. Indeed I did not only set down here the titles of my books, but also their size, incipits, contents, order, the utility of the division and order of the books, and what they contained of importance.[32]

Each of these versions consecrate to the *Aliza* only a few lines – and most of the time as a fringe topic – but we will see that we can anyway draw some conclusions.

In the lists of his works that Cardano makes in the different versions of his autobiography, the *Aliza* as a published book is of course only mentioned once, in 1576.[33] Most of the information concerning the *Aliza* comes collaterally to the description of the *Opus Arithmeticæ Perfectum* that can be found in the autobiography. This should have been a mathematical, encyclopedic work, composed by fourteen books, and probably conceived between the 1530s and the 1560s.[34] Unfortu-

[31] For more information on the chronology of Cardano's works, see Maclean's essay in the preface of [24].

[32] The English translation is by Maclean in [24, page 11], or "*[d]einde auxi, iterumque mutavi [...]. Iconem ergo quendam mihi in eo proposui omnium eorum quæ a me conscripta sunt, non solum ad memoriam confirmandam, eligendosque mihi libros quos prius absolverem et castigarem: sed ut doceam, quibus causis, temporibus, quoque ordine talia conscripserim, et ut vim numunis suo loco testarer [...]. Verum non hic solum docui librorum nomina, sed et magnitudinem, initium materiam pertractatam, ordinem divisionis, ordinisque librorum inter se utilitatem, quidque in se præcipuum continerent*" in [7, pages 23-24].

[33] See [17, page 40]. The other versions of Cardano's autobiography are all previous to 1570.

[34] See [93], [64], and [65].

nately, it has never been accomplished. It is quite difficult to gather some consistent information on the supposed table of contents of the *Opus Arithmeticæ Perfectum.*[35] It should have included calculations with integer, fractional, irrational, and denominated numbers, proportions, properties of numbers, commercial arithmetic, algebra, plane and solid geometry, and some arithmetical and geometrical problems. Veronica Gavagna[36] suggests that, though the general outfit follows the *Practica Arithmeticæ*'s structure, it also reveals an evident symmetry with Cardano's unpublished comment to the *Elements* (which will be later referred to as the *Nova Geometria*), suggesting that the *Opus Arithmeticæ Perfectum* should be considered as a sort of arithmetical counterpart. Moreover, it is likely that some of Cardano's printed mathematical treatises that are nowadays available should have filled up some of the books, or part of them, in the table of contents of the *Opus Arithmeticæ Perfectum.* Namely,[37] it is the case for the *Tractatus de Integris*, the *De Proportionibus*, and the *De Numerorum Proprietatibus*, which can be found in the fourth and tenth volumes of the *Opera Omnia.* The *Ars Magna Arithmeticæ*, the *Ars Magna*, and the *De Regula Aliza* also play a role in this game, but the matter is more ticklish, as we will see.

All the versions of Cardano's autobiography agree that Book X of the *Opus Arithmeticæ Perfectum* should have been on algebra and equations. In 1544, he says that "the tenth [book] is entitled Ars magna, it contains sixty-seven chapters",[38] whereas Books XIII and XIV "are assigned to arithmetical and geometrical problems".[39] This quotation raises the incidental question to know what was that "Ars magna" of which Cardano is speaking, since all the editions that we have with that title (handwritten or printed) contain only forty

[35]For a detailed discussion, see [24, pages 64-66].

[36]See [64, page 65].

[37]See below, Footnote 43, page 24.

[38]"*[D]ecimus inscribitur Ars magna, continet sexaginta septem capitula*", see [3, page 426].

[39]"*[T]ertiusdecimus ac quartusdecimus, quæstionibus Arithmeticis et Geometricis destinantur*", see [3, page 426].

chapters. Massimo Tamborini in [93, pages 178-179], Ian Maclean in [24, page 65], and Veronica Gavagna in [65] report on this incongruity, and I refer to them for an accurate discussion. In a nutshell, as we have seen and despite Maclean's interpretation, it seems more likely that both the *Ars Magna Arithmeticæ* and the *Ars Magna* (or, at least, their original common core on equations) should have concurred to Book X of the *Opus Arithmeticæ Perfectum.*[40] We moreover recall that, in Chapter XII of the 1545 *Ars Magna*, the "*aliza* problem" was coupled with a certain "book on geometrical problems" (see above, on page 19). Then, in 1544 the situation is the following. Book X of the *Opus Arithmeticæ Perfectum* is devoted to equations and possibly consists of some parts of the *Ars Magna Arithmeticæ* and *Ars Magna.* The *Aliza* is not explicitly mentioned, but is connected (at least, starting from 1545) to certain geometrical problems in the *Opus Arithmeticæ Perfectum*'s last book.

Later on, in 1550, Cardano devoted Book X to "all the chapters on the square together with the Aliza rule". Book XIV still concerned geometry, but was restricted to the "measure of figures".[41] Any reference to the "Ars magna" or to our *Ars Magna* had disappeared. Now, the *Aliza* was shifted from filling the last book to filling the tenth one and, accordingly, the last book of the *Opus Arithmeticæ Perfectum* was reduced in contents.

Afterwards, from 1557 on, the *Aliza* disappeared again, coming back to the same situation as in 1544.[42] This time Cardano gave the *incipit* and the lengths of the books in the *Opus Arithmeticæ Perfectum* that he has already written. In particular, he said that

---

[40]We also recall that, according to Veronica Gavagna, Book III of the *Opus Arithmeticæ Perfectum* (the *incipit* of which is "*Cum in radicibus quantitatum*") should have been covered by the first part of the *Ars Magna Arithmeticæ* on the arithmetisation of Book X of the *Elements*, see here page 13.

[41]"*In [decimo] omnia capitula supra quadratum cum regula Aliza*" and "*[i]n [quartodecimo] ad mensuram figurarum pertinentia, quæ geometrica vocantur*", see [23, page 9v].

[42]"*Decimus de regulis magnis, atque ideo ars magna vocatur: atque hic solus ex omnibus editus est*" and "*Tertiusdecimus quæstiones Arithmeticas, ut Quartusdecimus Geometricas*" in [6, pages 37-38] and [7, page 16].

Book X began with "*[h]æc ars olim a Mahomete*" and is made up of 83 *folia*. This enables[43] us to identify for sure Book X with the *Ars Magna*'s that we have nowadays available.

Finally, but almost twenty years later in 1576, Cardano told that in 1568 he joined the *Aliza* and the *De Proportionibus* to the *Ars Magna* and had them printed.[44]

We can now try to fix some temporal limits for the *Aliza*. A certain "*aliza*" was firstly mentioned in 1545 (but we can stretch this time limit of a year, up to 1544) and it sank into oblivion at least from 1557. It reappeared in 1568, but it was only a quick mention. We can therefore conjecture that the miscellany of writings that compose the *Aliza* (or a part of them) started to be conceived at worst between 1544 and 1545. Then, its proofreading could have lasted at best until 1557, when Cardano had already lost interest in it. Afterwards, it is likely that during the 1560s, due to his personal misfortune, Cardano desisted from the *Opus Arithmeticæ Perfectum* project and, in general, gave up all of his mathematical projects. Moreover, regarding at least the *casus irreducibilis*, the changes in the editorial project of the *Opus Arithmeticæ Perfectum* could also express Cardano's hope to find some new results. In particular, the lapse of time during which only the *Aliza* appears as a part of the *Opus Arithmeticæ Perfectum* could correspond to the lapse of time during which Cardano believed to manage to avoid the *casus irreducibilis*, namely from 1550 to (at best) 1557. But, when the *Aliza* was printed in 1570, its second title (just above 'Chapter I') recalled its origins as a problem, since it contains the word '*libellus*', which means 'pamphlet', 'booklet', 'small book'. No matter at that time to replace the *Ars Magna*.

We can finally address the sources that mention the *Aliza*, but that were not written by Cardano himself. I am only aware of one,[45] which

[43] By the way, this also enables us to stick the *Tractatus de integris* to the first book, the *De Proportionibus* to the fifth book, and the *De numerorum proprietatibus* to the sixth book of the *Opus Arithmeticæ Perfectum*, see [65].

[44] "*De proportionibus, et Aliza regula addidi anno MDLXVIII ad librum Artis magnæ et edidi*" in [17, page 41].

[45] I thank Veronica Gavagna for this reference.

is in [49, page 163]. Gian Luigi Betti relates about Ercole Bottrigari, printer, scholar, music theorist, and bibliophile, who lived in Bologna between 1531 and 1612. We do not know when he met Cardano for the first time, but he wrote in his *La Mascara*[46] that he questioned Cardano about the *Aliza* "two or three years" before 1570.[47] As Bottrigari testified, Cardano avoided to answer, but in doing so he dropped that the *Aliza* was going to be printed in Germany.[48] This is in accordance with Cardano's assertion in his 1576 autobiography.

Summing up, a certain "*aliza* problem", which, according to the *Ars Magna*, concerned the *casus irreducibilis* and which should have originally been linked to geometry, was already there in 1544 or 1545. We get the strong impression that, from that moment on, the problem at issue grew more and more in importance and attained its best around 1550. The reference to the "*aliza*" disappeared then in 1557. The *Aliza* (as we know it) was ready for press in 1568 and printed in 1570. Therefore, the core of the book on the *casus irreducibilis* had already been conceived a long time before its publication. At a certain point, a more or less sizeable number of pages on fringe topics could have been added and the name of the problem was handed on to the whole miscellany. In this way, what at the beginning was nothing but an "*aliza* problem" could have become an entire book – our *De*

[46]This is one of Bottigari's late works. Gian Luigi Betti found the autographic Manuscript MS B 45 in the library G. B. Martini at the Academy of music in Bologna.

[47]"*É invece certo che [Bottigari] gli [Cardano] abbia direttamente posto una questione 'sopra a quel suo capitolo Alizam algebratico [...] nella sua Arte magna, il qual da poi publicò due o tre anni, facendo ristampare essa Arte Magna insieme con il libro De propositionibus [sic in Betti] l'anno 1570'*" in [49, page 163].

[48]"*La risposta ricevuta al quesito proposto fu invero piuttosto singolare, poiché il Cardano avrebbe affermato : 'ch'egli non lo sapea, affermandomi che il suo genio era stato e non egli che lo aveva scritto, con soggiungermi che per ciò spesse volte egli stesso non sapea quello che avesse scritto, et che leggendolo non lo intendeva. Cosa che veramente mi scandalizzò molto a prima faccia, poi mi diè che molto a meravigliarmi. Finalmente mi spedì dicendomi ch'io aspettassi che tosto (come ho detto, che poi veramente fu) ei verrebbe in istampa da Alemagna'*" in [49, page 163] referring to the *Mascara*, page 84.

Regula Aliza. Obviously, giving a title to a miscellany of heterogeneous chapters does not automatically make the text consistent. Indeed, we see in Section 5.1 that the *Aliza* has not been proofread or proofread very quickly.

2.4. The readers of the *De Regula Aliza*

Due to the problems that I have detailed in the preceding section, the *Aliza* was, and still remains, a very unfamiliar book. During the centuries, it had nevertheless a handful of reader. In the following, I give an overview on the studies that have already been done on this book. Note that I do not mention the authors who simply quote the title.

According to the last sentence of the above quotation by Cossali,[49] one of the very first readers of the *Aliza* must have been Rafael Bombelli. Anyway, I did not manage to retrieve any comments in his *Algebra* (1572) about the *Aliza*. There is nevertheless an indirect link that deserves to be mentioned. It is Cardano himself that coupled in the first paragraph of the *Sermo de Plus et Minus* a reference to the *Aliza* and Bombelli's name.[50] What we know for sure is that Bombelli read the *Ars Magna*, since in the preface to the readers in his *Algebra* (1572), he criticised the obscure way in which the *Ars Magna* had been written.[51] This also shows that Cossali's last sentence in the above quotation is inaccurate.

Federico Commandino read for sure the *Aliza*, or at least some parts. In his edition of the *Elements* (1572), he briefly mentioned Cardano's alternative sign rule from *Aliza*, Chapter XXII (see below, Section 5.6 from page 378).[52]

[49]See above, on page 16.

[50]For more details, see below, Section 5.6.

[51]"*Hanno poi, e Barbari, e Italiani à nostri tempi scritto [. . .], oltre che il Cardano Mediolanese nella sua arte magna, ove di questa scientia assai disse, ma nel dire fu oscuro*", see [1, Agli Lettori, no page number].

[52]"*[C]oguntur dicere si minus per minus multiplicetur produci plus. Quod verum non esse primus animadvertit Hieronymus Cardanus non solum mathematicus,*

Belonging to the next generation of mathematicians, Simon Stevin made reference to the *Aliza* in his *Arithmétique* (1585). While solving the equation $x^3 = 6x + 40$, Stevin shortly stated that Cardano also put some examples in his *Aliza*.[53]

Also Thomas Harriot made reference to the *Aliza*, and twice. We know that he read (at least) one chapter of the *Aliza*. In the British Library,[54] Rosalind Tanner has found[55] the following note to one of Harriot's colleagues, Walter Warner.

> Although Cardane in the beginning the first Chapter (pag. 6) of his 10th booke of Arithmeticke wold have $\sqrt{9}$ to be $+3$ or -3, yet in his Aliza being a latter worke he was of another opinion. I prey read his 32 Chapter being at the 42 page.[56]

The chapter of the *Aliza* to which Harriot was referring is actually Chapter XXII, which fairly starts at the *folio* 42 of the 1570 edition. In her two articles,[57] Tanner establishes a relation between Harriot's remarks on the fact that there is a choice behind the sign rule and the new sign rule proposed by Cardano in the *Aliza* (see Section 5.6).

As far as I know, there is only one scholar who studied the *Aliza* pretty in details. It is Pietro Cossali, born in Verona in 1748 and

sed et philosophus ac medicus prestantissimus, ut apparet in libro de regula aliza", see [26, Book X, Proposition 34, Theorem III, page 149].

[53]"*Cardane met aussi en son* Aliza *quelques exemples, servans à ceste matiere, mais pas generaux*", see [39, Book II, page 309].

[54]It is the Add. Ms. 6783, folio 121. The page has no date and Tanner remarks that no sufficient indication can be found to place it in time. Anyway, she remarks that all the dated pages in the manuscript are between 1589 and 1619. Note that Tanner says that the folio 121 is a stray sheet among a batch of unconnected papers, but in truth it is associated with other notes on the *Ars Magna*, see [38].

The second reference to the *Aliza* is in Add. Ms. 6785, folio 197. I thank Jackie Stedall for these connections.

[55]See [95].

[56]I have picked the quotation up from [95, page 144]. The '10th booke of Arithmeticke' is in truth the *Ars Magna*, which at a certain point should have become Book X of the *Opus Arithmeticæ Perfectum*, see above, Section 2.3.

[57]See [95] and [94].

deceased in Padua in 1815. He was a priest of the Teatini's order in Milan, calculus, physics, and astronomy professor at the university in Padua, member of the Società italiana delle scienze, and pensionnaire of the Reale istituto italiano di scienze, lettere ed arti.[58] He also was a historian of mathematics. In 1799 his history of algebra in Renaissance Italy, the *Origine, trasporto in Italia, primi progressi in essa dell'Algebra*[59] was published in two volumes. Moreover, since 1966 we have available the commented transcription by Romano Gatto of Cossali's manuscript *Storia del caso irriducibile*,[60] which has a considerable number of pages in common with the history of algebra.

In the *Storia del caso irriducibile* and mostly in the second volume of the history of algebra, Cardano played a central role and the *Aliza* was handled on the same footing as the *Ars Magna*, as well as the *Ars Magna Arithmeticæ*. Nevertheless, it must be said that most of the time Cossali's accuracy as a historian is not up to standards. He seldom gave the exact references, had no hesitation in integrating or completely rewriting Cardano's proofs, and paid no attention to the interconnections between Cardano's treatises. In short, he was doing mathematics from a historical starting point rather than history of mathematics.

Actually, Cossali was a mathematician, and I believe that this could partially explain his attitude as a historian. Around this period, mathematicians started to study in depth the *casus irreducibilis* for cubic equations. In 1781 the Academy of Padua announced a competition to prove whether it is possible to free the cubic formula from imaginary numbers, but finally the prize was not assigned. Then, in 1799, Ruffini began to get involved in the study of equations and in 1813 he published an incomplete proof of the fact that, when a cubic equation undergoes the *casus irreducibilis*, none of its solutions can be expressed only by real radicals.[61] Cossali was very interested in the topic of the *casus irreducibilis*. His very first work, the *Particularis*

[58]For a more detailed biography, see [60, page 3]
[59]See [57].
[60]See [60].
[61]See [36] and [37].

methodi de cubicarum æquationum solutione a Cardano luci traditæ. Generalis posteriorum analystarum usus ex cap. I De Regula Aliza ipsius Cardani vitio luculentissime evictus. Atque mysterium casus irreducibilis post duo sæcula prorsus retecta causa sublatum specimen analyticum primum (1799) was devoted to it, and the *Disquisizione sui varj metodi di eliminazione con il componimento di uno nuovo* (1813) ranged around the same issue.[62] He wanted to take part in the 1781 competition, but he could not finish on time his *Sul quesito analitico proposto all'Accademia di Padova per il premio dell'anno 1781 di una assoluta dimostrazione della irriducibilità del binomio cubico.*[63] There, he also started a long lasting controversy with Anton Maria Lorgna on some mistakes that he made while dealing with the *casus irreducibilis*. Briefly, the mathematician Cossali worked pretty much on equations, and in particular on cubic equations. Cossali's strong interest toward the *casus irreducibilis* is the reason why – I suggest – he struggled with the *Aliza* under a mainly mathematical viewpoint.

In our chronological review of the *Aliza*'s readers, we stumble at this point upon some scholars who wrote an overall history of mathematics and limited themselves to choose and explain a bunch of tiny mathematical techniques from the *Aliza*. They are Charles Hutton (*Tracts on mathematical and philosophical subjects* (1812), on pages[64] 219-224), Moritz Cantor (*Vorlesungen über die Geschichte der Mathematik* (1892), on pages[65] 532-537), and Gino Loria (*Storia delle matematiche* (1931), on pages[66] 298-299). In particular, concerning the overall flowing of the *Aliza*, Loria said that

> [w]hile [...] all of Cardano's writings do not shine out for clarity, this last [referring to the *Aliza*] is despairingly unclear. Nobody or a few people (one of these is the

[62]See [58] and [59].
[63]See [56].
[64]See [70].
[65]See [52].
[66]See [74].

> historian Cossali) struggled to understand it and its influence was poor or none. The one who goes through it follows Cardano while he tries one thousand ways to solve that distressing enigma. He sees him stopping, moving backwards, then resuming his way ahead, pushed by the longing to discover the limits of applicability of the solving method conceived by Tartaglia.[67]

Finally, we have the contemporary readers, like the already mentioned Rosalind Tanner (*The alien realm of the minus: deviatory mathematics in Cardano's writings* (1980), on pages[68] 166-168), Silvio Maracchia (*Storia dell'algebra* (2005), on pages[69] 277 and 331-335), and Jacqueline Stedall (*From Cardano's great art to Lagrange's reflections: filling a gap in the history of algebra* (2011), on page[70] 10). In a greater or lesser measure, they conform to their predecessors and give an episodic account of the *Aliza*.

It is also worthwhile to mention the new impulse that has been given in Milan from the early 90s to the studies of Cardano's writings thanks to the project of the edition of his works (see the web site `http://www.cardano.unimi.it/`).

2.5. The mathematical context: how we deal with equations nowadays

As we will see in the following sections, a focus of Cardano's interest (especially in the *De Regula Aliza*) is, rather than numbers themselves,

[67]"*Mentre [...] tutti gli scritti del Cardano non brillano per chiarezza, quest'ultimo è di una disperante oscurità; nessuno o pochi (uno di questi è lo storico Cossali) si sforzarono di comprenderlo, ove scarsa per non dire nulla ne fu l'influenza. Chi lo percorre segue Il Cardano mentre tenta mille vie per sciogliere quell'angoscioso enigma; lo vede arrestarsi, retrocedere per poi riprendere il cammino in avanti, sospinto dalla brama di scoprire i limiti di applicabilità del metodo di risoluzione ideato dal Tartaglia*", see [74, page 298].

[68]See [94].

[69]See [76].

[70]See [92].

how numbers can be expressed. This is what we could call the 'algebraic shapes' of numbers, that is expressions written in Cardano's stenography like $\sqrt{a}+b$ or $\sqrt[3]{a}+\sqrt[3]{b}$ where a, b are rational numbers such that respectively their square roots or cubic roots are irrational. When Cardano addressed the problem entailed by the *casus irreducibilis*, he turned to these algebraic shapes rather than to the numerical values of the solutions or to the structural relationships between them (as Lagrange, Ruffini, Abel, and Galois will subsequently do). In the end, he was trying for a general theory of the shapes under which certain numbers can be written. Indeed, a recurring strategy in Cardano's mathematical treatises is to establish whether these shapes can convey a solution of an equation or not (see below, Section 4.2.4 and Section 5.3.1). As such, Cardano's attempt is doomed to failure. Nevertheless, this study of the algebraic shapes is very up-to-date and marks a trend that will go on for about two centuries up to the 18th century.

It must be said, moreover, that Cardano mainly deals with procedures or algorithms rather than with formulae: in these terms, the most problematic of Cardano's algebraic treatises, that is the *Aliza*, should not be considered only as a failure, but also as a treasury of results.

Let us now try to understand why, from a nowadays mathematical viewpoint, Cardano had to fail. In the following sections, I recall our solving methods that lead to the formulae for quadratic, cubic, and quartic equations. I deal in particular with the cases in which the coefficients are real. Then, especially concerning cubic equations, I explain, using Galois theory, why imaginary numbers necessarily appear in the cubic formula when the equation falls into the *casus irreducibilis*. Finally, I provide some paradigmatic examples in the cubic case.

2.5.1. Solving quadratic equations

We consider the generic quadratic equation

$$\alpha_2 x^2 + \alpha_1 x + \alpha_0 = 0, \tag{2.5.1}$$

with $\alpha_2, \alpha_1, \alpha_0 \in \mathbb{C}$ and $\alpha_2 \neq 0$. We are looking for its solutions in $\mathbb{C}$.

Dividing by α_2, we get the monic equation

$$x^2 + \frac{\alpha_1}{\alpha_2}x + \frac{\alpha_0}{\alpha_2} = 0.$$

We rewrite the above equation in the following way

$$x^2 + \frac{\alpha_1}{\alpha_2}x + \left(\frac{(\alpha_1)^2}{4(\alpha_2)^2} - \frac{(\alpha_1)^2}{4(\alpha_2)^2}\right) = -\frac{\alpha_0}{\alpha_2}.$$

Thus, using the formula for the square of a binomial, we find a square on the left side of the equation

$$\left(x + \frac{\alpha_1}{2\alpha_2}\right)^2 = -\frac{\alpha_0}{\alpha_2} + \frac{(\alpha_1)^2}{4(\alpha_2)^2}.$$

Taking the square root and developing calculations, we obtain

$$x_1 = \frac{-\alpha_1 + \sqrt{(\alpha_1)^2 - 4\alpha_2\alpha_0}}{2\alpha_2} \quad \text{and} \quad x_2 = \frac{-\alpha_1 - \sqrt{(\alpha_1)^2 - 4\alpha_2\alpha_0}}{2\alpha_2}. \tag{2.5.2}$$

The relation (2.5.2) is called the 'quadratic formula' and $\Delta_2 = (\alpha_1)^2 - 4\alpha_2\alpha_0$ is called the 'discriminant' of a quadratic equation.

2.5.2. Quadratic equations with real coefficients

If α_2, α_1, α_0 in equation (2.5.1) are real, we can then discuss the nature of its solutions starting from the sign of its discriminant Δ_2. In the case $\Delta_2 > 0$, it is easily seen that we have two distinct real solutions. Analogously, in the case $\Delta_2 = 0$, we have two coincident real solutions. Finally, in the case $\Delta_2 < 0$, we have two complex conjugate solutions.

The main idea behind this method is the so-called completion of the square. It is exposed above as an algebraic trick, but whenever the coefficients are real we can interpret it in a geometrical way, according to the following diagram.

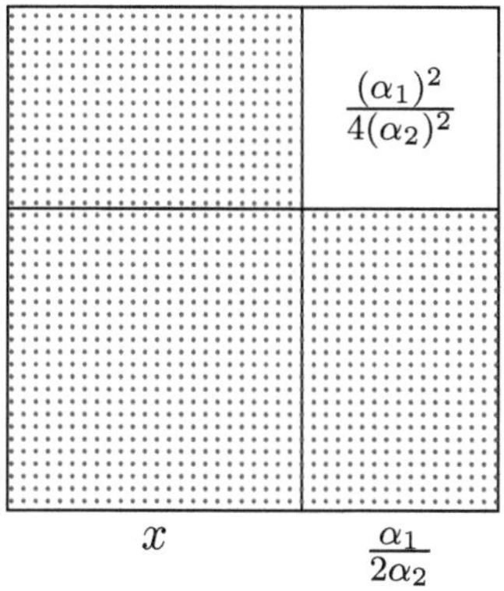

Figure 2.1. – Geometrical interpretation of the completion of the square for the equation $\alpha_2 x^2 + \alpha_1 x + \alpha_0 = 0$ with real coefficients.

The dotted part is given. If we add $\frac{(\alpha_1)^2}{4(\alpha_2)^2}$ to it, we obtain a square the side of which gives x. In this way, a support for intuition is provided.

2.5.3. Solving cubic equations

We consider the general cubic equation

$$\alpha_3 x^3 + \alpha_2 x^2 + \alpha_1 x + \alpha_0 = 0, \tag{2.5.3}$$

with $\alpha_3, \alpha_2, \alpha_1, \alpha_0 \in \mathbb{C}$ and $\alpha_3 \neq 0$. We are looking for its solutions in $\mathbb{C}$.

The substitution $x = y - \frac{\alpha_2}{3\alpha_3}$ yields to the so-called depressed cubic equation, which is monic and lacks in the second degree term. It is

$$y^3 + py + q = 0 \tag{2.5.4}$$

with

$$p = \frac{3\alpha_3\alpha_1 - (\alpha_2)^2}{3(\alpha_3)^2}, \tag{2.5.5}$$
$$q = \frac{2(\alpha_2)^3 - 9\alpha_3\alpha_2\alpha_1 + 27(\alpha_3)^3\alpha_0}{27(\alpha_3)^3}.$$

We introduce two variables u, v such that $y = u + v$. Substituting in equation (2.5.4), we get

$$u^3 + v^3 + (3uv + p)(u + v) + q = 0. \tag{2.5.6}$$

Let us notice that we can force on u, v (which have been arbitrarily chosen) another condition, for instance $3uv + p = 0$ suggested by the previous equation.[71] Therefore, we have $u^3v^3 = -\frac{p^3}{27}$ and equation (2.5.6)

$$u^3 + v^3 + q = 0.$$

becomes simpler. Thus, if we manage to solve the system coming from the two conditions on u, v

$$\begin{cases} u^3 + v^3 = -q \\ u^3v^3 = -\frac{p^3}{27} \end{cases}, \tag{2.5.7}$$

we can also solve equation (2.5.3). Let us notice that this system has two solutions in u^3, v^3 and eighteen solutions in u, v. We are interested in the sum $u + v$, which will give the three solutions of cubic equation (2.5.4), then of equation (2.5.3).

---

[71]In fact, the linear system given by

$$\begin{cases} y = u + v \\ 3uv + p = 0 \end{cases}, \quad \text{that is} \quad \begin{cases} y = u + v \\ uv = -\frac{p}{3} \end{cases},$$

always has solutions in $\mathbb{C}$.

We consider u^3 and v^3 as unknowns. Since we know their sum and their product, we associate[72] the system (2.5.7) with the equation

$$t^2 + qt - \frac{p^3}{27} = 0. \tag{2.5.8}$$

This is called the 'Lagrange resolvent' for the cubic equation (2.5.4). Now we use the quadratic formula to draw backwards in chain all the solutions of the other equations: we firstly find the two solutions t_1, t_2 of equation (2.5.8), from which we obtain the two couples of solutions

$$\begin{cases} u^3 = t_1 \\ v^3 = t_2 \end{cases} \quad \text{and} \quad \begin{cases} u^3 = t_2 \\ v^3 = t_1 \end{cases}$$

of system (2.5.7).

We consider the first couple. We want to draw u, v. The three cubic roots of $t_1, t_2 \in \mathbb{C}$ are respectively

$$\sqrt[3]{t_1},\ \omega\sqrt[3]{t_1},\ \omega^2\sqrt[3]{t_1} \quad \text{and} \quad \sqrt[3]{t_2},\ \omega\sqrt[3]{t_2},\ \omega^2\sqrt[3]{t_2},$$

with ω a primitive third root of unity[73]. Thus, there are nine rearrangements for $u + v$. At first, we arbitrarily choose u. Since the condition $uv = -\frac{p}{3}$ must hold, it follows that only three rearrangements out of nine fit, which are:

$$y_1 = \sqrt[3]{t_1}+\sqrt[3]{t_2}, \quad \text{or} \quad y_2 = \omega\sqrt[3]{t_1}+\omega^2\sqrt[3]{t_2}, \quad \text{or} \quad y_3 = \omega^2\sqrt[3]{t_1}+\omega\sqrt[3]{t_2}.$$

These are the three solutions of equation (2.5.4).

Considering then the second couple of solutions of system (2.5.7), where the values for u^3, v^3 are exchanged, we find the same values for

---

[72]Since

$$(\alpha - t)(\beta - t) = t^2 - (\alpha + \beta)t + \alpha\beta,$$

if we search two numbers of which we know their sum S and product P, it is enough to solve the quadratic equation $t^2 - St + P = 0$.

[73]That is, $\omega^3 = 1$ e $\omega^k \neq 1$ per $k = 1, 2$. In particular, if ω is a primitive third root of unity, then $\omega = -\frac{1}{2} + \imath\frac{\sqrt{3}}{2}$ e $\omega^2 = -\frac{1}{2} - \imath\frac{\sqrt{3}}{2}$.

the symmetric expression $u + v$.

Hence, the solutions of equation (2.5.3) are

$$x_i = y_i - \frac{\alpha_2}{3\alpha_3} \quad \text{for} \quad i = 1, 2, 3,$$

that is,

$$x_1 = \sqrt[3]{-\frac{q}{2} + \sqrt{\frac{q^2}{4} + \frac{p^3}{27}}} + \sqrt[3]{-\frac{q}{2} - \sqrt{\frac{q^2}{4} + \frac{p^3}{27}}} - \frac{\alpha_2}{3\alpha_3}, \tag{2.5.9}$$

$$x_2 = \omega \sqrt[3]{-\frac{q}{2} + \sqrt{\frac{q^2}{4} + \frac{p^3}{27}}} + \omega^2 \sqrt[3]{-\frac{q}{2} - \sqrt{\frac{q^2}{4} + \frac{p^3}{27}}} - \frac{\alpha_2}{3\alpha_3},$$

$$x_3 = \omega^2 \sqrt[3]{-\frac{q}{2} + \sqrt{\frac{q^2}{4} + \frac{p^3}{27}}} + \omega \sqrt[3]{-\frac{q}{2} - \sqrt{\frac{q^2}{4} + \frac{p^3}{27}}} - \frac{\alpha_2}{3\alpha_3},$$

explicitly written as function of the coefficients p, q in (2.5.5). This is called the 'cubic formula' for equation (2.5.3) and

$$\Delta_3 = \frac{q^2}{4} + \frac{p^3}{27}, \tag{2.5.10}$$

$$\text{where } p = \frac{3\alpha_3\alpha_1 - (\alpha_2)^2}{3(\alpha_3)^2}, q = \frac{2(\alpha_2)^3 - 9\alpha_3\alpha_2\alpha_1 + 27(\alpha_3)^3\alpha_0}{27(\alpha_3)^3},$$

is called[74] 'discriminant' of the cubic equation (2.5.3).

The main idea behind this method is to reveal the structure of a cubic equation introducing two new variables u, v linked by the linear condition $u + v = y$, and then to exploit the degree of freedom

[74]Note that, usually, the discriminant is defined from a higher viewpoint by $(x_1 - x_2)^2(x_2 - x_3)^2(x_3 - x_1)^2$, where x_1, x_2, x_3 are the roots of the cubic polynomial (see, for instance, [72, Chapter IV, Sections 6 and 8].). My definition differs from the usual one by the factor $-\frac{1}{108}$. More precisely, if Δ_3' is the usual discriminant, we have $\Delta_3 = -1/108\Delta_3'$. Nevertheless, I choose to use this definition, since in Cardano's text we explicitly find the value $\frac{q^2}{4} + \frac{p^3}{27}$ without the factor $-\frac{1}{108}$.

obtained in this way by imposing a further condition on the product uv. The method works because it takes advantage of symmetries in a fundamental way. In fact, the number of solutions of the system (2.5.7) decreases thanks to the symmetry in the expression $u + v$.

2.5.4. Cubic equations with real coefficients

We first remark that, if α_3, α_2, α_1, α_0 in equation (2.5.3) are real, then p, q in (2.5.5) are real too.

We can then discuss the nature of the solutions of equation (2.5.3) starting from the sign of its discriminant Δ_3. We know that each odd degree equation with real coefficients has at least one real solution, for instance as a simple consequence of the Intermediate values theorem.

It is convenient to neglect the term $\frac{\alpha_2}{3\alpha_3}$, which has no effect since it is real, and discuss the nature of the solutions of equation (2.5.4). In the case $\Delta_3 > 0$, we have one real solution and two complex conjugate solutions. In fact, we get that $\sqrt[3]{t_1}$, $\sqrt[3]{t_2}$ are real. Thus,

$$\begin{aligned} y_1 &= \sqrt[3]{t_1} + \sqrt[3]{t_2} \in \mathbb{R}, \\ y_2 &= \omega \sqrt[3]{t_1} + \omega^2 \sqrt[3]{t_2} = -\frac{1}{2}(\sqrt[3]{t_1} + \sqrt[3]{t_2}) + \imath\frac{\sqrt{3}}{2}(\sqrt[3]{t_1} - \sqrt[3]{t_2}) \in \mathbb{C}, \\ y_3 &= \omega^2 \sqrt[3]{t_1} + \omega \sqrt[3]{t_2} = -\frac{1}{2}(\sqrt[3]{t_1} + \sqrt[3]{t_2}) - \imath\frac{\sqrt{3}}{2}(\sqrt[3]{t_1} - \sqrt[3]{t_2}) \in \mathbb{C}, \end{aligned}$$

and y_2, y_3 are complex conjugate.

In the case $\Delta_3 = 0$, we have three real solutions, two of which coincide. In fact, we get $\sqrt[3]{t_1} = \sqrt[3]{-\frac{q}{2}} = \sqrt[3]{t_2}$. Thus,

$$\begin{aligned} y_1 &= 2\sqrt[3]{t_1} = -2\sqrt[3]{\frac{q}{2}} \in \mathbb{R}, \\ y_2 &= (\omega + \omega^2)\sqrt[3]{t_1} = -\sqrt[3]{-\frac{q}{2}} = \sqrt[3]{\frac{q}{2}} = (\omega^2 + \omega)\sqrt[3]{t_1} = y_3 \in \mathbb{R}. \end{aligned}$$

Finally, let $\Delta_3 < 0$. Then, we have three distinct real solutions. In fact, we get $\sqrt[3]{t_1}, \sqrt[3]{t_2} \in \mathbb{C}$. But, since the equality $\sqrt[3]{t_1}\sqrt[3]{t_2} = -\frac{p}{3} \in \mathbb{R}$ holds, $\sqrt[3]{t_1}, \sqrt[3]{t_2}$ are complex conjugate. Thus,

$$\begin{aligned}
y_1 &= \sqrt[3]{t_1} + \sqrt[3]{t_2} = \sqrt[3]{t_1} + \overline{\sqrt[3]{t_1}} \in \mathbb{R}, \\
y_2 &= \omega\sqrt[3]{t_1} + \omega^2\sqrt[3]{t_2} = \omega\sqrt[3]{t_1} + \overline{\omega\sqrt[3]{t_1}} \in \mathbb{R}, \\
y_3 &= \omega^2\sqrt[3]{t_1} + \omega\sqrt[3]{t_2} = \omega^2\sqrt[3]{t_1} + \overline{\omega^2\sqrt[3]{t_1}} \in \mathbb{R}.
\end{aligned}$$

We observe that $y_2+y_3 = (\omega+\omega^2)\sqrt[3]{t_1}+(\overline{\omega}+\overline{\omega^2})\overline{\sqrt[3]{t_1}} = -(\sqrt[3]{t_1}+\sqrt[3]{t_2}) = -y_1$.

More precisely (but only for equation (2.5.4)), when we write t_1, t_2 in trigonometric form, we have

$$\begin{aligned}
\sqrt[3]{t_1} &= \sqrt[3]{\rho}\sqrt[3]{\cos\theta + \imath\sin\theta} = \sqrt[3]{\rho}\left(\cos\frac{\theta}{3} + \imath\sin\frac{\theta}{3}\right), \\
\sqrt[3]{t_2} &= \overline{\sqrt[3]{t_1}} = \sqrt[3]{\rho}\left(\cos\frac{\theta}{3} - \imath\sin\frac{\theta}{3}\right),
\end{aligned}$$

with $\rho \in [0, 2\pi)$ and $\theta \in [0, +\infty)$, and then

$$\begin{aligned}
y_1 &= \sqrt[3]{t_1} + \overline{\sqrt[3]{t_1}} = 2\sqrt[3]{\rho}\cos\frac{\theta}{3}, \\
y_2 &= \omega\sqrt[3]{t_1} + \overline{\omega\sqrt[3]{t_1}} = 2\sqrt[3]{\rho}\cos\left(\frac{\theta}{3} + \frac{2}{3}\pi\right), \\
y_3 &= \omega^2\sqrt[3]{t_1} + \overline{\omega^2\sqrt[3]{t_1}} = 2\sqrt[3]{\rho}\cos\left(\frac{\theta}{3} + \frac{4}{3}\pi\right).
\end{aligned}$$

We remark that the three angles in the cosine function argument are at the distance $\frac{2}{3}\pi$ one from the other, so that one of them necessarily lies in the first or fourth quadrant where the cosine function is positive. Another of them necessarily lies in the second or third quadrant where the cosine function is negative. On the contrary, the remaining one varies in such a way that one cannot establish *a priori* the sign of its cosine. Hence, we obtain a positive and a negative solution, whereas the sign of the third solution is changeable.

This last case – when a cubic equation has three distinct real solutions, or $\Delta_3 < 0$ – is called the '*casus irreducibilis*'.[75] Note that $\Delta_3 < 0$ means that imaginary numbers appear in the cubic formula (see below, Section 2.5.6).

Let us notice that, in the cubic case with real coefficients, no geometric interpretation provides a good support for intuition as the completion of the square was in the quadratic case. In other words, the completion of the cube does not have an analogous heuristic value. Actually, it does not work at all. In fact, if we write equation (2.5.3) as

$$x^3 + \frac{\alpha_2}{\alpha_3}x^2 + \frac{\alpha_1}{\alpha_3}x = -\frac{\alpha_0}{\alpha_3}$$

and we complete the cube, we find the condition $\alpha_1 = \frac{(\alpha_2)^2}{3}$.

Suppose then $\alpha_1 \neq \frac{(\alpha_2)^2}{3}$. We can try to write equation (2.5.3) as

$$x^3 + \frac{\alpha_2}{\alpha_3}x^2 = -\frac{\alpha_1}{\alpha_3}x - \frac{\alpha_0}{\alpha_3},$$

and add and subtract the quantity $\frac{\alpha_2^2}{\alpha_3^2}\frac{x}{3} + \frac{\alpha_2^3}{27\alpha_3^3}$ in order to obtain the cube of a binomial

$$\left(x + \frac{\alpha_2}{3}\right)^3 = -\frac{\alpha_1}{\alpha_3}x - \frac{\alpha_0}{\alpha_3} + \frac{\alpha_2^2}{\alpha_3^2}\frac{x}{3} + \frac{\alpha_2^3}{27\alpha_3^3}.$$

But then we can in no way write the right side of the equation as a cube of a polynomial one term of which is x, since x^3 does not appear on the right side. Then, one needs to follow another way and introduce the new variables u, v.

[75]This is a late terminology that we do not find in Cardano's works. Up to now, the earliest reference to the *casus irreducibilis* that I have found is in the 1781 contest by the Accademia in Padova, where they asked to prove that Cardano's formula necessarily entails imaginary numbers when the equation falls into the *casus irreducibilis*.

2.5.5. Cubic equations with real coefficients solved in a trigonometrical way

In the preceding section, we have remarked that, when a cubic equation has three distinct real solutions, imaginary numbers appear in the cubic formula, since $\Delta_3 < 0$ and the discriminant is under square roots. Nevertheless, there is a way to obtain a real expression for the solutions. One can achieve this by exploiting the triple-angle formula

$$\cos 3\theta = 4\cos^3\theta - 3\cos\theta.$$

Consider the depressed equation 2.5.4 and set $y = u\cos\theta$. Then, $u^3\cos^3\theta + pu\cos\theta + q = 0$. We want to choose u in order to compare the equation with the triple-angle formula. We multiply both sides of the equation by $\frac{4}{u^3}$ and we have $4\cos^3\theta + \frac{4p}{u^2}\cos\theta + \frac{4q}{u^3} = 0$. There, we are allowed to choose $u = \sqrt{-\frac{4}{3}p}$, so that we get $4\cos^3\theta - 3\cos\theta + \frac{3}{2}\frac{q}{p}\sqrt{-\frac{3}{p}} = 0$. Note that u is well-defined, since $\Delta_3 < 0$ implies that $p < 0$. Then, by comparison, we have

$$-\cos 3\theta = \frac{3}{2}\frac{q}{p}\sqrt{-\frac{3}{p}}$$

and, since $y = u\cos\theta$, we finally have that

$$y_1 = \sqrt{-\frac{4}{3}p}\cos\left(\frac{1}{3}\arccos\left(\frac{3}{2}\frac{q}{p}\sqrt{-\frac{3}{p}}\right)\right),$$

$$y_2 = \sqrt{-\frac{4}{3}p}\cos\left(\frac{1}{3}\arccos\left(\frac{3}{2}\frac{q}{p}\sqrt{-\frac{3}{p}}\right) + \frac{2}{3}\pi\right),$$

$$y_3 = \sqrt{-\frac{4}{3}p}\cos\left(\frac{1}{3}\arccos\left(\frac{3}{2}\frac{q}{p}\sqrt{-\frac{3}{p}}\right) + \frac{4}{3}\pi\right).$$

The arccosine is defined only between -1 and 1, which gives the condition on the discriminant $27q^2 + 4p^3 \leq 0$. We remark that, as a matter of fact, the above formulae do not contain imaginary numbers.

Anyway, these formulae are not as interesting as the algebraic ones in Section 2.5.3, since they do not give as much information. Indeed, it is still out of reach to exactly calculate the cosine and arccosine functions. The above formulae look very easy, but in fact they hide the complex structure of the equation in an artificially introduced function (and its almost inverse function) that has a particular relationship with a certain cubic polynomial (the triple-angle formula). Moreover, when $\Delta_3 > 0$, one is compelled to use hyperbolic functions.

2.5.6. What Galois theory can say about cubic equations

There[76] is a deeper reason that explains why imaginary numbers necessarily appear in the cubic formula, and it surfaces using Galois theory. Galois' original definition of the group of an equation easily leads to impracticable calculations when the degree of the equation is greater than four. Nevertheless, his approach has the advantage of giving a very simple proof of the impossibility of overcoming the use of imaginary numbers to solve an irreducible cubic equation with real coefficients that falls into the *casus irreducibilis*.

Firstly, I will quickly sum up how to calculate the Galois group in the general case of an nth degree equation. Let $f(x) = 0$ be the equation and denote its (simple) solutions by $x_1, x_2, \ldots, x_n$. Then, Galois states[77] – as obvious – that it is possible to find some integers $A_1, A_2, \ldots, A_n$ such that the function of the solutions (the so-called Galois resolvent)

$$V(x_{i_1}, x_{i_2}, \ldots, x_{i_n}) = A_1 x_{i_1} + A_2 x_{i_2} + \ldots + A_n x_{i_n}$$

takes $n!$ different values on the set of the permutations of the solutions.

Once a particular value of this function, for instance

$$V_1 = A_1 x_1 + A_2 x_2 + \ldots + A_n x_n$$

[76] I thank Massimo Galuzzi for this section.

[77] See [35, page 110]. I have slightly modified Galois' original notation.

obtained by the identical permutation, is given, we have, in modern terms, a primitive element. All the solutions of $f(x) = 0$ may be expressed as rational functions of V_1, with the coefficients that are in the same field as $f(x)$.

Due to the fundamental theorem of symmetric polynomials, the coefficients of the polynomial

$$\varphi(X) = \prod_{i_1, i_2, \ldots, i_n} (X - (A_1 x_{i_1} + A_2 x_{i_2} + \ldots + A_n x_{i_n}))$$

are in the same field as $f(x)$. Let $\Phi(X)$ be the irreducible factor of $\varphi(X)$ that contains the root[78] V_1. The Galois group is given by the permutations that correspond to the roots of $\Phi(X)$.

What does it mean, in this approach, to solve an equation? It amounts to "reduce its group successively to the point where it does not contain more than a single permutation".[79] This reduction is, of course, accomplished by adjoining quantities that can be supposed as known.

Let us now detail this approach in the particular case of cubic equations. We consider the irreducible equation

$$x^3 + px + q = 0, \tag{2.5.11}$$

with p, q real, and let x_1, x_2, x_3 be its solutions. In this case, a Galois resolvent may be given by[80]

$$V(x_i, x_j, x_k) = x_i - x_k.$$

[78] Of course, in the general case, we have $\Phi(X) = \varphi(X)$. The polynomial $\Phi(x)$ is the minimal polynomial of V_1.

[79] See [35, page 121].

[80] There are two kinds of tests to make. First, if $x_1 - x_2 = x_2 - x_1$, then $x_1 = x_2$. Second, if $x_1 - x_2 = x_2 - x_3$, then $2x_2 = x_1 + x_3$, that is $3x_2 = x_1 + x - 2 + x_3 = 0$. The other cases are similar.

By means of this resolvent, the polynomial $\varphi(X)$ becomes

$$\varphi(X) = \prod_{x_i \neq x_j} (X - (x_i - x_j)) = X^6 + 6pX^4 + 9p^2X^2 + 4p^3 + 27q^2,$$

or

$$\varphi(X) = (X^3 + 3pX)^2 + 4p^3 + 27q^2.$$

In this calculation, we remark that the quantity $4p^3 + 27q^2$ has a real structural value and is not a mere device to make the demonstration work.

The result, namely that imaginary numbers are indispensable tools to solve a cubic equation when it falls into the *casus irreducibilis*, has been proved, during the 19th century, by Paolo Ruffini in 1813 (see [36] for an incomplete proof, and [37]), Pierre Laurent Wantzel in 1843 (see [42]), Vincenzo Mollame in 1890 (see [33] and [34]), Otto Hölder in 1891 (see [30]), Adolf Kneser in 1892 (see [31]), and Leopold Gegenbauer in 1893 (see [27]). I plan to develop the study of this topic in a forthcoming work.

2.5.7. Paradigmatic examples for cubic equations

We have already touched on the fact that Cardano is interested in the algebraic shapes for the solutions of a cubic equation. In particular, let us have a look at the family of equations $x^3 = a_1x + a_0$, which is fundamental in Cardano's arguments. We will try to deduce some information on the shapes of its solutions.

Since for most of the time Cardano uses rational coefficients and since in his treatises there is the well-identifiable topic of equations with rational coefficients, let us assume that a_1, a_0 are rational. In this case, consider a, b positive, rational. Cardano states that the (positive) irrational solutions go either under the shape $\sqrt{a} + b$, with a such that $\sqrt{a}$ is irrational, or under the shape $\sqrt[3]{a} + \sqrt[3]{b}$, with a, b such that $\sqrt[3]{a}$, $\sqrt[3]{b}$ are irrational and $\sqrt[3]{ab}$ is rational. Indeed, it turns out that these two shapes can never be solution of the same cubic equation, since, if a rational equation has a solution of the first shape, then it has $\Delta_3 < 0$ and, if it has a solution of the second shape, then it has $\Delta_3 > 0$.

In fact, we firstly consider a rational, irreducible polynomial of degree three and we suppose that it exists a, b positive, rational such that $x_1 = \sqrt{a} + b$ is irrational and is one of its roots. Then, by Galois theory, we have that the polynomial is irreducible over $\mathbb{Q}$, that is, if x_2 is another root of the polynomial, then it is rational. Moreover, $x_3 = -\sqrt{a} + b$. Since we get three real roots, the discriminant is negative.

We consider then a rational, irreducible polynomial of degree three and we suppose that it exists a, b positive, rational such that $x_1 = \sqrt[3]{a} + \sqrt[3]{b}$ is irrational and is one of its roots. Then, again by Galois theory, we have that, if x_2 is another root of the polynomial, then $x_2 = \omega^r \sqrt[3]{a} + \omega^s \sqrt[3]{b}$, with ω a primitive third root of unity and $r, s \in \mathbb{Z}/3\mathbb{Z}$ such that $r, s \not\equiv 0 \bmod 3$. If we moreover assume that $\sqrt[3]{ab}$ is rational, then $r + s \equiv 0 \bmod 3$. We can assume that $r = 1$ and $s = 2$ and $x_3 = \omega \sqrt[3]{a} + \omega^2 \sqrt[3]{b}$ (up to interchanging x_2 and x_3). It is easily seen that x_2, x_3 cannot be real, so that the discriminant of the equation is in the end positive.

For instance, consider the equation $x^3 = 6x + 4$, which is in *Aliza*, Chapters XXIV and LIX. It has $\Delta_3 < 0$ and its real solutions are $x_1 = 1 + \sqrt{3}$, $x_2 = -2$, and $x_3 = 1 - \sqrt{3}$. Instead, the equation $x^3 = 6x + 6$ (which is in *Ars Magna Arithmeticæ*, Chapters XXIII, XXVII, XXXII, in *Ars Magna*, Chapter XII, and in *Aliza*, Chapters XVIII, XIX, XXV, LVIII) has $\Delta_3 > 0$ and its real solution is $\sqrt[3]{4} + \sqrt[3]{2}$.

Moreover, there is the case in which an equation of the family $x^3 = a_1 x + a_0$ has $\Delta_3 > 0$ and the real solution is rational. For instance, 4 is the real solution of $x^3 = 6x + 40$, which is in *Ars Magna*, Chapter XII and *Aliza*, Chapters XLIX, LIII, LIX.

A consequence of the method in Section 2.5.3 is that solving cubic equations is often brought back to cubic root calculations, which is not always easy to carry out. For instance, when we want to draw a solution of $x^3 + 6x = 40$ using the cubic formula, $\sqrt[3]{20 + \sqrt{392}} + \sqrt[3]{20 - \sqrt{392}}$ is returned. In truth, this solution is 4, since the terms under the cubic roots are two cubes, namely $20+\sqrt{392}=(2+\sqrt{2})^3$ and $20-\sqrt{392}=(2-\sqrt{2})^3$, but this is not evident at all.[81] Indeed, in *Ars Magna*, Chapter XII Cardano does not know that the terms under the cubic roots are two cubes. In *Aliza*, Chapters XLIX and LIII he knows that a solution is 4, but there is no mention of the cubic formula. Instead, in *Aliza*, Chapter LIX he knows it, and he explicitly states it in the third corollary.

Much more interesting – in the context of Cardano's inquiries – is the case in which we use the cubic formula to try to solve $x^3=6x+4$. In fact, the formula returns $\sqrt[3]{2+\sqrt{-4}}+\sqrt[3]{2-\sqrt{-4}}$, which in truth is real and is $1+\sqrt{3}$, but Cardano can in no way remark the equality.

2.5.8. Solving quartic equations

We consider the general quartic equation

$$\alpha_4x^4+\alpha_3x^3+\alpha_2x^2+\alpha_1x+\alpha_0=0, \tag{2.5.12}$$

with $\alpha_4,\alpha_3,\alpha_2,\alpha_1,\alpha_0\in\mathbb{C}$ and $\alpha_4\neq 0$. We are looking for its solutions in $\mathbb{C}$.

The substitution $x=y-\frac{\alpha_3}{4\alpha_4}$ yields to the so-called depressed quartic equation, which is monic and lacks in the third degree term

$$y^4+py^2+qy+r=0, \tag{2.5.13}$$

[81] In short, if one wants to know if $\sqrt{a}+b$, with a,b rational, is a cube in $\mathbb{Q}(\sqrt{a})$, one has to search for two rational x,y such that $\sqrt[3]{\sqrt{a}+b}=\sqrt{x}+y$ holds. For more details about the calculation, see for instance [57, Capo V.4, p. 291] or [76, pages 242-244].

with

$$p = \frac{-3(\alpha_3)^2 + 8\alpha_4\alpha_2}{8(\alpha_4)^2}, \tag{2.5.14}$$

$$q = \frac{(\alpha_3)^3 - 4\alpha_4\alpha_3\alpha_2 + 8(\alpha_4)^2\alpha_1}{8(\alpha_4)^3},$$

$$r = \frac{-3(\alpha_3)^4 + 16\alpha_4(\alpha_3)^2\alpha_2 - 64(\alpha_4)^2\alpha_3\alpha_1 + 256(\alpha_4)^3\alpha_0}{256(\alpha_4)^4}.$$

We rewrite equation (2.5.13) in the following way

$$y^4 + py^2 = -qy - r,$$

in order to obtain two perfect squares on the left and right side of the equation.

First, we exploit the formula to expand the square of a binomial. Adding the identity $(y^2+p)^2 - y^4 - 2py^2 = p^2$ to equation (2.5.13) yields to

$$(y^2+p)^2 = py^2 - qy + p^2 - r. \tag{2.5.15}$$

Then, in short, we introduce a variable u such that the perfect square on the left side of equation (2.5.15) is preserved (but it will be the square of a trinomial rather than of a binomial). We will force then a further condition on u in order to have a perfect square on the right side. We will finally exploit the formula to expand the square of a trinomial. Following this plan, we add the identity $(y^2+p+u)^2 - (y^2+p)^2 = 2uy^2 + 2pu + u^2$ to equation (2.5.15). This yields to

$$(y^2+p+u)^2 = (p+2u)y^2 - qy + (p^2 + 2pu + u^2 - r). \tag{2.5.16}$$

We choose a value for u such that the right side of equation (2.5.16) becomes a perfect square. For that, we consider the second degree polynomial in y on the right side of equation (2.5.16), and we force its discriminant

$$\Delta_2 = q^2 - 4(p + 2u)(p^2 + 2pu + u^2 - r)$$

to be zero.

In this way, we want to solve the nested cubic equation in u

$$u^3 + \frac{5}{2}pu^2 + (2p^2 - r)u + (\frac{1}{2}p^3 - \frac{1}{2}pr - \frac{1}{8}q^2) = 0 \qquad (2.5.17)$$

in order to solve equation (2.5.16). We apply to it the method in Section 2.5.3 and we get the depressed nested cubic equation

$$v^3 + Pv + Q = 0, \qquad (2.5.18)$$

with

$$P = -\frac{p^2}{12} - r,$$
$$Q = -\frac{p^3}{108} + \frac{pr}{3} - \frac{q^2}{8},$$

that is $P, Q \in \mathbb{C}$ depend on the coefficients $\alpha_4, \alpha_3, \alpha_2, \alpha_1, \alpha_0$ of equation (2.5.12). The solutions of equation (2.5.17) are

$$u_1 = \sqrt[3]{-\frac{Q}{2} + \sqrt{\frac{Q^2}{4} + \frac{P^3}{27}}} + \sqrt[3]{-\frac{Q}{2} - \sqrt{\frac{Q^2}{4} + \frac{P^3}{27}}} - \frac{5}{6}p,$$
$$u_2 = \omega\sqrt[3]{-\frac{Q}{2} + \sqrt{\frac{Q^2}{4} + \frac{P^3}{27}}} + \omega^2\sqrt[3]{-\frac{Q}{2} - \sqrt{\frac{Q^2}{4} + \frac{P^3}{27}}} - \frac{5}{6}p,$$
$$u_3 = \omega^2\sqrt[3]{-\frac{Q}{2} + \sqrt{\frac{Q^2}{4} + \frac{P^3}{27}}} + \omega\sqrt[3]{-\frac{Q}{2} - \sqrt{\frac{Q^2}{4} + \frac{P^3}{27}}} - \frac{5}{6}p.$$

Each one of the above solutions gives a polynomial with a double root on the right side of equation (2.5.16), allowing us to factorise it in the following way.

Using such a u, we have that

$$(p+2u)y^2 - qy + (p^2 + 2pu + u^2 - r)$$

is a perfect square of the form

$$(\alpha y + \beta)^2 = \alpha^2 y^2 + 2\alpha\beta y + \beta^2.$$

By comparison, we get $\alpha = \sqrt{p+2u}$ and $\beta = \frac{-q}{2\sqrt{p+2u}}$. Equation (2.5.16) can be then rewritten as

$$(y^2 + p + u)^2 = \left((\sqrt{p+2u})y + \frac{-q}{2\sqrt{p+2u}}\right)^2.$$

Taking the square root and collecting like powers of y, we obtain

$$y^2 + (\mp\sqrt{p+2u})y + \left(p + u \pm \frac{-q}{2\sqrt{p+2u}}\right) = 0. \qquad (2.5.19)$$

These are two quadratic equations. We apply the method in Section 2.5.1 to get the solutions

$$y_1, y_2 = \frac{\sqrt{p+2u} \pm \sqrt{(p+2u) - 4\left(p + u + \frac{-q}{2\sqrt{p+2u}}\right)}}{2},$$

$$y_1', y_2' = \frac{-\sqrt{p+2u} \pm \sqrt{(p+2u) - 4\left(p + u - \frac{-q}{2\sqrt{p+2u}}\right)}}{2}.$$

The solutions of equation (2.5.12) are then

$$x_i = y_i - \frac{\alpha_3}{4\alpha_4} \quad \text{with} \quad i = 1, 2, 3, 4,$$

that is

$$x_1 = \frac{\sqrt{p+2u} + \sqrt{(p+2u) - 4\left(p+u+\frac{-q}{2\sqrt{p+2u}}\right)}}{2} - \frac{\alpha_3}{4\alpha_4},$$

$$x_2 = \frac{\sqrt{p+2u} - \sqrt{(p+2u) - 4\left(p+u+\frac{-q}{2\sqrt{p+2u}}\right)}}{2} - \frac{\alpha_3}{4\alpha_4}, \tag{2.5.20}$$

$$x_3 = \frac{-\sqrt{p+2u} + \sqrt{(p+2u) - 4\left(p+u-\frac{-q}{2\sqrt{p+2u}}\right)}}{2} - \frac{\alpha_3}{4\alpha_4} = x_1',$$

$$x_4 = \frac{-\sqrt{p+2u} - \sqrt{(p+2u) - 4\left(p+u-\frac{-q}{2\sqrt{p+2u}}\right)}}{2} - \frac{\alpha_3}{4\alpha_4} = x_2',$$

explicitly written as function of the coefficients p, q, r in (2.5.14).

2.5.9. Quartic equations with real coefficients

We firstly remark that, if α_4, α_3, α_2, α_1, α_0 in equation (2.5.12) are real, then p, q, r in (2.5.14) are real too, that is the nested cubic equation (2.5.17) has real coefficients.

We know that equation (2.5.17) has at least one real solution. We take u to be such a real solution.

We can assume that $u > -\frac{p}{2}$, so that $p + 2u > 0$ and $\sqrt{p+2u} \in \mathbb{R}$. We evaluate the polynomial $R(v) = v^3 + Pv + Q$ on the right side of equation (2.5.18) at $v = \frac{p}{3}$ and we get

$$R\left(\frac{p}{3}\right) = -\frac{q^2}{8} < 0.$$

But

$$\lim_{v \to +\infty} R(v) = +\infty.$$

This means that equation (2.5.18) has a real solution $v > \frac{p}{3}$, so that equation (2.5.17) has a real solution $u > v - \frac{5}{6}p = -\frac{p}{2}$.

Using such a u in equation (2.5.19), we get $\sqrt{p+2u} \in \mathbb{R}$ and the two quadratic equations have real coefficients.

Therefore, if α_4, α_3, α_2, α_1, α_0 in equation (2.5.12) are real, we can choose u such that equation (2.5.19) has real coefficients.

Afterwards, let us consider equation (2.5.12). For the sake of ease, we take $\alpha_4 = 1$, since we can always easily reduce any equation to its monic form

$$x^4 + \alpha_3 x^3 + \alpha_2 x^2 + \alpha_1 x + \alpha_0 = 0. \tag{2.5.21}$$

The coefficients of the depressed quartic equation (2.5.13) become then

$$\begin{aligned}
p &= \frac{-3(\alpha_3)^2 + 8\alpha_2}{8}, \\
q &= \frac{(\alpha_3)^3 - 4\alpha_3\alpha_2 + 8\alpha_1}{8}, \\
r &= \frac{-3(\alpha_3)^4 + 16(\alpha_3)^2\alpha_2 - 64\alpha_3\alpha_1 + 256\alpha_0}{256}.
\end{aligned}$$

Developing calculations in (2.5.20), we get four huge formulae that deliver the four solutions x_1, x_2, x_3, x_4 depending only on the coefficients of (2.5.12). There, a special role is played by

$$\begin{aligned}
\Delta_4 = -4\Big(\alpha_2^2 - 3\alpha_3\alpha_1 + 12\alpha_0\Big)^3 + \\
\Big(2\alpha_2^3 - 9\alpha_3\alpha_2\alpha_1 + 27\alpha_1^2 + 27\alpha_3^2\alpha_0 - 72\alpha_2\alpha_0\Big)^2 \\
= -4B^3 + A^2,
\end{aligned}$$

which is called the 'discriminant' of a quartic equation.

As in the case of quadratic and cubic equations, we would like to discuss the nature of the solutions of equation (2.5.12) starting from the sign of its discriminant Δ_4. Let us consider the quartic monic polynomial on the left side of equation (2.5.21)

$$x^4 + \alpha_3 x^3 + \alpha_2 x^2 + \alpha_1 x + \alpha_0. \tag{2.5.22}$$

The discriminant Δ_4 of a quartic polynomial is defined as the square of the product of the differences of the roots.[82] Let x_i with $i = 1, 2, 3, 4$ be the roots. Then

$$\Delta_4 = (x_1 - x_2)^2 (x_1 - x_3)^2 (x_1 - x_4)^2 (x_2 - x_3)^2 (x_2 - x_4)^2 (x_3 - x_4)^2.$$

The discriminant is well defined, in the sense that it does not depend on the order of the roots. Indeed, rearranging the x_i boils down to a permutation of the factors of the product, since the differences are raised to the square.

Using this formula, it is easily seen that if (at least) two of the roots coincide, then $\Delta_4 = 0$.

Let us suppose now that all the roots are distinct. Since the polynomial factors as the product of two quadratic polynomials, there are three possibilities for its roots. We can have four real roots, or two real roots and two complex conjugate roots, or four complex roots conjugate in pairs.

If the four roots are real, it is easily seen that $\Delta_4 > 0$, since it is the product of squares of real numbers.

Let us suppose that two roots are distinct and real, say $x_1, x_2 \in \mathbb{R}$ with $x_1 \neq x_2$, and two roots are complex conjugate, say $x_3, x_4 \in \mathbb{C}$ with $x_4 = \overline{x_3}$. We write $x_3 = a + \imath b$ and $x_4 = a - \imath b$. Then, the factor $(x_1 - x_2)^2 > 0$, since it is the square of a real number. We remark that four differences out of six occur as pairs of complex conjugate numbers, so that their products are real, and thus positive. In fact, considering for example $(x_1 - x_3)^2 (x_1 - x_4)^2$, we get $((x_1 - a - \imath b)(x_1 - a + \imath b))^2$. The remaining factor is $x_3 - x_4 = a + \imath b - (a - \imath b) = 2\imath b$, the square of which is negative. In this way, we get $\Delta_4 < 0$.

Finally, let us suppose that there are two pairs of complex conjugate roots, say x_1, x_2, x_3, $x_4 \in \mathbb{C}$ with $x_2 = \overline{x_1}$, and $x_4 = \overline{x_3}$. We write

[82] See [72, Chapter IV, Sections 6 and 8].

$x_1 = \alpha_1 + \imath b_1$, $x_2 = \alpha_1 - \imath b_1$, $x_3 = \alpha_3 + \imath b_3$, and $x_4 = \alpha_3 - \imath b_3$. As before, four of the six differences occur as pairs of complex conjugate numbers, so that their products are real, and then positive. The two remaining factors are $x_1 - x_2 = \alpha_1 + \imath b_1 - (\alpha_1 - \imath b_1) = 2\imath b_1$ and $x_3 - x_4 = \alpha_3 + \imath b_3 - (\alpha_3 - \imath b_3) = 2\imath b_3$, the square of which are negative. In this way, we get $\Delta_4 > 0$.

In particular, we see that $\Delta_4 < 0$ if and only if the polynomial (2.5.22) has exactly two real roots and two complex conjugate roots, and $\Delta_4 > 0$ if and only if the polynomial (2.5.22) has four real roots or four complex roots that are not real. This result can also be obtained by brute force calculations using the above formula for the roots x_1, x_2, x_3, x_4.

Therefore, we observe that in the quartic case the sign of the discriminant does not provide complete information on the nature of the roots. In particular, if $\Delta_4 \geq 0$, there is some ambiguity.

It is worth noting that if we have a polynomial of degree five or higher it is impossible to have an algebraic formula for the roots. This is a deep result of Galois theory.[83]

[83] See [68, Theorem 5.7.3].

3. Cardano solves equations in the *Ars Magna*

In this section, I give an insight of the *Ars Magna*. It is well-known that this is Cardano's most outstanding work, since it contains the most advanced results that he reached concerning equations – which were also the most advanced at that time.

There are[1] three editions of the *Hieronymi Cardani, præstantissimi mathematici, philosophi, ac medici, Artis magnae sive de regulis algebraicis, lib. unus, qui et totius operis de arithmetica, quod opus perfectum inscripsit, est in ordine decimus*, also shortly called *Ars Magna*. The first dates back to 1545 and was printed in Nuremberg by Iohannes Petreium. The second is a joint publication of the *Ars Magna* together with the *De Proportionibus* and the *De Regula Aliza* in 1570 in Basel by Oficina Henricpetrina. This edition was made during Cardano's lifespan and contains some additions. The third is posthumous and can be found in the fourth volume of Cardano's *Opera Omnia*, printed in 1663 in Lyon by Ioannis Antonii Huguetan and Marci Antonii Ravaud. For an extensive explanation of the reference to the *Opus Perfectum* in the *Ars Magna*'s title, see above Section 2.3, page 18. Moreover, there is the German translation *Roth, Petrus, Arithmetica philosophica, oder schöne künstliche Rechnung der Coss oder Algebrae. Im 1. Theil werden Herrn D. H. Cardani dreyzehn Reguln esetzt. Dessgleichen noch drey andere Reguln, zu den ersten drey cubiccosischen aequationibus gehörig*, made in Nuremberg in 1608 by Peter Roth. We also have available the Manuscript Plimpton 510/1700: s.a., which is entitled *L'algebra*, at the Columbia University Library in New York. Finally, we recall that we have available, with

[1]See [44].

assets and defects, the English compared translation [22] of the 1545, 1570 and 1663 editions of the *Ars Magna* made by Richard Witmer in 1968 and the French translation [25] of the 1545 *Ars Magna* made by Jean Peyroux in 2006.[2]

With relation to the topics ordering, the *Ars Magna* is a tricky writing. Unlike the algebra treatises of that time, Cardano did not start with the usual considerations on the nature of numbers, but rather with some miscellaneous considerations on the structure of equations and on relations between them (Chapters I-X). This is a first noteworthy clue to the novelty of this book. Afterwards, we find the block formed by Chapters XI-XXIII, where Cardano dealt uniquely with cubic equations. Finally, in Chapters XXIV-XL, some usual algebraic rules, such as the golden rule or the rule of equal position together with the cubic formulae, are applied to solve problems that in the majority involve equations.

In the following, my main concern is to analyse the methods, rules, techniques, and procedures at Cardano's disposal to deal with equations. I firstly present in Section 3.1 a selection of topics from the chapters in the first and last block of the *Ars Magna*. Note that they do not specially concern cubic equations. In Section 3.1.1, we follow Cardano while solving linear, quadratic, and other simple kinds of equations. One the one hand, this recalls the usual way to approach these subjects at that time. On the other hand, it bears witness to Cardano's early encounter with the so-called "sophistic" quantities or square roots of negative quantities. Then, Cardano shows some techniques to transform equations (see Section 3.1.2). Thanks to them, one can establish some relationships between two equations by the means of their coefficients and solutions. In this context, it sometimes happens that Cardano also considers negative solutions. These transformations of equations will prove to be one of the preferred and most useful methods to enquire on equations. All along the *Ars Magna*, we find the contrast between "particular" and "general", either concerning the solutions themselves or the methods employed.

[2]In this section, we mainly make reference to Witmer's work.

I report on this in Section 3.1.4. Cardano's efforts are obviously "general"-oriented. Anyway, we find some "particular" techniques to put a patch where the (expected) "general" method turns out to fail. The "general" method provided by Cardano to translate the problem of solving the cross-party class of equations $x^q + a_0 = a_p x^p$, where $q > p$, into the problem of solving a certain system is particularly noteworthy (see Section 3.1.3). Cardano also states the formula for the cube of a binomial (see Section 3.1.5). Proving this last formula, Cardano makes some important remarks about geometry. While analysing these passages – as for all the following, I particularly take care of the way in which Cardano proves his statements. In the *Ars Magna*, the most widespread pattern is to settle first of all a geometrical environment, where one can interpret the given equation or the equation which comes from the given problem. In this environment one can also draw some inferences exploiting the known results on geometrical quantities. In Sections 3.2, 3.3, and 3.4, we analyse the solving formulae for all the families of cubic equations. The proofs of these formulae follow the usual pattern. Moreover, in eleven cases out of thirteen, they rely on drawing back through transformations to an already solved equation. In the course of these proofs, the arguments employed become more and more repetitive until the point where they lose their link with the possible geometrical arguments that underlie their justification. Cardano soon runs into the major inconvenience of these formulae, the so-called *casus irreducibilis*, which prevents them from being as general as Cardano hoped. I show in Section 3.5, that the solving methods for the cubic equations are eventually all linked one to the other, even if sometimes in a twisted way. Finally, in Section 3.6, I give an overview of Cardano and Ferrari's method to solve quartic equations.

3.1. Miscellaneous considerations regarding equations

3.1.1. Solving linear, quadratic, and other simple equations. Coming across the square roots of negative numbers

Starting from the very beginning, I spend a few words on the solution of linear, quadratic, and other simple kinds of equations. In the *Ars Magna* aiming to be a compendium on equations, we naturally also find a basic explanation of these topics.

In **Chapter I** "On double solutions in certain types of cases [*De duabus æquationibus in singulis capitulis*]", after some introductory considerations, Cardano starts recalling briefly how to solve equations of type $x^n = a_0$ by taking the (real) n^{th}-roots, with n natural. We pay particular attention to the fact that here Cardano accepts also negative solutions: he states that the values of x in $x^2 = 9$ are 3 and -3 and that the values of x in $x^4 = 81$ are 3 and -3.

Then, Cardano speaks about the so-called "derivative cases [*derivativa capitula*]", where one can use substitutions such as $y = x^n$, with n natural, to go back to equations that one already knows how to solve. To quote an example, we mention $x^4 + 3x^2 = 28$.

In **Chapter V** "On finding the solution for equations composed of minors [*Ostendit æstimationem capitulorum compositorum minorum, quæ sunt quadratorum, numeri, et rerum*]", Cardano solves quadratic equations by radicals.

AM V.4-6. *Let us consider the three families of quadratic equations:*

- *if* $x^2 + a_1x = a_0$*, then* $x = \sqrt{a_0 + \left(\frac{a_1}{2}\right)^2} - \frac{a_1}{2}$*;*
- *if* $x^2 = a_1x + a_0$*, then* $x = \sqrt{a_0 + \left(\frac{a_1}{2}\right)^2} + \frac{a_1}{2}$*;*
- *if* $x^2 + a_0 = a_1x$ *and* $\left(\frac{a_1}{2}\right)^2 - a_0 > 0$*, then* $x = \frac{a_1}{2} \pm \sqrt{\left(\frac{a_1}{2}\right)^2 - a_0}$*.*

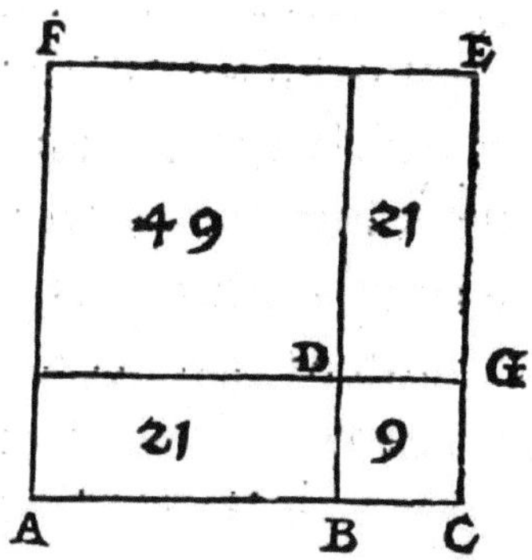

Figure 3.1. – Diagram from *Ars Magna* V.1. Kind courtesy of Biblioteca Virtual del Patrimonio Bibliográfico and Universidad de Sevilla.

For each rule above, Cardano gives a proof. We consider only the first one, since the others are similar.

AM V.1 — *Proof.* Cardano[3] shows that, given the equation $x^2 + 6x = 91$, a solution is $x = \sqrt{91 + \left(\frac{6}{2}\right)^2} - \frac{6}{2} = 7$. Considering the diagram in Figure 3.1, Cardano takes $\overline{AB} = x$ and $\overline{BC} = \frac{6}{2} = 3$. Then, he considers $\overline{AB} + \overline{BC}$ and draws the square on it, getting in this way four objects, two of which are equal by *Elements* I.43. They are $\overline{AB}^2$, $\overline{BC}^2$, and $\overline{AB}\,\overline{BD} = \overline{DG}\,\overline{GE}$.

By the given equation, Cardano gets then the equality $\overline{AB}^2 + 2(\overline{AB}\,\overline{BD}) = 91$. Completing the square, he obtains

$$\overline{AB}^2 + 2(\overline{AB}\,\overline{BD}) + \overline{BC}^2 = 91 + 9, \tag{$\star$}$$

that is $\overline{AC}^2 = 100$, then $\overline{AC} = 10$ [that is, $x = 7$]. □

We observe that Cardano presents the very standard technique of the completion of the square to solve quadratic equations. He provides a

[3]See [4, Chapter V, Paragraph 1, pages 9v-10r].

geometrical environment where he interprets the given equation, that is, where he identifies the quantities x and $\frac{a_1}{2}$ with the measure of some segments. This interpretation enables him to use some known results on geometrical quantities such as *Elements* I.43. Then, he only needs to exploit the relationships between the objects in his model to be done. In term of the general equation $x^2 + a_1x = a_0$, the relation ($\star$) says that $(x + \frac{a_1}{2})^2 = a_0 + (\frac{a_1}{2})^2$, which gives the formula once the square root is taken.

We remark that, in this chapter, Cardano hardly considers square roots of negative numbers. They could appear in the family of equations $x^2 + a_0 = a_1x$, where the quadratic formula is $x = \frac{a_1}{2} \pm \sqrt{(\frac{a_1}{2})^2 - a_0}$. Nevertheless, Cardano implicitly adds the condition $(\frac{a_1}{2})^2 - a_0 > 0$ saying that:

> [i]f that subtraction of the number from the square of the half of the number of the things cannot be done, the question itself is false and that that has been proposed cannot be[4]

However, Cardano harks back on the topic of square roots of negative numbers at the end of the *Ars Magna*. In **Chapter XXXVII** "On the rule for assuming a false [condition] [*De regula falsum ponendi*]" Cardano deals again with the solutions of quadratic equations. According to him, there are three ways to "assume a false": either admitting negative solutions ("*purum m:*"), or admitting solutions where the square root of a negative quantity appears ("*R m:* or *sophisticum m:*"), or finally admitting negative solutions where the square root of a negative quantity appears. In particular, the second and third ways occur for the family of equations $x^2 + a_0 = a_1x$ when $(\frac{a_1}{2})^2 - a_0 < 0$.

[4]See [22, page 217] or "*[q]uod si detractio ipsa numeri a quadrato dimidii numeri rerum fieri nequit, quæstio ipsa est falsa, necesse potest quod proponitu*", see [4, Chapter V, Paragraph 6, Rule III, page 11v].

Concerning the first way, Cardano says that he looks for solutions "which can at last be verified in the positive".[5] Indeed, he remarks that the solutions of $x^2 = a_1x + a_0$ are opposite in sign to the solutions of $x^2 + a_1x = a_0$. Then, a possible negative solution is in some sense admissible, since it corresponds to a positive solution of another family of equations where the coefficient of the first degree term is opposite in sign. Cardano states that, "[i]f the case is impossible [to solve] with either a positive or a negative, the problem is a false one. If it is a true [problem] with a positive solution in one, it will be a true [problem] with a negative solution in the other".[6]

Then come the "completely false [*omnino falsa*]" cases. Concerning the second way, Cardano only suggests one example, namely to solve the system

$$\begin{cases} 10 = x + y \\ 40 = xy \end{cases},$$

which is equivalent to the equation $x^2 + 40 = 10x$. Its solutions are $x = 5 \pm \sqrt{-15}$. It is easily seen that, in this way, we get two couples (x, y) that are both solutions, but by the symmetry of the system they reduce to

$$\begin{cases} x = 5 + \sqrt{-15} \\ y = 5 - \sqrt{-15} \end{cases}$$

alone. Cardano tries to justify the expressions $5 \pm \sqrt{-15}$ using the geometrical representation of the quadratic formula and this leads him to the incoherence of using a negative area to represent $\sqrt{-15}$. Hence, he verifies that $5 \pm \sqrt{-15}$ are the real demanded quantities, multiplying them as follows

$$(5 + \sqrt{-15})(5 - \sqrt{-15}) = 25 - (-15) = 40.$$

[5] See [22, page 217] or "*Primo igitur quærimus quæstionum solutiones, quæ per p: verificari minime possunt*", see [4, Chapter XXXVII, rule I, page 65r].

[6] See [22, page 219], or "*si casus est impossibilis, in utroque quæstio falsa est, per p: et per m: et si vera est, per p: in uno, erit vera per m: in alio*" see [4, Chapter XXXVII, rule I, page 65v].

There, Cardano is compelled to admit that $(\sqrt{-15})^2 = -15$. Concerning the third way, the only example provided by Cardano suffers a similar shortcoming.

Cardano says that "this [referring to the geometrical quantity that represents $\sqrt{-15}$] is truly sophisticated, since with it one cannot carry out the operations one can in the case of a pure negative and other [numbers]".[7] For example – he continues – one cannot use the quadratic formula. This is how far Cardano's researches on square roots of negative numbers could go. He tried for an arithmetic of square roots of negative quantities, which requires that $(\sqrt{-a})^2 = -a$. This means that he could perform basic, correct calculations with them, and in general he was able to use them. Nevertheless, a proper justification was missing, which explains his resistance in openly using square roots of negative numbers.

3.1.2. Transformations of equations

Let us now turn to cubic equations. One of the most used methods, which fills the whole of Cardano's research on equations, is what I call 'transformation' of equations. With this term I point at relationships between two equations, which are performed by a substitution that involves either some of the coefficients of the two equations or their solutions.

In **Chapter I** Cardano explains some relations that hold between the solutions of certain families of equations, depending on the sign of a quantity which, in the end, is the discriminant. This leads to an investigation of the number of "true [*vera*]" (that is, positive) and "fictitious [*ficta*]" (that is, negative) solutions of an equation. We take into account only the parts concerning cubic equations.

AM I.4. *Consider the equation* $x^3 + a_1x = a_0$. *Then, there is only a positive [real] solution.*

---

[7] "*[Q]uævere est sophistica, quoniam per eam, non ut in puro m: nec in alijs operatione exercere licet*", see [4, Chapter XXXVII, rule II, proof, page 66r].

More generally, Cardano observes that "[w]here several [odd] powers, or even a thousand of them, are compared with a number, there will be one true solution and none that is fictitious".[8]

AM I.5-6. *Consider the equation $x^3 + a_0 = a_1 x$. Then, its solutions correspond up to the sign to the solutions of $y^3 = a_1 y + a_0$.*
More precisely,

- *if [a condition[9] equivalent to $\Delta_3 = 0$ holds], $x_1 < 0$ and $x_2 = x_3 > 0$ are solutions of $x^3 + a_0 = a_1 x$. Moreover, $x_2 = \sqrt{\frac{a_1}{3}}$ and $x_1 = -(x_2 + x_3) = -2\sqrt{\frac{a_1}{3}}$. Then $y_1 = -x_1 > 0$ is a solution of $y^3 = a_1 y + a_0$;*
- *if [a condition equivalent to $\Delta_3 < 0$ holds], $x_1 < 0$ and $x_2, x_3 > 0$ are solutions of $x^3 + a_0 = a_1 x$. Moreover, $x_1 = -(x_2 + x_3)$. Then $y_1 = -x_1 > 0$ is a solution of $y^3 = a_1 y + a_0$;*
- *If [a condition equivalent to $\Delta_3 > 0$ holds], $x_1 < 0$ is a solution of $x^3 + a_0 = a_1 x$. Then $y_1 = -x_1 > 0$ is a solution of $y^3 = a_1 y + a_0$.*

In Cardano's explanation of the first point, we find *in nuce* the idea of multiple solutions, when he says that "the remaining fictitious solution of which we spoke in the other example is the sum of two true ones, but since the true ones are equal to each other, the fictitious one is twice the true one".[10]

Since the positive solutions are to be preferred to the negative ones, Cardano does not feel the need to draw the negative solutions of

[8] See [22, page 11] or "*ubi plures denominationes numero comparantur, etiamsi mille forent, una erit æstimatio rei vera, et nulla ficta*", see [4, Chapter I, Paragraph 4, page 4r].

[9] In all these cases, Cardano discusses the sign of $\frac{2}{3} a_1 \sqrt{\frac{a_1}{3}} - a_0$, which is clearly equivalent to discussing the sign of $\Delta_3 = \frac{\alpha_0^2}{4} + \frac{\alpha_1^3}{27}$. We remark that the condition on the discriminant Δ_3 is explicitly given by Cardano in the particular case $\alpha_3 = 1$, $\alpha_2 = 0$, $\alpha_1 = \pm a_1$, and $\alpha_0 = \pm a_0$ (see here page 33).

[10] See [22, page 12] or "*et ita relique ficta, de qua diximus, in alio exemplo, aggregatur ex duabus veris, sed quia vere sunt invicem æquales, ideo ficta semper dupla est veræ*", see [4, Chapter I, Paragraph 5, page 4r].

$y^3 = a_1 y + a_0$. Anyway, for symmetry reasons we can complete the above proposition as follows.

AM I.5-6 supplemented. *Consider the equation $x^3 + a_0 = a_1 x$. Then, its solutions correspond up to the sign to the solutions of $y^3 = a_1 y + a_0$.*
More precisely,

- *if [a condition equivalent to $\Delta_3 = 0$ holds], $x_1 < 0$ and $x_2 = x_3 > 0$ are solutions of $x^3 + a_0 = a_1 x$. Moreover, $x_2 = \sqrt{\frac{a_1}{3}}$ and $x_1 = -(x_2 + x_3) = -2\sqrt{\frac{a_1}{3}}$. Then $y_1 = -x_1 > 0$ and $y_2 = y_3 = -x_2 < 0$ are solutions of $y^3 = a_1 y + a_0$;*

- *if [a condition equivalent to $\Delta_3 < 0$ holds], $x_1 < 0$ and $x_2, x_3 > 0$ are solutions of $x^3 + a_0 = a_1 x$. Moreover, $x_1 = -(x_2 + x_3)$. Then $y_1 = -x_1 > 0$, $y_2 = -x_2 < 0$, and $y_3 = -x_3 < 0$ are solutions of $y^3 = a_1 y + a_0$;*

- *If [a condition equivalent to $\Delta_3 > 0$ holds], $x_1 < 0$ is a solution of $x^3 + a_0 = a_1 x$. Then $y_1 = -x_1 > 0$ is a solution of $y^3 = a_1 y + a_0$.*

Note that obviously only the real solutions are taken into account.

Concerning cubic equations lacking in the first degree term, Cardano observes the following.

AM I.8 i. *Consider the equation $x^3 = a_2 x^2 + a_0$. Then, there is only a positive [real] solution.*

AM I.8 ii. *Consider the equation $x^3 + a_2 x^2 = a_0$. Then, its solutions correspond up to the sign to the solutions of $y^3 + a_0 = a_2 y^2$.*
More precisely,

- *if [a condition*[11] *equivalent to* $\Delta_3 < 0$ *holds],* $x_1 > 0$ *and* $x_2, x_3 < 0$ *are solutions of* $x^3 + a_2x^2 = a_0$*. Moreover,* $(x_2 + x_3) - x_1 = a_2$*. Then* $y_1 = -x_1 < 0$ *and* $y_2 = -x_2, y_3 = -x_3 > 0$ *are solutions of* $y^3 + a_0 = a_2y^2$*;*

- *If [a condition equivalent to* $\Delta_3 > 0$ *holds],* $x_1 > 0$ *is a solution of* $x^3 + a_2x^2 = a_0$*. Then* $y_1 = -x_1 < 0$ *is a solution of* $y^3 + a_0 = a_2y^2$*.*

Cardano does not consider at all the case $\Delta_3 = 0$.

Then, Cardano moves on studying the complete cubic equations in a similar way.

AM I.9 i. *Consider the equation* $x^3 + a_1x = a_2x^2 + a_0$*. Then, there are no negative [real] solutions.*

Cardano rather affirms that, in this case, "there may be three solutions, all true and none fictitious". Anyway, his justification says that "the cube of a minus quantity is a minus, and thus [if there were a negative solution] a minus would equal a plus, which cannot be",[12] which shows that the equation $x^3 + a_1x = a_2x^2 + a_0$ cannot have negative solutions.[13]

AM I.9 ii. *Consider the equation* $x^3 + a_2x^2 + a_1x = a_0$*.*

[11] We remark as before that Cardano discusses the sign of $\frac{a_2}{3}\left(\frac{2}{3}a_2\right)^2 - a_0$, which is equivalent to a condition imposed on $\Delta_3 = \frac{\alpha_0}{108}\left(27\alpha_0 + 4\alpha_2^3\right)$. For instance, in the case $x^3 + a_2x^2 = a_0$, we get

$$\Delta_3 > 0 \Longleftrightarrow \frac{-a_0}{108}\left(-27a_0 + 4a_2^3\right) < 0 \Longleftrightarrow -27a_0 + 4a_2^3 < 0 \Longleftrightarrow \frac{4}{27}a_2^3 < a_0,$$

which is Cardano's condition. The condition on the discriminant Δ_3 is explicitly given by Cardano in the particular case $\alpha_3 = 1$, $\alpha_2 = \pm a_2$, $\alpha_1 = 0$, and $\alpha_0 = \pm a_0$ (see above, on page 33).

[12] See [22, page 17] or "*[i]n his autem capitulis, quæ duplici denominatione, impari et una pari ac numero constant, si cubus et res, æquales sint, quadratis et numero, æquationes possunt esse tres, et omnes veræ, et nulla ficta, quia ut dictum est, minus cum ad solidus deducitur, fit minus, et ita minus æquales esset plus, quod esse non potest*", see [4, Chapter I, Paragraph 9, page 5v].

[13] Actually, no cubic equation with real coefficients can have three positive solutions, see here Section 2.5.4.

- *If* $x_1 \geq 0$, $x_2, x_3 \leq 0$ *are its [real] solutions, then* $y_1 = -x_1 \leq 0$ *and* $y_2 = x_2, y_3 = x_3 \leq 0$ *are solutions of* $y^3 + a_1 y + a_0 = a_2 y^2$.

- *If* $x_1 \geq 0$ *is its [real] solution, then* $y_1 = -x_1 \leq 0$ *is a solution of* $y^3 + a_1 y + a_0 = a_2 y^2$.

The vice versa *also holds.*

AM I.10. *Consider the equation* $x^3 = a_2 x^2 + a_1 x + a_0$.

- *If* $x_1 \geq 0$, $x_2, x_3 \leq 0$ *are its [real] solutions, then* $y_1 = -x_1 \leq 0$ *and* $y_2 = x_2, y_3 = x_3 \leq 0$ *are solutions of* $y^3 + a_2 y^2 + a_0 = a_1 y$.

- *If* $x_1 \geq 0$ *is its [real] solution, then* $y_1 = -x_1 \leq 0$ *is a solution of* $y^3 + a_2 y^2 + a_0 = a_1 y$.

The vice versa *also holds.*

AM I.11. *Consider the equation* $x^3 + a_2 x^2 = a_1 x + a_0$.

- *If* $x_1 \geq 0$, $x_2, x_3 \leq 0$ *are its [real] solutions, then* $y_1 = -x_1 \leq 0$ *and* $y_2 = x_2, y_3 = x_3 \leq 0$ *are solutions of* $y^3 + a_0 = a_2 y^2 + a_1 y$.

- *If* $x_1 \geq 0$ *is its [real] solution, then* $y_1 = -x_1 \leq 0$ *is a solution of* $y^3 + a_0 = a_2 y^2 + a_1 y$.

The vice versa *also holds.*

No mention of $x^3 + a_1 x = a_2 x^2 + a_0$ is done. The explanation is the following: the equation the solutions of which are opposite in sign to the ones of $x^3 + a_1 x = a_2 x^2 + a_0$ is $x^3 + a_2 x^2 + a_1 x + a_0 = 0$, and we have seen (on page XIII) that this equation is not admissible for Cardano.

We remark that Cardano often gives no justification of his statements, or at best he sometimes adds a numerical example.

Cardano's results on the relations between equations through their solutions also provide an overview of his knowledge on the number of (real) solutions of a cubic equation. This is particularly relevant in the case of depressed equations and of equations lacking in the first degree

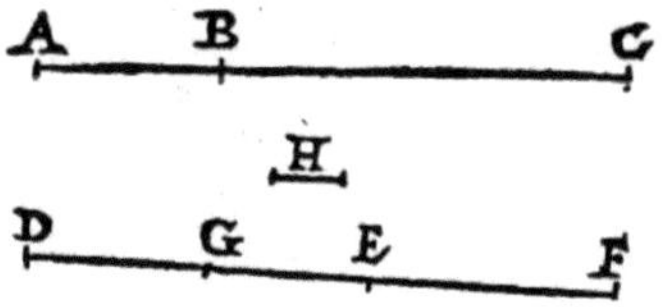

Figure 3.2. – Diagram from *Ars Magna* I.13. Kind courtesy of Biblioteca Virtual del Patrimonio Bibliográfico and Universidad de Sevilla.

term, where he manages to list (almost) all the possibilities depending on the discriminant. In order to do that, Cardano necessarily has to mention negative solutions. This chapter is one of the few places where it happens. So, despite the bad reputation that the negative, "fictitious" solutions have in Cardano's text (and his terminology reflects this aspect), it is anyway a matter of fact that he knows and uses them.

Up to this point in the *Ars Magna*, Cardano gave no proofs of his statements. But at the end of the chapter, he says that "it is meet that we should show these very wonderful things by a demonstration, as we will do throughout this whole book, so that, beyond mere experimental knowledge, reasoning may reinforce belief in them".[14] Then, we find a proof for one of the above cases, namely for the first point in (AM I.11).

AM I.13 — *Proof*. Cardano[15] considers the equation $x^3 + a_0 = a_2x^2 + a_1x$, where, considering the diagram in Figure 3.2, $x^3 = \overline{AB}$, $a_0 = \overline{BC}$, $a_2x^2 = \overline{DE}$, and $a_1x = \overline{EF}$. He takes $\overline{(H)} > 0$ to be a solution [and wants to show that $-\overline{(H)}$ is a solution of $x^3 + a_2x^2 = a_1x + a_0$.] The equation gives the equality $\overline{AB} + \overline{BC} = \overline{DE} + \overline{EF}$. Since $\overline{AB}+\overline{BC} = \overline{AC}$ and $\overline{DE}+\overline{EF} = \overline{DF}$, Cardano gets $\overline{AC} = \overline{DF}$.

[14]See [22, page 20] or "*[e]t iam oportunum est, ut ostendamus hæc demonstratione, quod etiam in toto hoc libro facturi sumus, ut rebus tam admirabilibus, ultra experientiam, fidei ratio accedat*", see [4, Chapter I, Paragraph 13, page 6v].

[15]See [4, Chapter I, Paragraph 13, page 6v].

Then, he marks G on DF such that $\overline{AB} = \overline{DG}$ [and as a consequence $\overline{BC} = \overline{GF}$]. Then, the following equalities hold

$$\begin{cases} \overline{DE} = \overline{AB} + \overline{GE} \\ \overline{BC} = \overline{GE} + \overline{EF} \end{cases},$$

that is

$$\begin{cases} \overline{DE} - \overline{AB} = \overline{GE} \\ \overline{BC} - \overline{FE} = \overline{GE} \end{cases}.$$

Taking $-\overline{(H)} < 0$ instead of $\overline{(H)}$, we have $x^3 = -\overline{AB} < 0$ and $a_1 x = -\overline{EF} < 0$, but $a_0 = \overline{BC} > 0$ and $a_2 x^2 = \overline{DE} > 0$. Then

$$\begin{cases} \overline{DE} - (-\overline{AB}) = \overline{GE} \\ \overline{BC} - (-\overline{FE}) = \overline{GE} \end{cases},$$

that is $\overline{DE} + \overline{AB} = \overline{BC} + \overline{EF}$, that is $x^3 + a_2 x^2 = a_1 x + a_0$. □

We observe that this is a common way to show that a certain quantity is the solution of an equation. One takes the hypothesis on the coefficients and on the solution, and shows by substitution that the equality is verified. Cardano settles a geometrical environment where he interprets the hypotheses as geometrical objects. This fixation of reference enables him to deal with these objects as if they were numbers, since he only considers their measures. We moreover observe that no dimensional considerations enter at this step since, for instance, x^3 is represented by the segment AB. Then, Cardano has only to take care to move on the other side of the equal the negative quantities and to translate back from the geometrical language to the language of equations.

In **Chapter VI** "On the methods for solving new cases [*De modi inveniendi capitula nova*]", Paragraph 3, Cardano states that there is a method

> which is called [the method of] similitude. It is four fold: [1] From the nature of the equation, as when the [solution in the] case of the cube equal to the first power and number is derived from the [solution in the] case of the cube and first power equal to the number. [2] From an augmentation of solutions, and thus we discovered the nonuniversal case of the fourth power, first power and number. [3] By the conversion of equations into [others] equivalent in nature, as we will show hereafter. [4] By the method of proceeding to solutions by the creations of cube and squares, or by proportion (as half or double), or by addition and subtraction, these being three variant methods within a whole.[16]

Later on, we will see more in detail what is the relationship between $x^3 = a_1x + a_0$ in (AM XII) and $x^3 + a_1x = a_0$ in (AM XI). It is in fact not clear what the second and fourth fold is.

In the subsequent paragraph, we have a hint on the third fold, namely on the "conversion [*conversio*]"[17] of equations into others equivalent in nature. There, Cardano says that there is also a "way of transmutation" by which he "discovered many general rules before [their] demonstration"[18] and, among others, those of $x^3 = a_2x^2 + a_0$ and $x^3 + a_2x^2 = a_0$. Cardano suggests the following example. He considers the problem to "find two numbers the sum of which is equal

[16]See [22, page 51] or "*qui similitudinis dicitur, atque hic quadruplex. A natura æquationis, velut cum capitulum cubi æqualis rebus et numero, extrahitur ex capitulo cubi et rerum æqualium numero. Ab augmentis æquationum, sicque capitulo non universalia invenimus quod quadrati, rerum, ac numeri. A conversione æquationum in natura ei æquivalentem, ut exponemus infra. A modo pro ædendi ad æquationes per cuborum vel quadratorum generationem, aut per proportionem ut dupli vel dimidij, aut per additionem vel diminutionem, tres enim sunt modi variandi in universum*", see [4, Chapter VI, Paragraph 3, page 15v].

[17]This is what I call here a 'transformation' of equations.

[18]See [22, page 51] or "*[e]st etiam transmutationis via, qua ante demonstrationem universalia capitula multa inveni*", see [4, Chapter VI, Paragraph 4, pages 15v-16r].

to the square of the second and the product of which is 8 [*duos invenias numeros, quoru aggregatum æquale sit alterius quadrato, et ex uno in alterum ducto, producatur 8*]", that is, to find two real numbers x, y such that the system

$$\begin{cases} x + y = y^2 \\ xy = 8 \end{cases}$$

holds. Depending on the substitution in the second line that Cardano chooses, he obtains two cubic equations

$$y^3 = y^2 + 8$$

taking $x = \frac{8}{y}$ or

$$x^3 + 8x = 64$$

taking $y = \frac{8}{x}$. Then, being able to solve one of the two equations, Cardano will also know the solution of the other *via* one of the substitutions $x = \frac{8}{y}$ or $y = \frac{8}{x}$. Jacqueline Stedall comments on this point (in her own translation of Cardano's text):

> [t]his is the first example we have in the history of algebra of the transformation of an equation by an operation on the roots. It is of fundamental importance. Cardano's optimism at this point shines through in his writing: "Transform problems that are by some ingenuity understood to those that are not understood", he wrote, "and there will be no end to the discovery of rules".[19]

As a matter of fact, this passage could give a hint on how Cardano had the idea of the series of transformations that we have seen in Chapter I and that we will see in Chapter VII.

In **Chapter VII** "On the transformation of equations [*De capitulorum transmutatione*]", Cardano gives two rules about the relation that

[19]See [92, page 9]. In the quotation we find Stedall's own translation of Cardano's text: "*[q]uæstiones igitur alio ingenio cognitas ad ignotas transfer positiones, nec capitulorum inventio finem est habitura*", see [4, Chapter VI, Paragraph 3, page 16r].

holds between the solutions of some equations, in a way similar to what he did in Chapter I. But in contrast to Chapter I, he no longer considers the sign of the discriminant, rather allowing some coefficients to change. He says that these rules are "general [*generalis*]". The first rule is two-fold.

AM VII.2-3. *Let us consider the following transformations of equations:*

$$x^q + a_0 = a_p x^p \xleftrightarrow{y=\frac{(\sqrt[q]{a_0})^{p'}}{x}} y^q + a_0 = a_{p'} y^{p'},$$

$$x^q = a_p x^p + a_0 \xleftrightarrow[y=\frac{(\sqrt[q]{a_0})^{p'}}{x}]{} y^q + a_{p'} y^{p'} = a_0$$

with q, p, p' *naturals such that* $p' = q - p > 0$ *and*

$$a_{p'} = \begin{cases} \frac{a_p}{(\sqrt[q]{a_0})^{p'-p}}, & \text{if } p < p' < q \\ (\sqrt[q]{a_0})^{p-p'} a_p, & \text{if } p' < p < q \end{cases}$$

The arrow means 'correspond to', that is, if x is the given solution of the equation on the left side of the arrow and if we take such a coefficient $a_{p'}$ as above, then $y = \frac{(\sqrt[q]{a_0})^2}{x}$ is the solution of the equation on the right side. The transformation works in both directions.

In particular, let us firstly take $q = 3$ and $p = 1$. Then, we get

$$\text{(AM XIII) } x^3 + a_0 = a_1 x \xleftrightarrow{y=\frac{(\sqrt[3]{a_0})^2}{x}} y^3 + a_0 = \frac{a_1}{\sqrt[3]{a_0}} y^2 \text{ (AM XVI)}$$

$$\text{(AM XII) } x^3 = a_1 x + a_0 \xleftrightarrow[y=\frac{(\sqrt[3]{a_0})^2}{x}]{} y^3 + \frac{a_1}{\sqrt[3]{a_0}} y^2 = a_0 \text{ (AM XV).}$$

Figure 3.3. – Diagram from *Ars Magna* VII.5. Kind courtesy of Biblioteca Virtual del Patrimonio Bibliográfico and Universidad de Sevilla.

Taking then $q = 3$ and $p = 2$, we get

$$\text{(AM XVI) } x^3 + a_0 = a_2x^2 \xleftrightarrow{y=\frac{(\sqrt[3]{a_0})^2}{x}} y^3 + a_0 = (\sqrt[3]{a_0}a_2)\, y \text{ (AM XIII)}$$
$$\text{(AM XIV) } x^3 = a_2x^2 + a_0 \xleftrightarrow[y=\frac{(\sqrt[3]{a_0})^2}{x}]{} y^3 + (\sqrt[3]{a_0}a_2)\, y = a_0 \text{ (AM XI).}$$

Cardano provides a demonstration for the first line of (AM VII.2-3) in the particular case $q = 3$ and $p = 2$. I will only sketch the proof.

AM VII.5 (idea) — *Proof*. Cardano[20] wants to show that $x^3 + a_0 = a_2x^2$ corresponds to $y^3 + a_0 = a_1y$, provided that $a_1 = \sqrt[3]{a_0}\, a_2$ and $y = \frac{(\sqrt[3]{a_0})^2}{x}$. First of all and as usual, Cardano translates his problem in a geometrical fashion. Considering the diagram in Figure 3.3, he assumes $a_2 = \overline{AM}$ and $x = \overline{AD}$ to be given. Rewriting the equation under the form $a_0 = x^2(a_2 - x)$, he draws a_0 depending on a_2 and x, that is $a_0 = \overline{AD}^2\, \overline{DM}$. Then, he assumes that there exist $\overline{FH}[= y]$, $\overline{(E)}[= \sqrt[3]{a_0}]$, and $\overline{FK} = \overline{FH} + \overline{HK}$ such that

$$\frac{\overline{DM}}{\overline{FH}} = \frac{\overline{FH}}{\overline{(E)}} = \frac{\overline{(E)}}{\overline{AD}} \quad \text{[that is, } y = \frac{(\sqrt[3]{a_0})^2}{x}\text{]}$$

[20]See [4, Chapter VII, Paragraph 5, pages 18v-19r].

and

$$\overline{FH\,FK} = \overline{(E)\,AM} \quad [\text{that is, } a_1 = \sqrt[3]{a_0}a_2].$$

Now Cardano has only to work out the relations given by the above proportions. [This corresponds to drawing the value of x from $y = \frac{(\sqrt[3]{a_0})^2}{x}$, substituting it in the equation $y^3 + a_0 = a_1 y$, and verifying to have obtained the good value for a_1]. □

We remark that, if we complete the bi-dimensional diagrams with the appropriate heights, we get a geometrical model, into which Cardano interprets the given hypothesis. As a result, he is enabled to use the language of proportions and their properties to derive the solution. More precisely, in the end, Cardano gets to the relation $\overline{FH}^3 + \overline{FH}^2\,\overline{HK} = (\overline{FK}\,\overline{FH})\overline{FH}$, which represents the equation $y^3 + a_0 = a_1 y$. The core of the proof, where the proportions are used as a calculation tool, is to show that $\overline{FH}^2\,\overline{HK} = \overline{AD}^2\,\overline{DM}$. Note that Cardano's assumptions respect a homogeneity bond concerning dimension, that is, only bodies of the same dimension are summed. Indeed, a_2 in the equation $x^3 + a_0 = a_2 x^2$ has dimension one (it is the segment AM) and a_1 in the equation $y^3 + a_0 = a_1 y$ has dimension two (it is the parallelogram $FK\,FH$) such that both $a_2 x^2$ and $a_1 y$ has dimension three. Moreover, a_0 has always dimension three (it is the parallelepipeds AD^2DM and FH^2HK).

The second rule is also two-fold.

AM VII.6-7. *Let us consider the following transformations of equations:*

$$x^q + a_0 = a_p x^p \xleftrightarrow{y=\sqrt[p]{a_p - x^{p'}}} y^q + a_0' = a_p y^{p'},$$

with $p' = q - p > 0$ *and* $a_0' = (\sqrt[p]{a_0})^{p'}$.

$$x^q = a_p x^p + a_0 \xleftrightarrow[y=\sqrt[p]{x^{p'} - a_p}]{} y^q + a_p y^{p'} = a_0',$$

with q, p, p' *naturals such that* $p' = q - p > 0$ *and* $a_0' = (\sqrt[p]{a_0})^{p'}$.

Here, the coefficient that changes is no longer the coefficient of the term of degree p, but rather the coefficient of degree zero. From the modern viewpoint, these last substitutions are not real changes of variable since the maps $v \mapsto v^2$ and $v \mapsto \sqrt{v}$ are not bijections. This is sometimes the case in Cardano's substitutions. Anyway, if we consider that he favoured positive numbers (and in particular he often considered only the positive roots while taking the $2n^{\text{th}}$ roots), his substitutions are invertible and indeed, in the examples that he provided, he used them in both directions.

In particular, considering firstly $q = 3$ and $p = 1$, we get

$$\text{(AM XIII) } x^3 + a_0 = a_1 x \xleftrightarrow{y=a_1-x^2} y^3 + a_0^2 = a_1 y^2 \text{ (AM XVI)}$$
$$\text{(AM XII) } x^3 = a_1 x + a_0 \xleftrightarrow[y=x^2-a_1]{} y^3 + a_1 y^2 = a_0^2 \text{ (AM XV).}$$

Considering then $q = 3$ and $p = 2$, we get

$$\text{(AM XVI) } x^3 + a_0 = a_2 x^2 \xleftrightarrow{y=\sqrt{a_2-x}} y^3 + \sqrt{a_0} = a_2 y \text{ (AM XIII)}$$
$$\text{(AM XIV) } x^3 = a_2 x^2 + a_0 \xleftrightarrow[y=\sqrt{x-a_2}]{} y^3 + a_2 y = \sqrt{a_0} \text{ (AM XI).}$$

Cardano gives a demonstration for the first line of (AM VII.6-7) in the particular case $q = 3$ and $p = 1$. I skip it since it is similar to the preceding one.

At the end of this chapter, Cardano gives two other rules without any further explication. We will only glance at them.

AM VII.12. *Let us consider the following transformation of equations:*

$$x^4 + a_2 x^2 + a_0 = a_3 x^3 \xrightarrow[y=\frac{\sqrt{a_0}}{x}]{} y^4 + a_2 y^2 + a_0 = (\sqrt{a_0}\, a_3)\, y.$$

AM VII.13. *Let us consider the following transformation of equations:*

$$x^3 + a_0 = a_2 x^2 \xrightarrow[y=x]{} y^3 + \left(\frac{a_2}{2}\right)^2 y = a_2 y^2 + \frac{a_0}{8}.$$

Note nevertheless that this last proposition is not true since the families of equations $x^3 + a_0 = a_2 x^2$ and $y^3 + \left(\frac{a_2}{2}\right)^2 y = a_2 y^2 + \frac{a_0}{8}$ do not have the same solutions.

3.1.3. Cardano solves "middle power equal to the highest power and a number" in "general"

In **Chapter VIII** "On the (general) solution for a middle power equal to the highest power and a number [*Docetur æquatio generaliter mediæ denominationis æqualis extremæ et numero*]", Cardano provides, on the one hand, some clarifications on the opposition of "particular" *versus* "general" (see below, Section 3.1.4, page 79) and, on the other hand, gives a method to find (or better to guess) a solution for a large class of equations (not only the cubic ones). He gives the following rule.

AM VIII.2. *Let $x^q + a_0 = a_p x^p$, with $0 < p < q$, and assume that two positive, real numbers f, g exist such that*

$$\begin{cases} a_p = f + g \\ a_0 = f\left(\sqrt[q-p]{g}\right)^p \end{cases}.$$

Then $x = \sqrt[q-p]{g}$.

We remark that solving the above system is equivalent either to solving an equation of degree $\frac{q}{p}$ or an equation of degree $\frac{q}{q-p}$. Indeed, the system is equivalent either to the following

$$\begin{cases} g = a_p - f \\ g = \left(\frac{a_0}{f}\right)^{\frac{q-p}{p}} \end{cases} \iff \begin{cases} g = a_p - f \\ f^{\frac{q}{p}} - a_p f^{\frac{q-p}{p}} + a_0^{\frac{q-p}{p}} = 0 \end{cases}$$

or to

$$\begin{cases} f = a_p - g \\ f = \frac{a_0}{g^{\frac{p}{q-p}}} \end{cases} \iff \begin{cases} f = a_p - g \\ g^{\frac{q}{q-p}} - a_p g^{\frac{p}{q-p}} + a_0 = 0 \end{cases}$$

In the particular case $q = 3$ and $p = 1$, that is when $x^3 + a_0 = a_1 x$, we want to solve the system

$$\begin{cases} a_1 = f + g \\ a_0 = f\sqrt{g} \end{cases}$$

in order to find a solution. Then $x = \sqrt{g}$. Let us consider Cardano's example $x^3 + 3 = 10x$. Taking $f = 1$ and $g = 9$, a solution is $x = 3$. The above example is the only cubic equation that Cardano suggests in this chapter. We add that in the case $q = 3$ and $p = 2$, that is when $x^3 + a_0 = a_2 x^2$, we want to solve the system

$$\begin{cases} a_2 = f + g \\ a_0 = fg^2 \end{cases}$$

in order to find a solution of the considered equation. Then $x = g$. We remark that, in both cases, solving the system is equivalent to solving some cubic equations. Indeed, in the first case, we get $f^3 - a_1 f^2 + a_0^2 = 0$ and $g\sqrt{g} - a_1\sqrt{g} + a_0 = 0$. In the second case, we get $g^3 - a_2 g^2 + a_0 = 0$ and $f\sqrt{f} - a_2\sqrt{f} + \sqrt{a_0} = 0$. By a simple substitution and considering only positive solutions, we remark that the above equations, which are not of degree three, can be reduced to cubic equations.

Here follows a remark from a modern viewpoint to make Cardano's proposition more understandable. Consider $x^3 + \alpha_1 x + \alpha_0 = 0$, with α_0, α_1 real, and let x_1, x_2, x_3 be its solutions. Then, by Vieta's formulae,

$$\begin{cases} 0 = -x_1 - x_2 - x_3 \\ \alpha_1 = x_1x_2 + x_2x_3 + x_1x_3 \\ \alpha_0 = -x_1x_2x_3 \end{cases},$$

that is

$$\begin{cases} x_3 = -x_1 - x_2 \\ \alpha_1 = -x_1^2 - x_2^2 - x_1x_2 \\ \alpha_0 = x_1x_2(x_1 + x_2) \end{cases}.$$

Let us call $G = x_1$ and $F = x_1x_2 + x_2^2$. Then, the above system is equivalent to

$$\begin{cases} -\alpha_1 = F + G^2 \\ \alpha_0 = FG \end{cases}. \tag{AM VIII.2 i}$$

Note that, given the system (AM VIII.2 i), we obtain the equivalent system

$$\begin{cases} F = -\alpha_1 - G^2 \\ G^3 + \alpha_1 G + \alpha_0 = 0 \end{cases}.$$

Therefore, solving the system is completely equivalent to solving the equation $x^3 + \alpha_1 x + \alpha_0 = 0$, The solutions are $x_1 = G$, together with the solutions of $x_2^2 + Gx_2 - F = 0$. In the end, we get

$$x_1 = G, \quad x_2 = -\frac{G}{2} + \sqrt{\frac{G^2}{4} + F}, \quad \text{and} \quad x_3 = -\frac{G}{2} - \sqrt{\frac{G^2}{4} + F}.$$

Note that if x_1, x_2, x_3 are all real, that is $\Delta_3 < 0$, we find that $\frac{G^2}{4} + F$ must be positive. We moreover remark that Cardano takes $g = G^2$, that is $G = \pm\sqrt{g}$, and $f = F$. From a modern viewpoint, the substitution $g = G^2$ is not a real change of variable, since the map $z \mapsto z^2$ is not a bijection. However, in the simple cases considered by Cardano, we have that each equation, or each system in G, is equivalent to two equations, or to two systems in g, where G is replaced by $\sqrt{g}$ or $-\sqrt{g}$. Namely, the system (AM VIII.2 i) is equivalent to either

$$\begin{cases} -\alpha_1 = f + g \\ \alpha_0 = f\sqrt{g} \end{cases} \quad \text{or} \quad \begin{cases} -\alpha_1 = f + g \\ -\alpha_0 = f\sqrt{g} \end{cases},$$

and the associated solutions are

$$x_1 = \pm\sqrt{g}, \quad x_2 = \mp\frac{\sqrt{g}}{2} + \sqrt{\frac{g}{4} + f}, \quad \text{and} \quad x_3 = \mp\frac{\sqrt{g}}{2} - \sqrt{\frac{g}{4} + f},$$

where the sign $+$ or $-$ is chosen according to the fact that (g, f) is a solution either of the first or of the second system. The first system has positive, real solutions f, g if and only if the equation $x^3+\alpha_1 x+\alpha_0=0$ has a positive, real solution x_1. The second system has positive, real solutions f, g if and only if the equation $x^3+\alpha_1 x+\alpha_0=0$ has a negative, real solution x_1. When the discriminant of $x^3+\alpha_1 x+\alpha_0=0$ is negative, both systems admit a solution, since there is always a positive solution of the equation and a negative one. Moreover,[21] x_1 is opposite in sign either to x_2 or to x_2. The sign of x_2, respectively of x_3, depends instead on the coefficients (see above, Section 2.5.4, page 37). Depending on the sign of x_1, Cardano sometimes chooses $G=\sqrt{g}$, sometimes $G=-\sqrt{g}$. Remark that, if we take $G=\sqrt{g}$, $F=f$, $\alpha_1=-a_1$, and $\alpha_0=a_0$, with a_1, a_0 positive real, we get the first system, which is Cardano's system in (AM VIII.2) for $x^3+a_0=a_1 x$.

Similarly, consider $x^3+\alpha_2 x^2+\alpha_0=0$, with α_0, α_2 real, and let x_1, x_2, x_3 be its solutions. Then,

$$\begin{cases} \alpha_2 = -x_1 - x_2 - x_3 \\ 0 = x_1x_2 + x_2x_3 + x_1x_3 \\ \alpha_0 = -x_1x_2x_3 \end{cases},$$

that is

$$\begin{cases} \alpha_2 = -x_1 - (x_2 + x_3) \\ x_2x_3 = -x_1(x_2 + x_3) \\ \alpha_0 = x_1^2(x_2 + x_3) \end{cases},$$

that is

$$\begin{cases} \alpha_2 = -x_1 - \frac{x_2^2}{x_1+x_2} \\ x_3 = -\frac{x_1x_2}{x_1+x_2} \\ \alpha_0 = x_1^2\left(\frac{x_2^2}{x_1+x_2}\right) \end{cases}.$$

[21]Namely, $\begin{cases} -\frac{G}{2} + \sqrt{\frac{G^2}{4}+F} > 0 & \text{if and only if } G<0 \text{ or } G>0 \text{ and } F>0 \\ -\frac{G}{2} - \sqrt{\frac{G^2}{4}+F} > 0 & \text{if and only if } G<0 \text{ and } F<0 \end{cases}.$

Let us call $G = x_1$ and $F = \frac{x_2^2}{x_1+x_2}$. Then, the above system is equivalent to

$$\begin{cases} -\alpha_2 = F + G \\ \alpha_0 = FG^2 \end{cases}. \qquad \text{(AM VIII.2 ii)}$$

The solutions of $x^3 + \alpha_2 x^2 + \alpha_0 = 0$ are $x_1 = G$, together with the solutions of $x_2^2 - Fx_2 - FG = 0$. In the end, we get

$$x_1 = G, \quad x_2 = \frac{F}{2} + \sqrt{\frac{F^2}{4} + FG}, \quad \text{and} \quad x_3 = \frac{F}{2} - \sqrt{\frac{F^2}{4} + FG}.$$

Similarly to the above, note that if x_1, x_2, x_3 are all real, that is $\Delta_3 < 0$, we find that $\frac{F^2}{4} + FG$ must be positive.[22] We moreover remark that, if we take $f = F$, $g = G$, $\alpha_2 = -a_2$, $\alpha_0 = a_0$, with a_2, a_0 positive real, we get Cardano's system in (AM VIII.2) for $x^3 + a_0 = a_2 x^2$.

In the end, we observe that solving the systems is not simpler than finding the solutions of the corresponding equations, since the systems are nothing else but another way to rewrite these equations. Nevertheless, my opinion is that rewriting the equations under the system form is to be considered as a way to help intuition to guess a solution. We will see in detail how the systems can be exploited in a very special case (where guessing a solution is of the utmost importance, since the cubic formula runs into troubles) dealing with Chapter XXV, on page 84.

Cardano gives a demonstration of this proposition in the case $q = 6$ and $p = 1$.

[22] About the sign of the solutions, we can only say that

$$\begin{cases} \frac{F}{2} + \sqrt{\frac{F^2}{4} + FG} > 0 & \text{if and only if } F > 0 \text{ or } F < 0 \text{ and } G < 0 \\ \frac{F}{2} - \sqrt{\frac{F^2}{4} + FG} > 0 & \text{if and only if } F > 0 \text{ and } G < 0 \end{cases}.$$

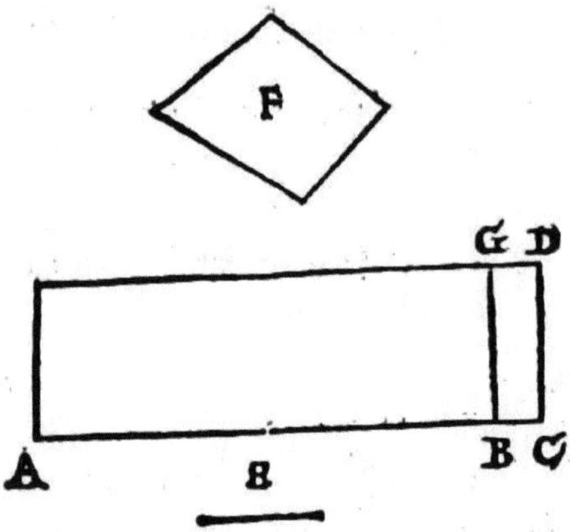

Figure 3.4. – Diagram from *Ars Magna* VIII.1. Kind courtesy of Biblioteca Virtual del Patrimonio Bibliográfico and Universidad de Sevilla.

AM VIII.1 (idea) — *Proof*. He[23] considers the equation $x^6+a_0 = a_1 x$ such that

$$\begin{cases} a_1 = f+g \\ a_0 = f\left(\sqrt[5]{g}\right) \end{cases}$$

holds. [We need to assume f, g real.] Cardano wants to show that $x = \sqrt[5]{g}$.

Again, he considers the diagram in Figure 3.4 and translates the problem in a geometrical fashion, taking $a_1 = (\overline{AB} + \overline{BC})\,\overline{CD}$, $f = \overline{AB}\,\overline{CD}$, $g = \overline{BC}\,\overline{CD}$, and calling $a_0 = \overline{(F)}$. Then, he only has to check that the equation $x^6 + a_0 = a_1 x$ holds when the above values for the coefficients are substituted and one takes $x = \sqrt[5]{g} = \overline{(E)}$. □

This proof clearly stands aside from the others. In fact, it is obvious that no dimensional argument can hold, as in the equation appears x^6 and the sixth dimension cannot be coherently represented in the same way as Cardano did before with the first three dimensions. Then, he has trickily to take $\overline{(E)}$ to represent the fifth root of a certain

[23]See [4, Chapter VII, Paragraph 1, page 21r].

quantity and $\overline{(F)}$ to represent the product of two quantities one of which is the fifth root of a certain quantity. As it will become more and more clear in the following, the distinctive feature of Cardano's geometry is not to use in an essential way the relationships given by the relative positions of the geometrical objects involved, but rather only to respect some dimensional constraints. In the most of the cases, there is a natural way to respect these constraints, which is implicit in the problem (for instance, when a cubic equation is involved). Then, Cardano's geometry can be seen as an almost one-to-one translation from arithmetic. But this cannot be the case here. It is indeed where the geometrical environment fails to imitate the arithmetical steps (as in the proof of (AM VIII.1)) that the real extent of Cardano's geometry comes to light.

We observe that the word 'general' does not appear anywhere else but in the title of the chapter. We remark that Cardano's proposition is really general in our sense. In fact, the conditions in the hypotheses on the coefficients are non-special ones, since it is always possible to find such f, g (by solving a cubic equation). Nevertheless, we will see that the sense of 'general' used by Cardano does not always match ours. I refer then to the next section (especially while commenting Chapter XXV) for further explanations.

Cardano ends this chapter observing that in the case $x^q + a_0 = a_p x^p$ "there are necessarily two solutions in all [cases] and most obviously so [in those] with a large constant".[24] No further explanation is given.

3.1.4. "Particular" *versus* "general"

Cardano introduces the contrast between "particular" and "general" for the first time in **Chapter IV** "On the general and particular solutions that follow [*De subiectis æquationibus generalibus et singularibus*]". He begins by stating that

[24]See [22, page 70] or "*in omnibus, præterque in maximo numero, duas æstimationes necessario habet*", see [4, Chapter VIII, Paragraph 3, page 21v].

> [s]olutions by which no class of cases can be completely resolved are called particular. Such are whole numbers or fractions or roots (square, cube, or what you will) of any number or, as I will put it, any simple quantity. Likewise, all *constantes* made up of two roots, either of which is a square root, a fourth root or, generally, any even root, and, therefore, those consisting of two [such] terms and their *apotomes* (or, as they are called, *recisa*) of the third and sixth classes are not suitable for a general solution.[25]

A "*constans*" or *binomium* is a (positive, irrational) quantity that can be written in the form $\alpha + \beta$, where[26] $\alpha^2 > \beta^2$. Moreover, when $\sqrt{\alpha^2 - \beta^2}$ is "commensurable [*commensurabilis*]" with α, Cardano says that it is of the

1st type if $\alpha = a$ is rational and $\beta = \sqrt{b}$ is irrational, that is $\alpha + \beta = a + \sqrt{b}$,

2nd type if $\alpha = \sqrt{a}$ is irrational and $\beta = b$ is rational, that is $\alpha + \beta = \sqrt{a} + b$,

3rd type if $\alpha = \sqrt{a}$ is irrational and $\beta = \sqrt{b}$ is irrational, that is $\alpha + \beta = \sqrt{a} + \sqrt{b}$,

and, when $\sqrt{\alpha^2 - \beta^2}$ is "incommensurable [*incommensurabilis*]" with α, we say that it is of the

---

[25] See [22, page 30] or "*[s]ingulares dicuntur æquationes, in quibus nullum capitulum perfecte potest absolvi, et tales sunt numerus integer, vel fractus, latus etiam omne numeri, su quadratum seu cubicum vel alterius generis, atque ut ita dicam, omnis simplex quantitas, item constantes ex duabus radicibus omnes, quarum altera sit quadrata, vel RR et generaliter radix par, unde quæ ex duobus constant nominibus, et apotome seu ut dicunt recisa tertij ac sexti generis, non apta sunt æquationi generali*", see [4, Chapter IV, Paragraph 1, page 9r].

[26] Cardano does not specify the definition here. Nevertheless, we find a detailed explanation in the *Ars Magna Arithmeticæ*, see [13, Chapter III, pages 307-308]. This comes from the first definitions in *Elements*, Book X. See also [9, Chapter IV, pages 8-13], where Cardano longerover explains some properties of the *binomia* and *recisa* and gives some rules to make calculations with them.

4th type if $\alpha = a$ is rational and $\beta = \sqrt{b}$ is irrational, that is $\alpha + \beta = a + \sqrt{b}$,

5th type if $\alpha = \sqrt{a}$ is irrational and $\beta = b$ is rational, that is $\alpha + \beta = \sqrt{a} + b$,

6th type if $\alpha = \sqrt{a}$ is irrational and $\beta = \sqrt{b}$ is irrational, that is $\alpha + \beta = \sqrt{a} + \sqrt{b}$,

where the square roots are irrational. We say that two real quantities x, y are "commensurable" if $\frac{x}{y}$ is rational and "incommensurable" if $\frac{x}{y}$ is irrational.[27] Likewise for an "*apotome*" or *recisum*, which is written in the form $\alpha - \beta$. Note that only positive numbers are taken into account (for instance, $a - \sqrt{b}$ has no place in this classification when $a < \sqrt{b}$).

Then, Cardano calls "particular [*particularis*]" solutions:

- a natural a,
- a positive rational a,
- an irrational $\sqrt[n]{a}$, with n natural,
- an irrational $\sqrt[n]{a} \pm \sqrt[n]{b}$, with n even natural (when $n = 2$, it is a "*binomium* or *recisum* of the 3rd or 6th type").

As a brief intermezzo, Cardano observes that

AM IV.2. *Every complete monic cubic equation can be reduced to another one by the substitution $y = x \pm \frac{a_2}{3}$, where a_2 is the coefficient of the second degree term.*

We remark that the reduced equation will be a depressed cubic equation.

Afterwards, Cardano observes that

AM IV.5. *A cubic* binomium *or* recisum *of the 3rd or 6th kind $x = \sqrt[3]{a} \pm \sqrt[3]{b}$ can be a "general" solution for $x^3 = a_1 x + a_0$ and $x^3 + a_1 x = a_0$ respectively.*

[27] See *Elements* X.5-8 in [32, pages 135v-137r].

In particular, this means that, if a cubic *binomium* is a solution of $x^3 = a_1x + a_0$, then its *recisum* is a solution of $x^3 + a_1x = a_0$.

Finally, Cardano observes that

AM IV.6. *The equation* $x^3 = a_2x^2 + a_0$ *has a solution of the form* $x = \sqrt[3]{a} + c + \sqrt[3]{b}$, *where* $\sqrt[3]{a} : c = c : \sqrt[3]{b}$, *[that is* $\sqrt[3]{ab} = c^2$*].*

The equation $x^3 + a_2x^2 = a_0$ *has a solution of the form* $x = \sqrt[3]{a} - c + \sqrt[3]{b}$, *where* $\sqrt[3]{a} : c = c : \sqrt[3]{b}$, *[that is* $\sqrt[3]{ab} = c^2$*].*

In short, Cardano calls "general [*generalis*]" solutions:

- $\sqrt[3]{a} \pm \sqrt[3]{b}$,
- $\sqrt[3]{a} \pm c + \sqrt[3]{b}$, where $\sqrt[3]{ab} = c^2$.

These (up to the signs) indeed are the shapes under which a real solution of a cubic equation can go, assuming that a, b are real. These shapes can be easily gathered simply by looking at the cubic formulae that Cardano provides further on. This chapter is quite short and contains no further explanations. It gives the impression of being a sort of an *a posteriori* melange between a condensed version of the study of the shapes of the irrational solutions of a rational cubic equation inspired by *Elements*, Book X (as it can be found in the *Ars Magna Arithmeticæ* or in the *De Regula Aliza*, see below, Sections 4.2.4 and 5.3.1 respectively on pages 196 and 291) and of the new formulae in the *Ars Magna*.

Let us now turn back to the quotation at the beginning. A first approximation of what Cardano means could be to think of "general solutions" as the most general shape under which the solution of an equation can go. In other words, this is the natural meaning that we would give to the term 'general' nowadays. This would imply that "general" or "particular" solutions should concern a class of equations, namely equations of a same degree. Then, for example, a "general" solution for a cubic equation is its cubic formula. This explains why the "simple quantities" that Cardano lists are "particular" in the sense that they are not "general" or obtained by further calculations starting from a "general" solution. Obviously, this does not mean that no solution of a cubic equation is among those listed in the above quotation. If "general" refers to the shape under which the solution of an equation always goes, then by opposition "particular" should be used for some special shapes under which the solution of an equation sometimes goes. Anyway, this interpretation does not explain why Cardano calls "general" solutions the quantities $\sqrt[3]{a} \pm \sqrt[3]{b}$ and $\sqrt[3]{a} \pm c + \sqrt[3]{b}$, where $\sqrt[3]{ab} = c^2$. We should then look for a weaker version.

In **Chapter XXV** "On imperfect and particular rules [*De capitulis imperfectis et particularibus*]", after having stated all the cubic formulae in Chapters XI-XXIII as we will see later on, Cardano harks back to the opposition of "particular" *versus* "general" saying that

> [t]he [foregoing] rules are called general for two reasons. The first is because the method itself is general, even though it is repugnant to the nature of a solution that it should be universal. Thus, if someone should say that the product of every number that is multiplied by itself is a square, that is a general rule. But it does not follow from this rule that I will know the square of every number, since it is impossible to know every number that is produced by multiplying another number by itself. A rule may also be called general because it exhausts a universal type of solution, although a solution does not exhaust the rule. Nevertheless there are also particular rules [and they are so called] since we cannot solve every given problem by them.[28]

[28]See [22, page 160] or "*[r]egulæ hæ, dicuntur generales, et hoc duabus de causis, prima, qua modus in se generalis est, quamque repugnet naturæ æstimationis, ut sit universalis, velut si quis dicat, omnis numerus productus ex aliquo in se ducto, quadratus est, regula est generalis, nec tamen sequitur, quod per hanc regulam, cognoscam omnem numerum quadratum, quia non licet cognoscere omnem numerum, qui e alio in se ducto producitur. Dicitur et generalis regula, quia exhaurit æstimationis genus universum, quamque æstimatio non exhauriat regulam particulares tamen sunt regulæ, quia non omnem propositam quæstionem per illa solvere posumus*", see [4, Chapter XXV, page 46r].

We find again the terms "general" and "particular", but referring to rules. Here, there are two possible meanings for "general" rules and, as a consequence, for "particular" rules. The second one seems to refer back to Chapter IV, since it concerns the shape under which a solution of an equation can go. A rule is "general" if it "exhausts a universal [that is, "general"] type of solution", and "particular" if not. In other words, a "general" rule is a rule that leads back to a "general" solution, otherwise it is a "particular" rule. The other meaning, instead, concerns the methods employed to find the solution of an equation. A rule is "general" if "the method itself is general", and "particular" if "we cannot solve every given problem" by it. In other words, a "general" rule is a rule that employs a "general" method, where "general" method is supposedly a method by which every problem can be solved. Otherwise, it is a "particular" rule.

These words by Cardano introduce a series of "particular" rules. Regarding the comment at the end of Chapter XII (see below, on page 114) and regarding the fact that almost all the examples by Cardano[29] have $\Delta_3 < 0$, we will assume that these rules should be especially useful in the case $\Delta_3 < 0$. Then, all the equations will have three real solutions. We take f, g, h to be positive real.

AM XXV.1. *Consider* $x^3 = a_1 x + a_0$.
If

$$\begin{cases} a_1 = f + g \\ a_0 = f\sqrt{g} \end{cases},$$

then $x = \sqrt{f + \frac{g}{4}} + \frac{\sqrt{g}}{2}$.

We remark that this is the system (AM VIII.2 i), with $G = -\sqrt{g}$, $F = f$, $\alpha_1 = -a_1$ and $\alpha_0 = -a_0$. The third solution $-\sqrt{f + \frac{g}{4}} + \frac{\sqrt{g}}{2}$ is always negative.

AM XXV.2. *Consider* $x^3 = a_1 x + a_0$.

[29] Except in (AM XXV.3), (AM XXV.5), and (AM XXV.6) in the second example.

If

$$\begin{cases} g = \sqrt{a_1 + f} \\ a_0 = fg \end{cases}, \text{ [that is } \begin{cases} a_1 = g^2 - f \\ a_0 = fg \end{cases}, \text{]}$$

then $x = \sqrt{a_1 + f}\,[= g = \sqrt{a_1 + \frac{a_0}{x}}]$.

Note that my addition in square brackets is necessary in order to preserve the truth of the statement. In fact, if one only considers the positive square root (as Cardano does), it is not true that $x = \sqrt{a_1 + \frac{a_0}{x}}$ follows from $x^3 = a_1 x + a_0$ (rather, the valid implication is the one in the opposite direction).

It is easy to observe that this proposition gives the negative solution $-\sqrt{g}$ in (AM XXV.1), and that the solution which was positive in (AM XXV.1) is now negative. We remark in fact that the system in the hypothesis of (AM XXV.2) is the system (AM VIII.2 i) with $G = g$, $F = -f$, $\alpha_1 = -a_1$ and $\alpha_0 = -a_0$. The other solutions $-\frac{g}{2} \pm \sqrt{f + \frac{g}{4}}$ are always negative.

In the 1570 and 1663 editions, a few words are added to justify some of the rules. In particular, in this case, Cardano observes that, if the hypothesis in (AM XXV.2) holds, then

$$g^3 = g^2 g = (a_1 + f)g, \quad \text{that is} \quad g^3 = a_1 g + a_0.$$

AM XXV.3. *Consider* $x^3 = a_1 x + a_0$.
If

$$\begin{cases} a_1 = f + g \\ \frac{a_0}{2} = \sqrt{f}\,g + f\sqrt{g} \end{cases}, \text{ [that is } \begin{cases} a_1 = f + g \\ a_0 = 2\sqrt{f}\,g + 2f\sqrt{g} \end{cases}, \text{]}$$

then $x = \sqrt{f} + \sqrt{g}$.

We remark that this is the system (AM VIII.2 i), with $G = \sqrt{f} + \sqrt{g}$, $F = -2\sqrt{fg}(\sqrt{f} + \sqrt{g})$, $\alpha_1 = -a_1$ and $\alpha_0 = -a_0$. The other solutions are always negative.

In the 1570 and 1663 editions, the proposition is justified remarking that
$$\frac{y^3+z^3+y^2z+y\,z^2}{2yz(y+z)}=\frac{y^2+z^2}{2y\,z},$$
with y, z positive real. We notice that this means that
$$y^3+z^3+y^2z+y\,z^2=(y^2+z^2)(y+z).$$
If we take $y=\sqrt{f}$ and $z=\sqrt{g}$, then we can rewrite the hypothesis in (AM XXV.3) as follows
$$\begin{cases} a_1=y^2+z^2 \\ \frac{a_0}{2}=y^2z+y\,z^2 \end{cases}. \qquad \text{(AM XXV.3 bis)}$$
In this case, we get $y^3+z^3+y^2z+y\,z^2=a_1(y+z)$. Assuming moreover $x=y+z$, we verify that the equation $x^3=a_1x+a_0$ holds, that is
$$(y+z)^3=(y^3+y^2z+y\,z^2+z^3)+(2y^2z+2y\,z^2).$$
Then, besides (AM VIII.2 i), we remark that it exists a second access point to this rule, namely the substitution $x=y+z$.

AM XXV.4. *Consider* $x^3=a_1x+a_0$.
If
$$\begin{cases} a_1=f+g+h \\ f:g=g:h \\ a_0=g(\sqrt{f}+\sqrt{h}) \end{cases}, \text{ [that is } \begin{cases} a_1=\left(\frac{f+g}{\sqrt{f}}\right)^2-g \\ a_0=g\left(\frac{f+g}{\sqrt{f}}\right) \end{cases}, \text{]}$$
then $x=\sqrt{f}+\sqrt{h}\,[=\frac{f+g}{\sqrt{f}}]$.

We remark that this is the system (AM VIII.2 i), with $G=\frac{f+g}{\sqrt{f}}$, $F=-g$, $\alpha_1=-a_1$ and $\alpha_0=-a_0$. The other solutions are always negative.

In the 1570 and 1663 editions, the proposition is justified remarking that

$$\frac{y^3+z^3+2y^2z+2y\,z^2}{yz(y+z)}=\frac{y^2+z^2+yz}{yz}\quad\text{(and }\ \frac{y^2}{yz}=\frac{yz}{z^2}\text{)}$$

with y, z positive real. We notice that this means that

$$y^3+z^3+2y^2z+2y\,z^2=(y^2+z^2+yz)(y+z).$$

If we take $\sqrt{f}=y$ and $\sqrt{h}=z$, then $g=\sqrt{fh}=yz$. Then, we can rewrite the hypothesis in (AM XXV.4) as follows

$$\begin{cases} a_1=y^2+z^2+yz \\ a_0=y^2z+y\,z^2 \end{cases}. \qquad \text{(AM XXV.4 bis)}$$

In this case, we get $y^3+z^3+2y^2z+2y\,z^2=a_1(y+z)$. Assuming moreover $x=y+z$ to be a solution, we verify that the equation $x^3=a_1x+a_0$ holds, that is

$$(y+z)^3=(y^3+2y^2z+2y\,z^2+z^3)+(y^2z+y\,z^2).$$

Then, besides (AM VIII.2 i), we remark that it exists a second access point to this rule, namely the substitution $x=y+z$.

AM XXV.5. *Consider* $x^3=a_1x+a_0$.
If

$$\begin{cases} f^2+g^2=a_1+fg \\ \frac{a_0}{3}=fg(f+g) \end{cases}, \ \textit{[that is} \begin{cases} a_1=(f+g)^2-3fg \\ a_0=3f^2g+3f\,g^2 \end{cases}, \textit{]}$$

then $x=f+g$.

We remark that this is the system (AM VIII.2 i), with $G=f+g$, $F=-3fg$, $\alpha_1=-a_1$ and $\alpha_0=-a_0$. The other solutions are always negative.

In 1570 (and 1663) editions, Cardano affirms that it is "the converse of a general rule". Consider the diagram in Figure 3.5 and let us call

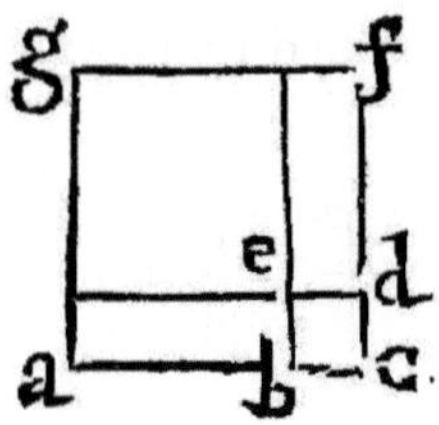

Figure 3.5. – Diagram from *Ars Magna* XXV.5. Kind courtesy of Biblioteca Virtual del Patrimonio Bibliográfico and Universidad de Sevilla.

$y = \overline{AB}$ and $z = \overline{BC}$. Taking $f = y$ and $g = z$, Cardano rewrites the hypothesis in (AM XXV.4) as follows

$$\begin{cases} a_1 = (y-z)^2 + yz \\ a_0 = 3z\,y^2 + 3y\,z^2 \end{cases} . \qquad \text{(AM XXV.5 bis)}$$

We notice that, assuming $x = f + g$, this means that

$$a_1 x = (f^2 + g^2 - fg)(f+g) = f^3 + g^3.$$

So, Cardano verifies that the equation $x^3 = a_1 x + a_0$, that is

$$(f+g)^3 = (f^3 + g^3) + (3f^2 g + 3f\,g^2)$$

holds. Then, besides (AM VIII.2 i), we remark that it exists a second access point to this rule, namely the substitution $x = y + z$.

AM XXV.6. *Consider $x^3 = a_1 x + a_0$.*
If $f^3 + a_0 = a_1 f$, then $x + f$ divides $x^3 + f^3$ and $a_1 x + a_0 + f^3$.
If $f^3 - a_0 = a_1 f$ [with $f^3 > a_0$], then $x - f$ divides $x^3 - f^3$ and $a_1 x + a_0 - f^3$.

We will try to get a deeper insight of this proposition. If we can find a common factor in both sides of the equation, then we manage to lower the degree of the equation. The easiest case is to consider a common factor of degree one $x - \alpha$, with α real. The only α such that

$x - \alpha$ divides x^3 is 0. But, since $a_0 \neq 0$, x does not divide $a_1 x + a_0$. Then, we add to both sides of the equation a constant such that the left side always turns out to be divisible by $x - \alpha$. In this way, we will only have to check the divisibility on the right side. Let us take $-\alpha^3$. Then,

$$\frac{x^3 - \alpha^3}{x - \alpha} = \frac{a_1 x + a_0 - \alpha^3}{x - \alpha}.$$

Since

$$a_1 x + a_0 - \alpha^3 = a_1(x - \alpha) + (a_1 \alpha + a_0 - \alpha^3),$$

we force $a_1 \alpha + a_0 - \alpha^3 = 0$. By the way, we remark that this means that α is a solution of $x^3 = a_1 x + a_0$ (this is nothing else but Ruffini's theorem). Performing the division, we get the quadratic equation $x^2 + \alpha x + \alpha^2 - a_1 = 0$. Its solutions are $-\frac{\alpha}{2} \pm \sqrt{a_1 - \frac{3}{4}\alpha^2}$. We conclude that the solutions of the equation $x^3 = a_1 x + a_0$ are

$$\alpha, \quad -\frac{\alpha}{2} + \sqrt{a_1 - \frac{3}{4}\alpha^2} \quad \text{and} \quad -\frac{\alpha}{2} - \sqrt{a_1 - \frac{3}{4}\alpha^2}.$$

We know[30] that, when the coefficients are real and $\Delta_3 < 0$, the equation $x^3 = a_1 x + a_0$ always has one positive and one negative solution. Cardano takes $\alpha = \pm f$.

We remark then that, once one already knows – no matter how – a solution of the cubic equation, this proposition helps in finding the two remaining real solutions. In this case, then, we can avoid to use the cubic formula, which will instead contain imaginary numbers.

We moreover remark that this proposition is another access point to (AM XXV.1) and (AM XXV.2).

AM XXV.7. *Consider* $x^3 = a_1 x + a_0$.

Then $a_0 = \left(\frac{\alpha}{2} + \sqrt{a_1 - \frac{3}{4}\alpha^2}\right)\left(\frac{\alpha}{2} - \sqrt{a_1 - \frac{3}{4}\alpha^2}\right)^2 + \left(\frac{\alpha}{2} - \sqrt{a_1 - \frac{3}{4}\alpha^2}\right)\left(\frac{\alpha}{2} + \sqrt{a_1 - \frac{3}{4}\alpha^2}\right)^2$,

where α *is a solution of the considered equation.*

---

[30]See above, on page 37.

Let us try to understand where this proposition comes from. We know by (AM VIII.2 i) that the solutions of $x^3 = a_1x + a_0$ are G and $-\frac{G}{2} \pm \sqrt{\frac{G^2}{4} + F}$ where F, G satisfy the system

$$\begin{cases} a_1 = F + G^2 \\ -a_0 = FG \end{cases}.$$

Since $F = a_1 - G$, we can rewrite the solutions as G and $-\frac{G}{2} \pm \sqrt{a_1 - \frac{3}{4}G^2}$. Another possible rewriting of the same solutions is $-\left(\left(-\frac{G}{2} + \sqrt{a_1 - \frac{3}{4}G^2}\right) + \left(-\frac{G}{2} - \sqrt{a_1 - \frac{3}{4}G^2}\right)\right)$ (see also Cardano's remark after proposition (AM I.5-6), above on page 61), and then $-\frac{G}{2} \pm \sqrt{a_1 - \frac{3}{4}G^2}$. Since we suppose $\Delta_3 < 0$, all these solutions are real. Moreover, we know that a_0 is the product of the three solutions, so that

$$a_0 = \left(-\frac{G}{2} + \sqrt{a_1 - \frac{3}{4}G^2}\right)\left(-\frac{G}{2} - \sqrt{a_1 - \frac{3}{4}G^2}\right)$$
$$\left(\left(\frac{G}{2} - \sqrt{a_1 - \frac{3}{4}G^2}\right) + \left(\frac{G}{2} + \sqrt{a_1 - \frac{3}{4}G^2}\right)\right),$$

which gives the equality in (AM XXV.7) taking $G = \alpha$. Then, this proposition is basically derived from the fact that a_0 is the product of the three solutions of $x^3 = a_1x + a_0$ and again from (AM VIII.2 i).

AM XXV.8. *Consider* $x^3 = a_1x + a_0$.
Then $\frac{a_0}{2} = \left(\frac{\alpha}{2} + \sqrt{\frac{a_0}{2\alpha} + a_1 - \frac{3}{4}\alpha^2}\right)\left(\frac{\alpha}{2} - \sqrt{\frac{a_0}{2\alpha} + a_1 - \frac{3}{4}\alpha^2}\right)^2$
$+\left(\frac{\alpha}{2} - \sqrt{\frac{a_0}{2\alpha} + a_1 - \frac{3}{4}\alpha^2}\right)\left(\frac{\alpha}{2} + \sqrt{\frac{a_0}{2\alpha} + a_1 - \frac{3}{4}\alpha^2}\right)^2$,
where α *is a solution of the considered equation.*

We remark that this proposition is a variation of the above one, thus also deriving from the fact that a_0 is the product of the three solutions of $x^3 = a_1x + a_0$ and from (AM VIII.2 i).

AM XXV.9. *Consider* $x^3 + a_0 = a_1x$.

If

$$\begin{cases} [a_1 = -f + (a_1 + f)] \\ a_0 = f\sqrt{a_1 + f} \end{cases},$$

then $x = \frac{\sqrt{a_1+f}}{2} \pm \sqrt{a_1 - \frac{3}{4}(a_1 + f)}$.

Note that in this case both solutions are positive, since $\frac{\sqrt{a_1+f}}{2} \geq \sqrt{a_1 - \frac{3}{4}(a_1 + f)}$.

We remark that this is the system (AM VIII.2 i), with $G = -\sqrt{a_1 + f}$, $F = -f$, $\alpha_1 = -a_1$ and $\alpha_0 = a_0$.

Moreover, we remark that this proposition corresponds to (AM XXV.1). Indeed, if we write a_1', a_0' for the coefficients and f', g' for the functions of the coefficients in (AM XXV.1), then it is enough to take $f' = -f$ and $\sqrt{g'} = \sqrt{a_1 + f}$ in order to verify that $a_1 = a_1'$ and $a_0 = -a_0'$.

AM XXV.10. *Consider* $x^3 + a_0 = a_1 x$.
If $f = a_1\sqrt[3]{f - a_0}$, *then* $x - \sqrt[3]{f - a_0}$ *divides* $x^3 + a_0 - f$ *and* $a_1 x - f$.

We remark that a reasoning similar to (AM XXV.6) is present in this proposition. In this case, Cardano takes $\alpha = \sqrt[3]{a_0 - f}$. We remark that, if it holds $f = a_1\sqrt[3]{f - a_0}$, then $\sqrt[3]{f - a_0}$ is a solution of $x^3 + a_0 = a_1 x$. I do not have any hints on why Cardano does not also consider the divisibility by $x + \sqrt[3]{f - a_0}$.

We moreover remark that this proposition is another access point to (AM XXV.9).

AM XXV.11. *Consider* $x^3 + a_0 = a_1 x$.
If

$$\begin{cases} a_1 = f + g + h \\ f : g = g : h \\ \frac{a_0}{3} = g(\sqrt{f} - \sqrt{h}) \end{cases}, \text{ [that is } \begin{cases} a_1 = \left(\frac{f-g}{\sqrt{f}}\right)^2 + 3g \\ a_0 = 3g\left(\frac{f-g}{\sqrt{f}}\right) \end{cases},]$$

then $x = \sqrt{f} - \sqrt{h}\,[= \frac{f-g}{\sqrt{f}}]$.

Note that $f > g$, since we assume that Cardano is admitting only positive solutions.

We remark that this is the system (AM VIII.2 i), with $G = \frac{f-g}{\sqrt{f}} > 0$, $F = 3g$, $\alpha_1 = -a_1$ and $\alpha_0 = a_0$. Among the other solutions, there is a negative one and a positive one, which is $x = \frac{g-f}{2\sqrt{f}} + \sqrt{\frac{(f-g)^2}{4f} + 3g}$.

Cardano remarks that this proposition corresponds to (AM XXV.4). We observe in fact that, if we write a'_1, a'_0 for the coefficients and f', g' for the functions of the coefficients in (AM XXV.4), then we can express f', g' depending on f, g (even if this cannot be done in an easy way) in order to verify that $a_1 = a'_1$ and $a_0 = -a'_0$. Moreover, we can try to justify (AM XXV.11) in a way similar to Cardano's 1570 (and 1663) editions. Let us call $y = \sqrt{f}$ and $z = \sqrt{h}$. Then, we can rewrite the hypothesis in (AM XXV.11) as follows

$$\begin{cases} a_1 = y^2 + z^2 + yz \\ a_0 = 3y^2 z - 3yz^2 \end{cases}. \qquad \text{(AM XXV.11 bis)}$$

In this case, we get $y^3 - z^3 = a_1(y - z)$. Assuming moreover $x = y - z$ to be a solution, we verify that the equation $x^3 + a_0 = a_1 x$ holds, that is

$$(y - z)^3 + (3y^2 z - 3yz^2) = y^3 - z^3.$$

Then, besides (AM VIII.2 i), we remark that it could exist a second access point to this rule, namely the substitution $x = y + z$.

AM XXV.12. *Consider* $x^3 + a_0 = a_1 x$.
If

$$\begin{cases} \frac{a_1}{\sqrt[3]{a_0}} = f + g \\ a_0 = fg^2 \end{cases},$$

then $\sqrt[3]{a_0} : x = x : f$, *[that is* $x = \sqrt{f \sqrt[3]{a_0}} = \sqrt[6]{f^4 g^2}$*].*

We remark that this is the system (AM VIII.2 i), with $G = \sqrt{f \sqrt[3]{f\, g^2}}$, $F = g\sqrt[3]{f\, g^2}$, $\alpha_1 = -a_1$ and $\alpha_0 = a_0$. Among the other solutions, there is a negative one and a positive one, which is $x = \frac{\sqrt{5}-1}{2}\sqrt[6]{f^4g^2}$.

AM XXV.13. *Consider* $x^3 + a_0 = a_1 x$.
If

$$\begin{cases} \frac{a_1}{3} = f + g \\ \frac{a_0}{2} = f\sqrt{f} + g\sqrt{g} \end{cases}, \text{ [that is } \begin{cases} a_1 = 3f + 3g \\ a_0 = 2f\sqrt{f} + 2g\sqrt{g} \end{cases}, \text{]}$$

then $x = \sqrt{f} + \sqrt{g}$.

We remark that this is the system (AM VIII.2 i), with $G = \sqrt{f} + \sqrt{g}$, $F = \frac{2f\sqrt{f}+2g\sqrt{g}}{\sqrt{f}+\sqrt{g}}$, $\alpha_1 = -a_1$ and $\alpha_0 = a_0$. Among the other solutions, there is a negative one and a positive one, which is $x = -\frac{\sqrt{f}+\sqrt{g}}{2} + \sqrt{9f + 6\sqrt{f}\sqrt{g} + 9g}$.

Cardano remarks that this proposition corresponds to (AM XXV.3). We observe in fact that, if we write a'_1, a'_0 for the coefficients and f', g' for the functions of the coefficients in (AM XXV.3), then we can express f', g' depending on f, g (even if this cannot be done in an easy way) in order to verify that $a_1 = a'_1$ and $a_0 = -a'_0$. Moreover, we can try to justify (AM XXV.13) in a way similar to Cardano's 1570 (and 1663) editions. Let us call $y = \sqrt{f}$ and $z = \sqrt{g}$. Then, we can rewrite the hypothesis in (AM XXV.13) as follows

$$\begin{cases} a_1 = 3y^2 + 3z^2 \\ a_0 = 2y^3 + 2z^3 \end{cases}. \qquad \text{(AM XXV.13 bis)}$$

In this case, we get $3(y^3 + y^2z + yz^2 + z^3) = a_1(y + z)$. Assuming moreover $x = y + z$ to be a solution, we verify that the equation $x^3 + a_0 = a_1 x$ holds, that is

$$(y + z)^3 + (2y^3 + 2z^3) = 3(y^3 + y^2z + yz^2 + z^3).$$

Then, besides (AM VIII.2 i), we remark that it could exist a second access point to this rule, namely the substitution $x = y + z$.

AM XXV.17. *Consider* $x^3 + a_0 = a_1 x$.

If

$$\begin{cases} \sqrt{a_1} = f + g \\ a_0 = 2f\,g^2 + f^2 g \end{cases}, \ \textit{[that is} \begin{cases} a_1 = g^2 + 2fg + f^2 \\ a_0 = g(2fg + f^2) \end{cases}, \textit{]}$$

then $x = g$.

We remark that this proposition corresponds to (AM XXV.2). Indeed, if we write a_1', a_0' for the coefficients and f', g' for the functions of the coefficients in (AM XXV.2), then it is enough to take $f' = -g$ and $g' = 2fg + f^2$ in order to verify that $a_1 = a_1'$ and $a_0 = -a_0'$.

Moreover, we remark that this is the system (AM VIII.2 i), with $G = g$, $F = 2fg + f^2$, $\alpha_1 = -a_1$ and $\alpha_0 = a_0$. Among the other solutions, there is a negative one and a positive one, which is $x = -\frac{g}{2} + \sqrt{\frac{g^2}{4} + 2fg + f^2}$.

AM XXV.14. *Consider* $x^3 + a_2 x^2 = a_0$.
If

$$\begin{cases} a_2 = f + g \\ a_0 = fg^2 \end{cases},$$

then $x = \sqrt{f\left(g + \frac{f}{4}\right)} - \frac{f}{2}$.

We remark that this is the system (AM VIII.2 ii), with $G = -g$, $F = -f$, $\alpha_2 = a_2$ and $\alpha_0 = -a_0$. The third solution is always negative.

AM XXV.18. *Consider* $x^3 + a_2 x^2 = a_0$.
If

$$\begin{cases} fg = \left(\frac{a_2}{3}\left(\sqrt[3]{f} - \sqrt[3]{g}\right)\right)^3 \\ a_0 = f - g \end{cases}, \ \textit{[that is} \begin{cases} a_2 = \dfrac{3\sqrt[3]{fg}}{\sqrt[3]{f} - \sqrt[3]{g}} \\ a_0 = f - g \end{cases}, \textit{]}$$

then $x = \sqrt[3]{f} - \sqrt[3]{g}$.

Note that $f > g$, since we assume that Cardano is only admitting positive solutions.

We remark that this is the system (AM VIII.2 ii), with $G = \sqrt[3]{f} - \sqrt[3]{g} > 0$, $F = -\frac{f-g}{(\sqrt[3]{f}-\sqrt[3]{g})^2}$, $\alpha_2 = a_2$ and $\alpha_0 = -a_0$. The other solutions are always negative.

AM XXV.15. *Consider* $x^3 + a_0 = a_2 x^2$.
If

$$\begin{cases} \frac{a_2}{4} < f < \frac{a_2}{3} \\ \frac{a_0}{f} = g^2 \\ \frac{g}{2} + a_2 = 4f \end{cases}, \text{ [that is} \begin{cases} \frac{a_2}{4} < f < \frac{a_2}{3} \\ a_0 = f\,g^2 \\ a_2 = 4f - \frac{g}{2} \end{cases}, \text{]}$$

then $x = 2f \pm \sqrt{4f(a_2 - 3f)}$.

Note that the condition $f < \frac{a_2}{3}$ in the hypothesis grants that $4f(a_2 - 3f) > 0$, that is that the solutions are all real. The condition $\frac{a_2}{4} < f$ grants that $2f - \sqrt{4f(a_2 - 3f)} > 0$.

We remark that this proposition corresponds to (AM XXV.14). Indeed, if we write a_2', a_0' for the coefficients and f', g' for the functions of the coefficients in (AM XXV.14), then it is enough to take $f' = -4f$ and $g' = \frac{g}{2}$ in order to verify that $a_2 = -a_2'$ and $a_0 = -a_0'$.

Moreover, we remark that this is the system (AM VIII.2 ii), with $G = -\frac{g}{2}$, $F = 4f$, $\alpha_2 = -a_2$ and $\alpha_0 = a_0$.

Cardano gives no further justifications for these rules, saying only that they are founded on the propositions in Chapter VI (on the cube of a binomial) and that they are self-evident to those familiar with his books on Euclid.[31]

Let the propositions (AM XXV.7), (AM XXV.8) stand aside. In the following I will refer only to the remaining ones. We observe that there are strong links between these propositions, as represented in Table 3.1. Note that only (non-complete) equations where it can be $\Delta_3 < 0$ are involved.

Moreover, except for (AM XXV.6) and (AM XXV.10) which are instead directly linked respectively the first to (AM XXV.1) and (AM

[31] See [22, page 168], [4, Chapter XXV, Note, page 48v]. The books on Euclid of which Cardano is speaking are the *Novæ Geometriæ*, see [24, page 58] and [75]. The *Novæ Geometriæ* was destroyed by Cardano himself in 1573.

Table 3.1. – Links between "particular" propositions in *Ars Magna*, Chapter XXV.

$x^3 = a_1x + a_0$		$x^3 + a_0 = a_1x$	$x^3 + a_2x^2 = a_0$		$x^3 + a_0 = a_2x^2$
(AM XXV.1)	$\longleftrightarrow$	(AM XXV.9)	(AM XXV.14)	$\longleftrightarrow$	(AM XXV.15)
(AM XXV.2)	$\longleftrightarrow$	(AM XXV.17)	(AM XXV.18)		
(AM XXV.3)	$\longleftrightarrow$	(AM XXV.13)			
(AM XXV.4)	$\longleftrightarrow$	(AM XXV.11)			
(AM XXV.5)		(AM XXV.12)			
(AM XXV.6)	$\longleftrightarrow$	(AM XXV.10)			

XXV.2), and the second to (AM XXV.9), they all can be drawn back either to (AM VIII.2 i) or to (AM VIII.2 ii), so that at the end they are all linked to each other. In some cases, I have directly highlighted the connections that seem to be the most natural.

From a modern viewpoint, none of the previous substitutions, which help to pass from a statement in Chapter XXV either to (AM VIII.2 i) or to (AM VIII.2 ii), or to another statement in the same Chapter XXV are real substitutions, since they are not invertible. But given a solution of the transformed system, one can always find at least one solution of the original system inverting in a naive way the substitution. The point is that we are interested in finding only one solution of a system (and not all the solutions), since it is enough to lower the degree of the equation. Anyway – as already observed, considering the systems in (AM VIII.2 i) or (AM VIII.2 ii) cannot help at all in finding the solutions of the corresponding equation, since finding the solution G of one of the systems is perfectly equivalent (and as difficult as) to solve the corresponding cubic equation.

Let us now turn back to Cardano's alleged viewpoint. It is true that, even if (AM VIII.2) or one of the above equivalent propositions are not a step in finding a cubic formula that works at every turn, at least they move towards the direction of bypassing the problem set by the *casus irreducibilis*. Indeed, once a solution is found, no matter how, one could lower the degree of the equation by polynomial division. By the way, this is why Cardano is probably not interested in listing all the positive, real solutions that he can deduce from the systems, but only one. Indeed, one solution is enough to lower the degree of the cubic equation. And once it happens to fall down on a quadratic equation, imaginary quantities would no longer appear. Indeed, we are interested in the case $\Delta_3 < 0$, where we know that all the solutions are real and quadratic equations do not bear the *casus irreducibilis*. The propositions (AM XXV.6) and (AM XXV.10) partially reveal how to perform polynomial division.[32] Under this perspective, it seems

[32]Even if similar propositions for cubic equations lacking in the first degree term exist, Cardano does not enunciate them.

natural to me to consider the other propositions as an aid to guess a solution. More precisely, if it is very hard to predict a solution looking only at the appearance of an equation, it could sometimes be easier to guess f, g in the system and from there to reconstruct a solution in the way taught by Cardano in his propositions. In this sense, the systems in the hypotheses of the above propositions provide assistance to intuition. Unfortunately, a considerable amount of uncertainty still remains. The coefficients of the equation have to be simple enough, or one has to be smart enough, to guess one of its solutions. In short, the propositions in Chapter XXV do not provide a safe algorithmic way to bypass the *casus irreducibilis.* More work still needs to be done in this direction. Moreover, there is another viewpoint from which the rules in Chapter XXV turn out to be useful. Indeed, one may use them to inquire the shape of a solution of an equation of a certain family, choosing adapted coefficients and starting from their composition in terms of f, g, h.

Then, we can stretch our interpretation of "general" and "particular" rules as referring to "general" and "particular" methods (see above, on page 84) in order to include the above propositions in it as "particular". Indeed, even if it is true that they work *a priori* for each equation (since, in the end, they are equivalent to solve a cubic equation),[33] in concrete terms they only work if one can find such f, g. This means that the above propositions provide a solution of a cubic equation where $\Delta_3 < 0$ only if one already found another solution, which is intended to be discovered case-by-case, since the cubic formula leads to imaginary numbers. In other words, the above rules prove to be *a priori* "general", but in practice Cardano manages to apply them only in some cases, and thus they are "particular" in the sense of the opening quotation.

[33] Note that [57] remarks that a "particular" rule is derived from a particular structure: "*oltre le generali regole per risolvere le equazioni di terzo grado insegna Cardano parecchie regole di scioglimenti speciali ricavati dai modi dalla struttura loro, dalla composizione dei termini ignoti, e dal termine noto, dai rapporti tra gli uni e l'altro*", see [57, Volume II, Chapter II, Soluzioni speciali, page 166].

Under this interpretation, also (AM VIII.2) becomes "particular", contradicting the title of the chapter. Indeed, this is a real inconsistency in Cardano's text, and one can only explain it resorting to the changeableness of the terms 'general' and 'particular'.

We moreover observe that there are other links between the above propositions. Indeed, there is a second access point to (AM XXV.1), (AM XXV.2), and (AM XXV.9) which is given by polynomial division in (AM XXV.6) and (AM XXV.10). There is also a second access point to (AM XXV.3), (AM XXV.4), (AM XXV.5), (AM XXV.11), and (AM XXV.13) which is given by the substitution $x = y + z$.

After the proposition (AM XXV.15), Cardano observes that "none of these rules can be [considered] general with respect to a solution" and that he will show this in one case for all.[34] This means that the above propositions are "particular" not only in terms of the methods employed, but also in terms of the solutions.

Cardano considers the rule given in (AM XXV.15), because – he says – this case is the more readily believable one, since he has already obtained some results concerning the family of equations $x^3 + a_0 = a_2x^2$ (see, for instance, (AM I.8 ii), (AM VII.2-3), (AM VII.6-7), and (AM VIII.2)).

AM XXV.16 — *Proof*. Cardano[35] shows that (AM XXV.15) is "particular" with respect to solutions.

He draws[36] an equation of the family $x^3 + a_0 = a_2x^2$, where a_0 is not integer and a_2 is integer, such that he knows a non-integer

[34]See [22, page 168] or "*Opræprecium fuerit nunc ostendere, quod hæ regulæ non possunt esse generales, respectu æstimationis, et modus in uno sufficiet ad ostendum in reliquis capitulis*", see [4, Chapter XXV, Paragraph 16, pages 48v-49r].

[35]See [4, Chapter XXV, Paragraph 16, page 48v-48r].

[36]Namely, he considers the equation $x^3 + a_0 = 7x^2$, where a_0 is left at first undetermined. Then, he assumes that the solution goes under the form $x = x' + \sqrt{x''}$. Taking $x' = \frac{8}{3}$, he finds $x'' = 16$, which implies by substitution that $x = \frac{20}{3}$, and $a_0 = \frac{400}{27}$.

solution x. He uses this example to justify[37] the statement that, if a_2 is integer and x non-integer, then a_0 is not integer.

Afterwards, Cardano remarks that $a_0 = x^2(a_2 - x)$. We notice that this also comes from the hypothesis of (AM XXV.15). Indeed, we get by the system that

$$a_0 = \left(-\tfrac{g}{2}\right)^2 \left(a_2 + \tfrac{g}{2}\right),$$

knowing by (AM VIII.2 ii) that $-\frac{g}{2}$ is a solution.

Then, Cardano assumes $a_2 = 7$ and moreover supposes x to be integer. He remarks that the integer part of the maximum of $x^2(a_2 - x)$ for $x \in [0, 7]$ is 50. Then, $a_0 = 6, 20, 36, 48, 50$. This means that there are only five equations of the family $x^3 + a_0 = a_2x^2$, where $a_2 = 7$ and a_0 natural, the solution of which is $x \in [0, 7]$ integer.

Cardano ends saying that therefore the rule "is particular and, indeed, exceedingly particular".[38] □

Cardano's proof is riddled with oversights or sometimes with real mistakes. I did not manage to use this poof to help in clarifying Cardano's statements about the fact that the above propositions should be "particular" with respect to a solution. By the way, [57] is of the opinion that 'general' and 'particular' rules was quite a common terminology on which Cardano's contemporaries did not agree.[39]

3.1.5. Cardano's "royal road"

In **Chapter VI** Cardano says that

> [w]hen, moreover, I understood that the rule that Niccolò Tartaglia handed to me had been discovered by him

[37] Actually, this is true by Gauss's lemma (in this case, by the fact that each rational root of a monic polynomial with integer coefficients is integer).

[38] See [22, page 169] or "*particularis igitur est, ac valde etiam particularis*", see [4, Chapter XXV, Paragraph 16, page 49r].

[39] "*Pare che l'idea delle accennate regole non fosse comunemente abbastanza chiara, e ben fissa, che per non distinguerne i diversi aspetti, chi errasse in voler loro negato ogni senso di generalità, e chi in attibuirlo alle medesime oltre dovere; e che Cardano mirato abbia a correggere e gli uni, e gli altri*", see [57, Volume II, Chapter II, Soluzioni speciali, page 183].

> through a geometrical demonstration, I thought that this would be the royal road to pursue in all cases.[40]

Without interruption, Cardano puts forward three "highly useful propositions", as they followed Tartaglia's geometrical "royal road". We consider only the first one about the cube of a binomial, the others being variations of it.

AM VI.6. *If a quantity is divided into two parts, the cube of the whole is equal to the cubes of the two parts plus three times the products of each and the square of the other.*[41]

AM VI.6 — *Proof*. [We consider the diagram in Figure 3.6. Let $\overline{AC}$ be divided into $\overline{AB}$ and $\overline{BC}$. Cardano wants to prove that $\left(\overline{AB}+\overline{BC}\right)^3 = \overline{AB}^3 + 3\overline{AB}^2\overline{BC} + 3\overline{AB}\,\overline{BC}^2 + \overline{BC}^3$.] Cardano[42] draws the cube on AC, obtaining eight parts. He observes then that, by *Elements* II.4, the following equalities hold

$$\overline{BC}^2 = \overline{(CD)}^2, \quad \overline{AB}^2 = \overline{(DF)}^2, \quad \text{and} \quad \overline{AB}\,\overline{BC} = \overline{(AD)} = \overline{(DE)}.$$

Then, since $\overline{AC}^2 = \overline{(CD)}^2 + \overline{(DF)}^2 + \overline{(AD)} + \overline{(DE)}$, it holds that

$$\overline{AC}^3 = \overline{AC}\,\overline{AC}^2 = \overline{AC}\left(\overline{(CD)}^2 + \overline{(DF)}^2 + \overline{(AD)} + \overline{(DE)}\right),$$

[40]See [22, page 52] or "*[c]um autem intellexissem capitulum, quod Nicolaus Tartalea mihi tradiderat, ab eo fuisse demonstrationem inventum Geometrica, cogitavi eam viam esse regiam, ad omnia capitula venanda*", see [4, Chapter VI, Paragraph 5, page 16r].

[41]See [22, page 52] or "*[s]i quantitas in duas partes dividatur, cubus totius aæqualis est, cubis ambarum partium, triploque productorum, uniuscuisque earum, vicissim in alterius quadratum*", see [4, Chapter VI, Paragraph 6, page 16r].

[42]See [4, Chapter VI, Paragraph 6, pages 16r-16v].

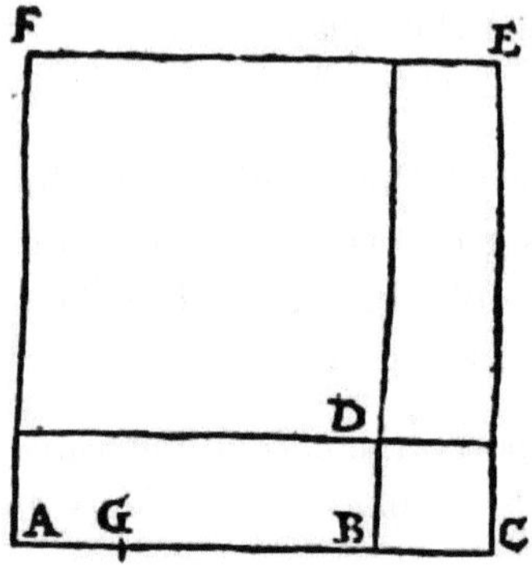

Figure 3.6. – Diagram from *Ars Magna* VI.6. Kind courtesy of Biblioteca Virtual del Patrimonio Bibliográfico and Universidad de Sevilla.

and, since $\overline{AC} = \overline{AB} + \overline{BC}$, it holds that

$$\overline{AC}^3 = \overline{AB}\left(\overline{(CD)}^2 + \overline{(DF)}^2 + \overline{(AD)} + \overline{(DE)}\right) + \\ \overline{BC}\left(\overline{(CD)}^2 + \overline{(DF)}^2 + \overline{(AD)} + \overline{(DE)}\right).$$

But, since "the bases and heights being the same", by *Elements* I.43, the following equalities hold

$$\overline{AB}\,\overline{(DF)}^2 = \overline{AB}^3, \quad \overline{BC}\,\overline{(CD)}^2 = \overline{BC}^3,$$
$$\overline{AB}\,\overline{(CD)}^2 = \overline{BC}\,\overline{(AD)} = \overline{BC}\,\overline{(DE)} = \overline{AB}\,\overline{BC}^2,$$
$$\overline{AB}\,\overline{(AD)} = \overline{AB}\,\overline{(DE)} = \overline{BC}\,\overline{(DF)}^2 = \overline{AB}^2\,\overline{BC}.$$

Then,

$$\overline{AC}^3 = \overline{AB}^3 + \overline{BC}^3 + 3\overline{AB}^2\,\overline{BC} + 3\overline{AB}\,\overline{BC}^2.$$

□

The proof flows in a quite easy way, since it is mainly based on remarking some equalities between measures (of the parts of the big solid that Cardano supposes to draw). As usual, he provides a geometrical environment to interpret the problem. The diagram acts as fixation of reference for the objects involved, so that one can think about the quantities in the hypotheses as measures associated to some geometrical objects.The references being fixed by the diagram, Cardano is enabled to use all the known results which are stated in terms of geometrical quantities. In particular, *Elements* II.4 and *Elements* I.43 are employed. Then, the amount and the kind (positional or not) of geometry in Cardano's proof depends in turn on the amount and the kind of geometry that one grants to those propositions in Euclid's *Elements*. In proofs of this kind, I have a penchant in bestowing an extremely narrow role to geometry, which means that I interpret the above proof mainly as an arithmetical calculation, once the references are fixed by the diagram. In the end, if it is true that the relative position of the geometrical objects involved is for sure useful in identifying objects that have the same measure (and which are the same), but are called with different names, like $\overline{AB(DF)^2} = \overline{AB^3}$, it is in the end not essential.

We finally remark that this very common kind of diagram is recurring – probably it is the most exploited – in the *Ars Magna*.

3.1.6. Summing up

We begin now to have a better overview of Cardano's knowledge on equations, in particular on cubic equations. The *Ars Magna* is a treasury of miscellaneous observations, rules, methods, and special cases. In the main, the ordering of the material is not always easy flowing. My impression is that the arrangement of the chapters does not match the chronology of the discoveries such that we often find repetitions, summaries, and in general a non-linear course.

In this section I have recalled some of the methods at Cardano's disposal to handle equations. According to the tradition, he knows how to solve linear, quadratic, and other simple equations such as $x^n = a_0$, with n natural. One of the most used method in the whole of Cardano's research on equations are the transformations of equations such as in Chapters I and VII. Outwardly, when Cardano wants to solve certain systems and finds more than one substitution that fits, he gets to different equations, which bear a relation to each other.

Cardano then shows a "general" way to deal with the cross-party family of equations $x^q + a_0 = a_p x^p$, with $q > p$. He is able to rewrite these equations under the form of certain systems. Anyway, finding the solutions of the systems is then equivalent to find the solutions of the equation (and as taught as it was for equations). This can nevertheless provide an aid to intuition to guess a solution when needed. Indeed, this strategy is especially useful in the "particular" cases of Chapter XXV, where the cubic formula contains imaginary numbers. All along this section, we have remarked Cardano's disposition towards the opposition of "general" *versus* "particular". At best, Cardano's aim is to find "general" rules to provide "general" solutions for an equation. But, if this terminology is intended in our modern terms, it could be quite misleading. We will look then for a better understanding in the following. Moreover, there is a second sense of this opposition applied to methods instead of to solutions. In this second sense, Cardano uses the term 'particular' to mean that a method cannot be applied in every case, where this impossibility is not only to be intended in a theoretical way, but also in practice.

Almost all the proofs follow a similar pattern. Basically, Cardano provides a geometrical environment where he interprets the problem. The diagrams serve for the fixation of reference. Then, Cardano can deal with the given quantities as if they were measures of certain geometrical quantities. In this way, he can apply the whole amount of known results that deal with geometrical quantities, such as the treasury of Euclid's *Elements*. Since the geometrical objects are considered only under their quantitative feature, their relative position becomes unimportant. This is a distinctive feature of Cardano's geometry. Then, the only constraint that one must respect in using such a geometry concerns the bonds established by the dimension. And it is exactly when Cardano needs to deal with dimensions higher than three that we can appreciate how much his geometry tends to mime arithmetic, and not the contrary.

3.2. Solving depressed cubic equations

In the middle of *Ars Magna*, there is a compact block, which is composed by Chapters XI–XXIII. These are entirely devoted to the study of cubic equations. Cardano assigns one chapter to each one of the thirteen families of cubic equations (see above, on page XIII).

All the chapters follow a similar pattern. They begin with a "proof [*demonstratio*]", from where is subsequently drawn a "rule [*regula*]" which is an algorithmic or step-by-step description of the operations that one has to perform on the coefficients of the given equation in order to get a solution. After that, we sometimes find an additional or alternative proof. Cardano always ends these chapters with a list of solved examples.

The so-called depressed cubic equations are considered easier by Cardano and are dealt with first. Afterwards, the cubic equations lacking in the first degree term and the complete ones follow. In the following paragraphs (each one devoted to a family of cubic equations), I resume in detail Cardano's treatment.

3.2.1. "On the cube and some things equal to a number"

In **Chapter XI** Cardano deals with the equation $x^3 + a_1x = a_0$. He says that

> Scipio del Ferro of Bologna well-nigh thirty years ago discovered this rule and handed it on to Antonio Maria Fior of Venice, whose contest with Niccolò Tartaglia of Brescia gave Niccolò occasion to discover it. He [Tartaglia] gave it to me in response to my entreaties, though withholding the demonstration. Armed with this assistance, I sought out its demonstration in [various] forms. This was very difficult. My version of it follows.[43]

[43]See [22, page 98] or "*Scipio Ferreus Bononiensis iam annis ab hinc triginta ferme capitulum hoc invenit, tradidit vero Anthonio Mariæ Florido Veneto, qui cum in certamen cum NicolaoTartales Brixellense aliquando venisset, occasionem dedit, ut Nicolaus invenerit et ipse, qui cum nobis rogantibus tradidisset, suppressa*

Cardano shows that:

AM XI. *Given the equation $x^3 + a_1 x = a_0$, a solution is*

$$x = \sqrt[3]{\sqrt{\left(\frac{a_0}{2}\right)^2 + \left(\frac{a_1}{3}\right)^3} + \frac{a_0}{2}} + \sqrt[3]{\sqrt{\left(\frac{a_0}{2}\right)^2 + \left(\frac{a_1}{3}\right)^3} - \frac{a_0}{2}}.$$

We observe that in this case we always have $\Delta_3 > 0$. Indeed, taking in the formula (2.5.10) on page 36 $\alpha_3 = 1$, $\alpha_2 = 0$, $\alpha_1 = a_1 > 0$, and $\alpha_0 = -a_0 < 0$, we get $\Delta_3 = \left(\frac{-a_0}{2}\right)^2 + \left(\frac{a_1}{3}\right)^3 > 0$.

AM XI (1545) — *Proof.* Let the equation [$x^3 + 6x = 20$ and, considering the diagram in Figure 3.7, $\overline{GH} = x$ be a solution, that is] $\overline{GH}^3 + 6\overline{GH} = 20$. Cardano[44] assumes that $\overline{BC} = \overline{CK}$ and $\begin{cases} 20 = \overline{AC}^3 - \overline{BC}^3 \\ \frac{6}{3} = \overline{AC}\,\overline{BC} \end{cases}$.

He wants to show that $\overline{AB} = \overline{GH}$, that is, that the equation $\overline{AB}^3 + 6\overline{AB} = 20$ holds. [This means that $\overline{AB} = x$ is also a solution of $x^3 + 6x = 20$.]

Cardano observes that the cubes $(DF)^3$, $(DC)^3$ respectively measure $\overline{AB}^3$, $\overline{BC}^3$ and that the parallelepipeds (AD), (DE) respectively measure $3\overline{AB}^2\,\overline{BC}$, $3\overline{AB}\,\overline{BC}^2$.

[Take $\overline{AB}$ to be a solution. We write $\overline{AB} = x$.] Since $\overline{AC}\,\overline{BC} = \frac{6}{3} = 2$, $3\overline{AC}\,\overline{BC} = 6$ follows, and then

$$(3\overline{AC}\,\overline{BC})\overline{AB} = 6x.$$

demonstratione, freti hoc auxilio, demonstrationem quæsivimus, eamque in modos, quod difficillimum fuit, redactam sic subijciemus", see [4, Chapter XI, page 29v].

[44] See [4, Chapter XI, pages 29v-30r].

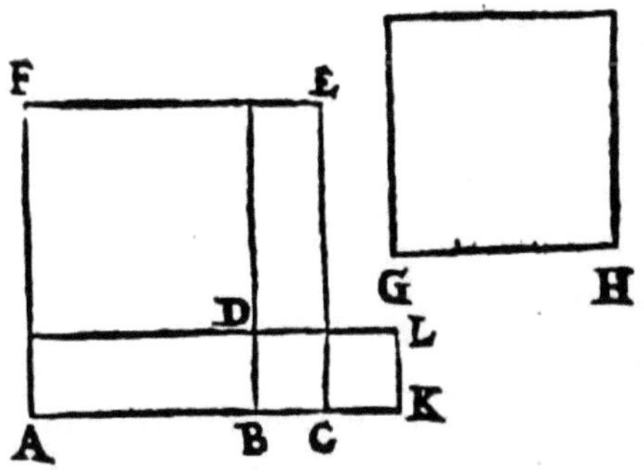

Figure 3.7. – Diagram from *Ars Magna* XI. Kind courtesy of Biblioteca Virtual del Patrimonio Bibliográfico and Universidad de Sevilla.

[Since $\overline{AC} = \overline{AB} + \overline{BC}$, I observe that we get $3\overline{AB}\,\overline{BC}^2 + 3\overline{AB}^2\,\overline{BC} = 6x$.] By $\overline{AC}^3 - \overline{BC}^3 = 20$, it follows that

$$\overline{AB}^3 + 3\overline{AB}^2\,\overline{BC} + 3\overline{AB}\,\overline{BC}^2 = 20,$$

applying (AM VI.6) to $\overline{AC} = \overline{AB} + \overline{BC}$.

Now, Cardano assumes that $\overline{BC} < 0$. He verifies in an analogous way that the last one of the above relations holds. Then, by *Elements* I.35 and by *Elements* XI.31, Cardano gets $\overline{AB} = \overline{GH}$. □

We observe that this kind of demonstration is opaque, meaning that the process of the discovery is not clearly explained. Indeed, no hints are given on how Cardano had the idea to assume the system

$$\begin{cases} 20 = \overline{AC}^3 - \overline{BC}^3 \\ \frac{6}{3} = \overline{AC}\,\overline{BC} \end{cases},$$

which is the core of the proof. But the system reminds of that part of Tartaglia's poem about $x^3 + a_1 x = a_0$, which says that

Quando che'l cubo con le cose appresso
    Se agguaglia à qualche numero discreto

Trovan dui altri differenti in esso.
Da poi terrai questo per consueto
Che 'l lor produtto sempre sia eguale
Al terzo cubo delle cose neto
El residuo poi suo generale
Delli lor lati cubi ben sottratti
Varra la tua cosa principale.

Indeed, we observe that Tartaglia and Cardano assume the same system up to changing variables. Indeed, Tartaglia suggests to solve the system

$$\begin{cases} y - z = a_0 \\ yz = (\frac{a_1}{3})^3 \end{cases}$$

and to draw the solution of the equation from the difference of the sides of two cubes $x = \sqrt[3]{y} - \sqrt[3]{z}$. Taking $y = Y^3$ and $z = Z^3$, we get

$$\begin{cases} Y^3 - Z^3 = a_0 \\ Y^3 Z^3 = (\frac{a_1}{3})^3 \end{cases},$$

which is Cardano's system

$$\begin{cases} Y^3 - Z^3 = a_0 \\ YZ = \frac{a_1}{3} \end{cases}.$$

Let us now glance at the link between the cubic formula and the proof, that is to say, why the above proof shows that the cubic formula given is a solution of the equation $x^3 + a_1 x = a_0$. The link is not explicitly given, even if it is easy to recover it. If we assume $\overline{AC} = \overline{AB} + \overline{BC}$ and the system

$$\begin{cases} a_0 = \overline{AC}^3 - \overline{BC}^3 \\ \frac{a_1}{3} = \overline{AC}\,\overline{BC} \end{cases},$$

then we can draw $\overline{AC}, \overline{BC}$ depending on the given coefficients a_1, a_0. In order to solve the system, we have to solve the equation $\overline{BC}^6 +$

$a_0\overline{BC}^3 - \frac{a_1^3}{27} = 0$, or equivalently $y^2 + a_0 y - \frac{a_1^3}{27} = 0$ taking $y = \overline{BC}^3$. It is easy to verify that the solutions of the system are

$$\begin{cases} \overline{BC} = \sqrt[3]{-\frac{a_0}{2} + \sqrt{\left(\frac{a_0}{2}\right)^2 + \left(\frac{a_1}{3}\right)^3}} \\ \overline{AC} = \sqrt[3]{\frac{a_0}{2} - \sqrt{\left(\frac{a_0}{2}\right)^2 + \left(\frac{a_1}{3}\right)^3}} \end{cases}, \begin{cases} \overline{BC} = \sqrt[3]{-\frac{a_0}{2} - \sqrt{\left(\frac{a_0}{2}\right)^2 + \left(\frac{a_1}{3}\right)^3}} \\ \overline{AC} = \sqrt[3]{\frac{a_0}{2} + \sqrt{\left(\frac{a_0}{2}\right)^2 + \left(\frac{a_1}{3}\right)^3}} \end{cases}.$$

We choose $x = \overline{AB}$ such that $\overline{AC} - \overline{BC} > 0$ and we get

$$x = \sqrt[3]{\frac{a_0}{2} + \sqrt{\left(\frac{a_0}{2}\right)^2 + \left(\frac{a_1}{3}\right)^3}} - \sqrt[3]{-\frac{a_0}{2} + \sqrt{\left(\frac{a_0}{2}\right)^2 + \left(\frac{a_1}{3}\right)^3}}.$$

We remark – as we previously did in the preceding section – that the proof is based on fixing the reference of the given quantities by the means of the diagram within the geometrical environment. This enables Cardano to deal with the given quantities as if they were the measures of some geometrical objects, and thus to apply to them results concerning geometrical objects. In particular, note that Cardano explicitly employs (AM VI.6), the proof of which makes in turn reference to *Elements* II.4. Then, assessing whether Cardano's proof makes an essential use of geometry or not (and in which sense geometry is possibly used) comes back to what interpretation one gives of *Elements* II.4 (concerning either geometrical objects or numbers, which are the measures of these geometrical objects). The same is also true concerning *Elements* I.35 and XI.31, which never generated such a great debate as *Elements*, Book II. Moreover, there is another issue related to this topic. One can ask whether the proof contains the recipe for drawing the solution segment AB, or only the instructions to calculate its measure. In my opinion, we should give credit on the second option. Indeed, one does not really need to pick up the ruler and the compass to draw AB. It is enough to settle the problem in a language in which we have certain propositions that act as inference rules and let the machinery work. Moreover, it seems clear from the above explanation of the relations between the cubic formula and the proof that the way of thinking is arithmetically tackled. Since very similar arguments also hold in the following, I will omit from now on these remarks.

In the 1570 and 1663 editions, this proof varies considerably. I will not present it here, as it is probably corrupted.[45] In the same editions, Cardano also explains how, taking two numbers to be a solution of $x^3 + a_1x = a_0$ and one of its coefficients, one can draw the equation, that is the remaining coefficient, using the fact that $a_0 = a_1x + x^3$ and $a_1 = \frac{a_0 - x^3}{x}$. Then, he starts from a depressed equation (like $x^3 + a_0 = a_1x$) of which he knows one solution, he takes this solution and one of its coefficients to be a solution and a coefficient of $x^3 + a_1x = a_0$, and draws the remaining coefficient. Finally, he says that "there may be a solution common to all types of cases [*æquatio hæc communis esse potest omnibus capitulis*]".

3.2.2. "On the cube equal to some things and a number"

In **Chapter XII** Cardano shows that

AM XII. *Given the equation* $x^3 = a_1x + a_0$, *a solution is*

$$x = \sqrt[3]{\frac{a_0}{2} + \sqrt{\left(\frac{a_0}{2}\right)^2 - \left(\frac{a_1}{3}\right)^3}} + \sqrt[3]{\frac{a_0}{2} - \sqrt{\left(\frac{a_0}{2}\right)^2 - \left(\frac{a_1}{3}\right)^3}}.$$

AM XII — *Proof*. Let the equation $x^3 = a_1x + a_0$. Considering the diagram in Figure 3.8, Cardano[46] assumes that

$$\begin{cases} a_0 = \overline{AB}^3 + \overline{BC}^3 \\ \frac{a_1}{3} = \overline{AB}\,\overline{BC} \end{cases}.$$

Then, he wants to show that $\overline{AC} = \overline{AB} + \overline{BC} = x$. [Take $\overline{AC}$ to be a

[45]See Witmer's comment in [22, Footnote 1, page 97].
[46]See [4, Chapter XII, page 31r].

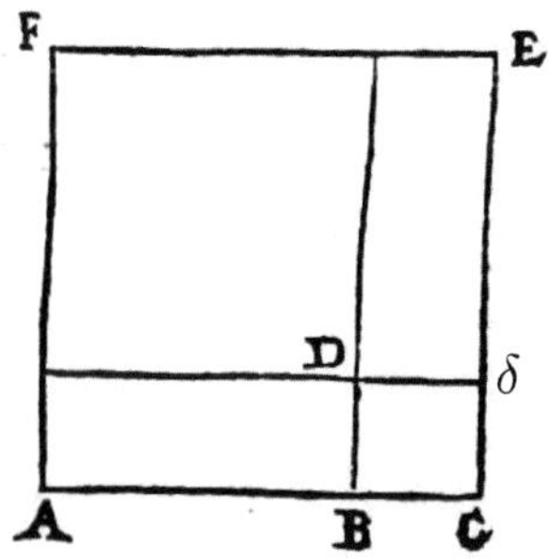

Figure 3.8. – Diagram from *Ars Magna* XII. Kind courtesy of Biblioteca Virtual del Patrimonio Bibliográfico and Universidad de Sevilla.

solution.] By $\overline{AB}\,\overline{BC} = \frac{a_1}{3}$, it follows that $3\overline{AB}\,\overline{BC} = a_1$, then

$$(3\overline{AB}\,\overline{BC})\overline{AC} = a_1\overline{AC}.$$

Cardano affirms that $(3\overline{AB}\,\overline{BC})\overline{AC} = 3\overline{AB}^2\,\overline{BC} + 3\overline{AB}\,\overline{BC}^2$. Note that he delays the proof of this statement. Then,

$$3\overline{AB}\,\overline{BC}^2 + 3\overline{AB}^2\,\overline{BC} = a_1\overline{AC}.$$

Therefore, by (AM VI.6),

$$3\overline{AB}^2\,\overline{BC} + 3\overline{AB}\,\overline{BC}^2 + \overline{AB}^3 + \overline{BC}^3 = \overline{AC}^3$$

and we know by hypothesis that $a_0 = \overline{AB}^3 + \overline{BC}^3$. Then, the equation $\overline{AC}^3 = a_1\overline{AC} + a_0$ holds.

It remains to show that $(3\overline{AB}\,\overline{BC})\overline{AC} = 3\overline{AB}^2\,\overline{BC} + 3\overline{AB}\,\overline{BC}^2$. Cardano rather shows that $(\overline{AB}\,\overline{BC})\overline{AC} = \overline{AB}^2\,\overline{BC} + \overline{AB}\,\overline{BC}^2$.

[The multiplication between measures of segments – that is, between numbers – being associative, then] $(\overline{AB}\,\overline{BC})\overline{AC} = \overline{AB}(\overline{BC}\,\overline{AC})$. Since "all sides are equal to all sides" (namely, $\overline{AC} = \overline{CE}$, $\overline{BC}$

unchanged), it holds that

$$\overline{AB}(\overline{BC}\,\overline{AC}) = \overline{AB}(\overline{BC}\,\overline{CE}),$$

that is

$$\overline{AB}(\overline{BC}\,\overline{AC}) = \overline{AB}(\overline{BC}(\overline{C\,\delta} + \overline{\delta\,E})) = \overline{AB}(\overline{BC}\,\overline{C\,\delta} + \overline{BC}\,\overline{\delta\,E}).$$

[The multiplication distributes over the sum] and again, since "all sides are equal to all sides" (namely, $\overline{C\,\delta} = \overline{BC}$, $\overline{\delta\,E} = \overline{AB}$), Cardano gets

$$\begin{aligned}\overline{AB}(\overline{BC}\,\overline{C\,\delta} + \overline{BC}\,\overline{\delta\,E}) &= \overline{AB}(\overline{BC}\,\overline{C\,\delta}) + \overline{AB}(\overline{BC}\,\overline{\delta\,E}) \\ &= \overline{AB}\,\overline{BC}^2 + \overline{AB}^2\,\overline{BC}.\end{aligned}$$

□

We observe that this proof follows the very same structure of the proof of (AM XI), with slightly variations. This is what we were expecting, since we recall that in Chapter VI (see above, on page 66) Cardano said that the solution of $x^3 = a_1x + a_0$ is derived from the solution of $x^3 + a_1x = a_0$ in Chapter XI.

Anyway, there is a difference. Indeed, Cardano additionally shows, as a sort of lemma, that

$$(3\overline{AB}\,\overline{BC})\overline{AC} = 3\overline{AB}^2\,\overline{BC} + 3\overline{AB}\,\overline{BC}^2,$$

whereas in the proof of (AM XI), where he also uses a similar consideration, he left this remark unsaid.

As before, no hints are given on how Cardano discovered the system in the hypotheses, which plays a central role. Again, it is the same as Tartaglia's one up to changing variables. Indeed, Tartaglia suggests to solve the system

$$\begin{cases} y + z = a_0 \\ yz = (\frac{a_1}{3})^3 \end{cases}$$

and to draw the solution of the equation from the sum of two cube's sides $x = \sqrt[3]{y} + \sqrt[3]{z}$. Taking $y = Y^3$ and $z = Z^3$, we get

$$\begin{cases} Y^3 + Z^3 = a_0 \\ Y^3 Z^3 = (\frac{a_1}{3})^3 \end{cases},$$

which implies Cardano's system

$$\begin{cases} Y^3 + Z^3 = a_0 \\ YZ = \frac{a_1}{3} \end{cases}.$$

As before, let us glance at the link between the cubic formula and proof. If we assume

$$\begin{cases} a_0 = \overline{AB}^3 + \overline{BC}^3 \\ \frac{a_1}{3} = \overline{AB}\,\overline{BC} \end{cases},$$

then we can draw $\overline{AB}, \overline{BC}$, and then $\overline{AC}$, depending on the given coefficients a_1, a_0. It is easy to verify that the solutions of the system are

$$\begin{cases} \overline{BC} = \sqrt[3]{\frac{a_0}{2} + \sqrt{(\frac{a_0}{2})^2 - (\frac{a_1}{3})^3}} \\ \overline{AB} = \sqrt[3]{\frac{a_0}{2} - \sqrt{(\frac{a_0}{2})^2 - (\frac{a_1}{3})^3}} \end{cases}, \begin{cases} \overline{BC} = \sqrt[3]{\frac{a_0}{2} - \sqrt{(\frac{a_0}{2})^2 - (\frac{a_1}{3})^3}} \\ \overline{AB} = \sqrt[3]{\frac{a_0}{2} + \sqrt{(\frac{a_0}{2})^2 - (\frac{a_1}{3})^3}} \end{cases}.$$

The resulting expression $\overline{AC} = \overline{AB} + \overline{BC}$ is symmetric, thus we get

$$x = \sqrt[3]{\frac{a_0}{2} + \sqrt{\left(\frac{a_0}{2}\right)^2 - \left(\frac{a_1}{3}\right)^3}} + \sqrt[3]{\frac{a_0}{2} - \sqrt{\left(\frac{a_0}{2}\right)^2 - \left(\frac{a_1}{3}\right)^3}}.$$

At the end of this chapter, Cardano makes a fundamental remark. He observes that, in order to apply this rule, the following condition has to be satisfied

> [w]hen the cube of one-third the coefficient of x is greater

> than the square of one-half the constant of the equation, which happens whenever the constant is less than three-fourths of this cube or when two-thirds of the coefficient of x multiplied by the square root of one-third the same number is greater than the constant of the equation,

that is

$$\left(\frac{a_0}{2}\right)^2 \geq \left(\frac{a_1}{3}\right)^3,$$

which[47] means that $\Delta_3 \geq 0$. If this condition does not hold (that is, if $\Delta_3 < 0$), we are in the *casus irreducibilis*, where a cubic equation has three distinct real solutions, but the cubic formula yields imaginary numbers.[48] This is why Cardano leaves this case out, going in the same direction of *Ars Magna*, Chapter V (see here, on page 60).

Then – he says[49] in the 1545 edition – one should better refer to the "Aliza problem [*quæstio Aliza*]" or to Chapter XXV of the *Ars Magna*. We remind that Chapter XXV deals with "particular" rules and solutions. There, Cardano explicitly states that the foregoing cases are "general" and in particular this is also referred to Chapter XII. But the fact that Chapter XII is "general" conflicts with Cardano's characterisation of "general" rules given in Chapter XXV, since indeed not every given problem (namely, the ones with $\Delta_3 < 0$) can be solved – for Cardano and his contemporaries – by the method in Chapter XII. Apart from this, what is more important is that Cardano's strategy[50] in Chapter XXV fails to provide a method safe enough to enable one

---

[47]See here 2.5.3, page 33.

[48]See here Section 2.5.4, page 37.

[49]"*At ut cubis tertiæ partis numeris rerum, excedat quadratum dimidij numeri, æquationis, quod accedit quandocunque numerus æquationis est minor $\frac{3}{4}$ cubi illius, vel ubi ex $\frac{2}{3}$ numerum rerum, producitur in R $\frac{1}{2}$ eiusdem numeri maior numerus numero æquationis, tunc hoc dissolvitur per quæstionem Alizam, de qua in libro de quæstionibus Geometricis dictum est, sed si libet tantam effugere difficultatem, plerumque capitulum 25m huius tibi satisfaciet*", see [4, Chapter XII, page 31v].

[50]See here Section 3.1.4, page 97.

to avoid imaginary numbers at all turns. At best, he manages to avoid to use the cubic formula, but this only happens in some lucky cases.

Later on, in the 1570 (and 1663) editions of the *Ars Magna*, Cardano makes reference[51] to the *De Regula Aliza* alone, which was firstly published together with the *Ars Magna* in 1570. I have already extensively dealt with this quotation from the *Aliza* in Section 2.3. I only recall that this quotation is a key point for my interpretation, since it explicitly links the *Aliza* with the *casus irreducibilis*.

3.2.3. "On the cube and a number equal to some things"

In **Chapter XIII** Cardano deals with the equation $x^3 + a_0 = a_1 x$. First of all, he says that its solving method "derives from the preceding ones [*ex præcedenti trahitur*]". We remind that in (AM I.5-6) Cardano explained that $x^3 + a_0 = a_1 x$ and $x^3 = a_1 x + a_0$ have the same solutions, but that they are opposite in sign.

Then, Cardano states that

AM XIII. *Given[52] the equation $x^3 + a_0 = a_1 x$, a solution is*

$$x = \frac{y}{2} \pm \sqrt{a_1 - 3\left(\frac{y}{2}\right)^2},$$

where y is a solution of $y^3 = a_1 y + a_0$.

Nothing guarantees that the above formula gives a real solution, except if we ask that $a_1 - 3\left(\frac{y}{2}\right)^2 > 0$. Moreover, despite the one example

[51] "*At ut cubis tertiæ partis numeris rerum, excedat quadratum dimidij numeri, æquationis, quod accedit quandocunque numerus æquationis est minor $\frac{3}{4}$ cubi illius, vel ubi ex $\frac{2}{3}$ numerum rerum, producitur in R $\frac{1}{2}$ eiusdem numeri maior numerus numero æquationis, tunc consules librum Alizæ hic adjectum*", see [12, Chapter XII, page 251].

[52] Expressed as before, the cubic formula is

$$x = \sqrt[3]{-\frac{a_0}{2} + \sqrt{\left(\frac{a_0}{2}\right)^2 - \left(\frac{a_1}{3}\right)^3}} + \sqrt[3]{-\frac{a_0}{2} - \sqrt{\left(\frac{a_0}{2}\right)^2 - \left(\frac{a_1}{3}\right)^3}}.$$

Cardano does not provide the formula in this form.

that Cardano gives (which is $x^3+3=8x$, the solutions of which are $-3, \frac{3}{2}-\frac{\sqrt{5}}{2}, \frac{3}{2}+\frac{\sqrt{5}}{2}$), nothing either guarantees that both the solutions in the formula are positive. And the last carelessness is that the equation has two (distinct) solutions only when $\Delta_3<0$, but it is not always the case.

This makes me think that the idea of the above formula originates in a completely different manner from (AM XI) or (AM XIII). As always, no hint is given on how Cardano discovered this formula. But the striking similarity with the formula in (AM XXV.1) on page 84 (or in (AM VIII.2 i) on page 75) stands out, even though these formulae concern the family $y^3=a_1y+a_0$. Indeed, in the *Ars Magna*, Cardano has available a path from the formula in (AM XXV.1) to the formula in (AM XIII), even though we have no evidence that he was aware of it. We assume that we are in the case $\Delta_3<0$ for $x^3+a_0=a_1x$. We consider $y^3=a_1y+a_0$, which accordingly also has $\Delta_3<0$. Suppose that one manages to find f,g positive such as in (AM XXV.1). Then, $y=\sqrt{f+\frac{g}{4}}+\frac{\sqrt{g}}{2}$, or $y=\sqrt{a_1-\frac{3}{4}g}+\frac{\sqrt{g}}{2}$ taking $f=a_1-g$, is a (positive) solution of $y^3=a_1y+a_0$. By (AM I.5-6), one now knows that $x=-y=-\sqrt{a_1-\frac{3}{4}g}-\frac{\sqrt{g}}{2}$ is a (negative) solution of $x^3+a_0=a_1x$. For the moment we keep it aside. Again, having found such f and g, $x=\sqrt{g}$ is a (positive) solution of $x^3+a_0=a_1x$ by (AM VIII.2). Then, $y=-x=-\sqrt{g}$ is a (negative) solution of $x^3+a_0=a_1x$. Substituting this value in $x=-\sqrt{a_1-\frac{3}{4}g}-\frac{\sqrt{g}}{2}$, we have one of the solutions in (AM XIII). Finally, we also know by (AM I.5-6) how to draw the third solution depending on the first two for each $x^3+a_0=a_1x$ and $y^3=a_1y+a_0$, from where by a similar reasoning we get to the other solution in (AM XIII). Clearly, this is only my supposition. But we observe that, even though we need to assume that $\Delta_3<0$, in order to have the three real solutions, it is automatically gained that $a_1-3\left(\frac{y}{2}\right)^2>0$, since we took f,g positive such that $a_1=f+g$ and $y=-\sqrt{g}$.

Anyway, even if my hypothesis holds, no mark is left in Cardano's proof.

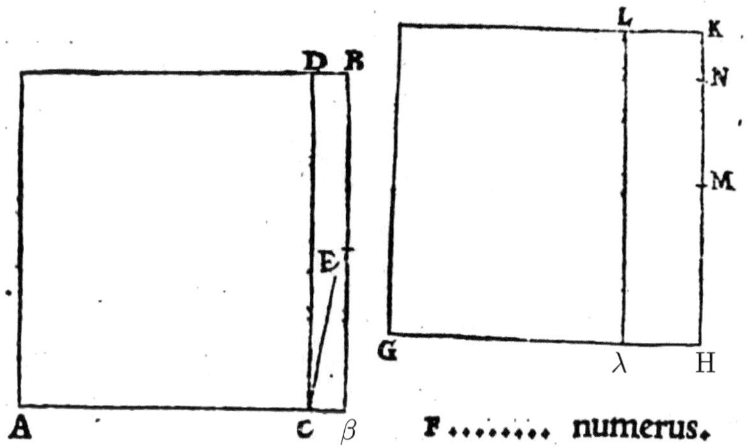

Figure 3.9. – Diagram from *Ars Magna* XIII. Kind courtesy of Biblioteca Virtual del Patrimonio Bibliográfico and Universidad de Sevilla.

AM XIII — *Proof***.** Consider the diagram in Figure 3.9 and let the equation $\overline{HK}^3 + F = \overline{A\beta}^2\,\overline{HK}$, where F is a [real] number. Cardano[53] assumes that

$$\overline{HK}^2 = 4\overline{LK}\,\overline{HK}, \quad \overline{HK} = 2\overline{MK},$$
$$\overline{HK}^2 - (\overline{LK}\,\overline{HK}) = \overline{G\lambda}\,\overline{\lambda L} = \overline{AC}\,\overline{CD},$$
$$\overline{MN} = \overline{CE}, \quad \sqrt{\overline{C\beta}\,\overline{\beta B}} = \overline{CE}.$$

Then, he wants to show that either $\overline{HN}$ is a solution of $x^3 + F = \overline{A\beta}^2 x$, that is that the equation $\overline{HN}^3 + F = \overline{A\beta}^2\,\overline{HN}$ holds, or $\overline{NK}$ is a solution of $x^3 + F = \overline{A\beta}^2 x$, that is that the equation $\overline{NK}^3 + F = \overline{A\beta}^2\,\overline{NK}$ holds. [We will write in the first case $\overline{HN} = x$ and in the second case $\overline{NK} = x$.]

[53]See [4, Chapter XIII, pages 31v-32r].

Cardano is looking for a writing of $\overline{A\beta}^2$ depending on the segments in the square on the right in the diagram

$$\begin{aligned}\overline{A\beta}^2 & [= (\overline{AC}\,\overline{CD}) + (\overline{C\beta}\,\overline{\beta B}) \\ &= \overline{HK}^2 - (\overline{LK}\,\overline{HK}) + \overline{CE}^2 \\ &= \frac{3}{4}\overline{HK}^2 + \overline{MN}^2] \\ &= 3\overline{MK}^2 + \overline{MN}^2.\end{aligned}$$

In the first case with respect to $\overline{HN}$, he forces $\overline{HN}$ to appear in the expression for $\overline{A\beta}^2$

$$\begin{aligned}\overline{A\beta}^2 - \overline{HN}^2 & [= 3\overline{MK}^2 + \overline{MN}^2 - \overline{HN}^2 \\ &= 2\overline{MK}(\overline{MK} - \overline{MN}) \\ &= 2\overline{MK}\,\overline{NK}] \\ &= \overline{HK}\,\overline{NK}.\end{aligned}$$

In this way, he is able to rewrite $\overline{HN}^3 = \overline{HN}\,\overline{HN}^2 = \overline{HN}(\overline{A\beta}^2 - \overline{HK}\,\overline{NK})$. Moreover, $\overline{A\beta}^2 = \overline{HN}^2 + \overline{HK}\,\overline{NK}$. Since the equation $\overline{HK}^3 + F = \overline{A\beta}^2\,\overline{HK}$ holds by assumption, Cardano looks for a writing of F depending on the segments in the second diagram, in particular depending also on $\overline{HN}$

$$\begin{aligned}F &= \overline{HK}(\overline{HK}^2 - \overline{A\beta}^2) \\ &= \overline{HK}((\overline{G\lambda}\,\overline{\lambda L} + \overline{LK}\,\overline{HK}) - (\overline{AC}\,\overline{CD} + \overline{C\beta}\,\overline{\beta B})) \\ &= \overline{HK}(\overline{LK}\,\overline{HK} - \overline{C\beta}\,\overline{\beta B}) \\ &= \overline{HK}(\overline{MK}^2 - \overline{MN}^2) \\ &= \overline{HK}(2\overline{MN}\,\overline{NK} + \overline{NK}^2 \\ &= \overline{HK}\,\overline{NK}\,\overline{HN}.\end{aligned}$$

Then,

$$\begin{aligned}\overline{HN}^3 + F &= \overline{HN}(\overline{A\beta}^2 - \overline{HK}\,\overline{NK}) + \overline{HK}\,\overline{NK}\,\overline{HN}\\ &= \overline{A\beta}^2\,\overline{HN},\end{aligned}$$

that is the equation $\overline{HN}^3 + F = \overline{A\beta}^2\,\overline{HN}$ holds.

In the second case with respect to $\overline{NK}$, Cardano shows in a very analogous way that, under his hypothesis, the equation $\overline{NK}^3 + F = \overline{A\beta}^2\,\overline{NK}$ holds. □

Let us glance at the link between the solution formula and the proof, that is to say why the above proof shows that the formula $x = \frac{y}{2} \pm \sqrt{a_1 - 3\left(\frac{y}{2}\right)^2}$, with y solution of $y^3 = a_1 y + a_0$, is a solution of the equation $x^3 + a_0 = a_1 x$. Cardano affirmed that this rule derives from (AM XII). Indeed, he needs to know how to solve $y^3 = a_1 y + a_0$ in order to solve $x^3 + a_0 = a_1 x$. By (AM I.5-6), Cardano knows that, if there are two positive solutions x_1 and x_2 of $x^3 + a_0 = a_1 x$ (namely, in the case $\Delta_3 < 0$), then $y_3 = x_1 + x_2$ is a solution of $y^3 = a_1 y + a_0$. Cardano himself makes this remark.[54] We take $y = \overline{HK}$. Then, the solutions that Cardano is looking for are $x_1 = \overline{HN} = \overline{HM} + \overline{MN} = \overline{MK} + \overline{MN}$ and $x_2 = \overline{NK} = \overline{MK} - \overline{MN}$, that is

$$x_1 = \frac{1}{2}\overline{HK} + \overline{MN} \quad \text{and} \quad x_2 = \frac{1}{2}\overline{HK} - \overline{MN},$$

where $\overline{MN} = \sqrt{\overline{A\beta}^2 + 3\overline{MK}^2} = \sqrt{\overline{A\beta}^2 + 3\left(\frac{\overline{HK}}{2}\right)^2}$. Since $\overline{A\beta}^2 = a_1$ and $\overline{HK} = y$, we find the formula we were looking for.

[54] "From this demonstration it is evident that the solution for the cube equal to the first power and number is equal to the sum of the solutions for the cube plus the same number equal to the first power with the same coefficient", see [22, page 106] or "*[e]x hac demonstratione patet, quod æquatio cubi æqualis rebus et numero, æqualis est ambabus æquationis cubi et eiusdem numeri æqualium totidem rebus simul iunctis*", see [4, Chapter XIII, page 32r]. Note that in the 1570 and 1663 editions this remark is shifted at the end of the chapter.

Note nevertheless that, in order to use the above formula for $x^3 + a_0 = a_1x$, one should be able to find a solution of $y^3 = a_1y + a_0$, which is not always the case because of the *casus irreducibilis.*

Afterwards, Cardano adds another proposition.

AM XIII bis. *Let us consider the equation* $x^3 + a_0 = a_1x$ *and assume* x_1 *to be a solution. Then,*

$$x_2 = \sqrt{a_1 - 3\left(\frac{x_1}{2}\right)^2} - \frac{x_1}{2}$$

is another solution.

We remark that again one needs to ask that $\Delta_3 < 0$ to guarantee the existence of the second real solution. Once that one agrees with my reasoning in (AM XIII) on page 116 and, as a consequence, he acknowledges that Cardano knows that, under certain conditions, two of the solutions of $x^3 + a_0 = a_1x$ can be written as $x_1 = \sqrt{g}$ and $x_3 = -\sqrt{a_1 - \frac{3}{4}g} - \frac{\sqrt{g}}{2}$, then the above formula is gathered by the opposite of the sum of these two solutions according to (AM I.5-6).

I will skip the proof in details, just sketching without calculations the idea behind it. At first sight, the proof is not so easy-reading, but it comes true that deep down it is very similar to (AM XVI bis).

AM XIII bis (idea) — *Proof*. Considering the diagram in Figure 3.10, Cardano[55] takes $x_1 = \overline{AD}$, $a_1 = \overline{AD^2} - \overline{AD^2}$, and $x_2 = \overline{FH}[= \sqrt{g}]$. He wants to show that such a $\overline{FH}$ is a solution of $x^3 + a_0 = a_1x$.

He knows that it exists $\overline{EF}$ such that $\overline{EF}^2 = a_1 - \overline{FH}[= f]$. Now, it remains to show that $\overline{EF}^2\,\overline{FH} = a_0$, in order to infer by (AM VIII.2) that $\overline{FH}$ is a solution. For that, Cardano considers $\overline{AD}$ which is a solution of $x^3 + a_0 = a_1x$ by hypothesis. Then,

[55] See [4, Chapter XIII, page 32v].

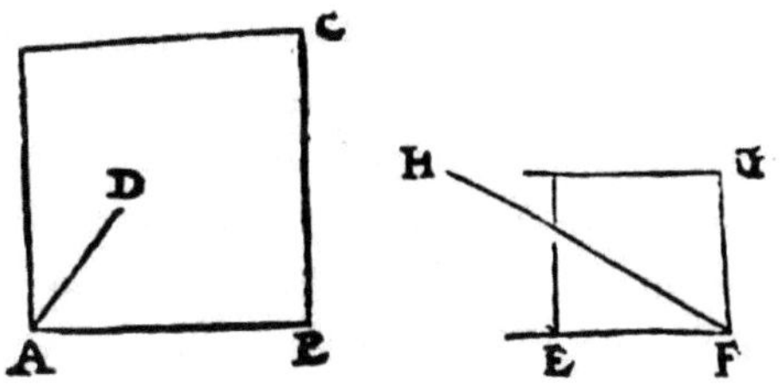

Figure 3.10. – Diagram from *Ars Magna* XIII bis. Kind courtesy of Biblioteca Virtual del Patrimonio Bibliográfico and Universidad de Sevilla.

$a_0 = a_1\overline{AD} - \overline{AD}^3 = (a_1 - \overline{AD}^2)\overline{AD} = \overline{AB}^2\,\overline{AD}$. He finally shows that $\overline{EF}^2\,\overline{FH} = \overline{AB}^2\,\overline{AD}$. This ends the proof. □

We observe that this proposition provides a formula to find a second solution of $x^3 + a_0 = a_1 x$, which bears no conditions to be satisfied. Anyway, one still has to be able to guess – no matter how – a first solution of the same equation. This could help in bypassing the *casus irreducibilis* at least for the second solution, but does not help at all in finding the first solution.

In the 1570 and 1663 editions, three corollaries are added. In the first, Cardano recalls that the (positive) solution of $x^3 = a_1 x + a_0$ is given by the sum of the (positive) solutions of $x^3 + a_0 = a_1$, which is a way to restate (AM I.5-6). The second corollary simply reminds (AM XIII bis). Finally, the third corollary is about the comparison of the shapes of the (positive) solutions of $x^3 = a_1 x + a_0$ and $x^3 + a_0 = a_1$. As usual through an example, Cardano shows that, if the (positive) solution of $x^3 = a_1 x + a_0$ is a *binomium* (of the 2nd or 5th kind), then (one of the positive) solutions of $x^3 + a_0 = a_1$ is its *recisum*.

3.2.4. Summing up

Cardano gives in (AM XI) and (AM XII) two cubic formulae, respectively for $x^3 + a_1 x = a_0$ and $x^3 = a_1 x + a_0$. The proofs are independent but strictly related. They are essentially what one expects. Indeed,

Cardano simply verifies that a certain segment, when substituted in a certain equation, gives a true equality. In order to do that, he has to write down an expression for the segment in term of the coefficients of the equation. Thanks to his interpretation of the hypotheses in a geometrical environment, it is easy for him to do that. The trick behind these proofs is to write $x = y \pm z$. Then, Cardano basically uses the binomial formula in the case of the third power and the properties of the operations on the measures of segments. As previously argued, my opinion is that Cardano's proofs do not deliver the instruction to draw the solution segment, but rather to calculate its measure.

We soon run into the major drawback of Cardano's formulae, the so-called *casus irreducibilis*. Except for the cubic formula in (AM XI), Cardano has to check not to have negative quantities under the square roots and this is made by forcing a condition on the coefficients. This means that, under certain circumstances, Cardano is not able to draw a solution from the formula for that family of cubic equations. Since the formula in (AM XIII) refers back to the one in (AM XII), it also has to meet the same condition and then it bears the same problem.

Concerning the equation $x^3 + a_0 = a_1 x$, Cardano also finds a way to get another solution, thus bypassing the cubic formula in (AM XII). It is interesting to remark how this alternative formula could have originated from "particular" arguments such as the ones in Chapter VII and XXV and by the transformations in Chapter I.

3.3. Solving cubic equation lacking in the first degree term

3.3.1. "On the cube equal to some squares and a number"

In **Chapter XIV** Cardano deals with the equation $x^3 = a_2 x^2 + a_0$. He says that

> [i]f the cube is equal to the square and constant, the equation can be changed into one of the cube equal to the first power and constant by the first method of conversion,

> which is from the whole to the part. (The second method is from the part to the whole, the third by the difference between parts, and the fourth by proportion).[56]

We recall[57] that in Chapter VI Cardano spoke of a method of "similitude" to solve new cases, the third part of which was "the conversion of equations into [others] equivalent in nature". He now better specifies what this "method of conversion" is, saying that it is divided into parts.

AM XIV. *Given the equation $x^3 = a_2x^2 + a_0$, a solution is*

$$x = \sqrt[3]{\left(\frac{a_2}{3}\right)^3 + \frac{a_0}{2} + \sqrt{\left(\left(\frac{a_2}{3}\right)^3 + \frac{a_0}{2}\right)^2 - \left(\frac{a_2}{3}\right)^6}}$$
$$+ \sqrt[3]{\left(\frac{a_2}{3}\right)^3 + \frac{a_0}{2} - \sqrt{\left(\left(\frac{a_2}{3}\right)^3 + \frac{a_0}{2}\right)^2 - \left(\frac{a_2}{3}\right)^6}} + \frac{a_2}{3}.$$

We observe that, in this case, we always have $\Delta_3 > 0$. Indeed, taking in the formula (2.5.10) on page 36 $\alpha_3 = 1$, $\alpha_2 = -a_2 < 0$, $\alpha_1 = 0$, and $\alpha_0 = -a_0 < 0$, we get $\Delta_3 = \frac{-a_0}{108}\left(-27a_0 - 4a_2^3\right) > 0$. This fact also appears in the above formula, since $\left(\left(\frac{a_2}{3}\right)^3 + \frac{a_0}{2}\right)^2 - \left(\frac{a_2}{3}\right)^6 = \frac{a_0^2}{4} + a_0\frac{a_2^3}{27} > 0$.

AM XIV — *Proof.* Let the equation $\overline{AC}^3 = 6\overline{AC}^2 + 100$. [From now on, the diagrams appear no longer in the proofs. This is because Cardano[58] makes reference to the diagram in Chapter XII, see the figure 3.8 on page 111.]

---

[56]See [22, page 110] or "*[q]uod, si cubus æqualis sit quadratis et numero, convertetur capitulum in cubum æqualem rebus et numero, primo conversionis modo, qui est a toto ad partem, nam secundus est a parte ad totum, tertius a differentia partium, quartus a proportione*", see [4, Chapter XIV, page 33r].

[57]See here, page 66.

[58]See [4, Chapter XIV, pages 33r-33v].

[Take $\overline{BC} = \frac{6}{3} = 2$. Then, Cardano shows that the equation $\overline{AC}^3 = 6\overline{AC}^2 + 100$ can be changed into $\overline{AB}^3 = 12\overline{AB} + 116$ *via* the substitution $\overline{AC} = \overline{AB} + 2$.]

[By *Elements* II.4,] Cardano knows that $\overline{AC}^2 = \overline{AB}^2 + 2\overline{AB}\,\overline{BC} + \overline{BC}^2 = \overline{AB}^2 + 4\overline{AB} + 4$. Then, by the assumed equation,

$$\begin{aligned}\overline{AC}^3 &= 6\overline{AC}^2 + 100\\ &= 6\overline{AB}^2 + 24\overline{AB} + 24 + 100.\end{aligned}$$

Using the other way, by (AM VI.6) Cardano gets

$$\begin{aligned}\overline{AC}^3 = (\overline{AB} + \overline{BC})^3 &= \overline{AB}^3 + 3\overline{AB}^2\,\overline{BC} + 3\overline{AB}\,\overline{BC}^2 + \overline{BC}^3\\ &= \overline{AB}^3 + 6\overline{AB}^2 + 12\overline{AB} + 8.\end{aligned}$$

Then, by comparison,

$$\begin{aligned}6\overline{AB}^2 + 24\overline{AB} + 24 + 100 &= \overline{AB}^3 + 6\overline{AB}^2 + 12\overline{AB} + 8,\\ \overline{AB}^3 &= 12\overline{AB} + 116.\end{aligned}$$

Now, Cardano knows how to solve the transformed equation in $\overline{AB}$ by (AM XII). The relation $\overline{AC} = \overline{AB} + \overline{BC}$ yields the solution of the equation in $\overline{AC}$. □

Let us glance at the link between the cubic formula and the proof. Call $\overline{AC} = x$ and $\overline{AB} = y$. We observe that, substituting $x = y + \frac{a_2}{3}$ in $x^3 = a_2x^2 + a_0$, we get the following transformation

$$x^3 = a_2x^2 + a_0 \xrightarrow{x=y+\frac{a_2}{3}} y^3 = \frac{a_2^2}{3}y + (\frac{2}{27}a_2^3 + a_0),$$

which is basically what Cardano performs in the proof. Then, if we apply the cubic formula in (AM XII) to the transformed equation $y^3 = \frac{a_2^2}{3}y + \frac{2}{27}a_2^3 + a_0$, we obtain the cubic formula of this chapter. We remark that this transformation is new compared to the ones in Chapters VI and VII for $x^3 = a_2x^2 + a_0$.

If we already knew how the following chapters deal with the remaining families of equations, we would realise that the common strategy is not to provide the formulae, but rather the transformations to get the formulae. Then, we can easily guess the transformation used here and restate the proposition as follows.

AM XIV restated. *Given the equation $x^3 = a_2x^2 + a_0$, a solution is*

$$x = y + \frac{a_2}{3},$$

where y is a solution of[59] $y^3 = \frac{a_2^2}{3}y + \frac{2}{27}a_2^3 + a_0$.

Recall that in Proposition (AM IV.2) (see above, on page 81) Cardano displays the above substitution, even if it is not directly referred to this equation. Then, my guess is not that imaginative (moreover, see below, Proposition (AMA XXI.9) on page 195 and the figure 4.4 on page 241).

Note that, despite the fact that the equation in x always has $\Delta_3 > 0$, this is not *a priori* the case for the transformed equation in y. Anyway, if we go and check the discriminant of the transformed equation, we get

$$\Delta_3 = \frac{a_0^2}{4} + \frac{1}{27}a_0a_2^3,$$

which is always positive, since we took a_0, a_2 positive.

3.3.2. "On the cube and some squares equal to a number"

In **Chapter XV** Cardano deals with the equation $x^3 + a_2x^2 = a_0$.

---

[59]Note that, despite the fact that the equation in x has always $\Delta_3 > 0$, this is not *a priori* the case for the transformed equation in y. Anyway, if we go and check the discriminant of the transformed equation, we get

$$\Delta_3 = \frac{a_0^2}{4} + \frac{1}{27}a_0a_2^3,$$

which is always positive since we took a_0, a_2 positive.

AM XV. *Given[60] the equation* $x^3 + a_2 x^2 = a_0$, *a solution is*

$$x = y - \frac{a_2}{3},$$

where y *satisfies the following conditions*

i. *if* $a_0' > 0$, *then* $y^3 + a_0' = a_1' y$,

ii. *if* $a_0' < 0$, *then* $y^3 = a_1' y + (-a_0)$,

iii. *if* $a_0' = 0$, *then* $y^3 = a_1' y$,

with $a_1' = 3\left(\frac{a_2}{3}\right)^2 \left[= \frac{a_2^2}{3}\right]$ *and* $a_0' = 2\left(\frac{a_2}{3}\right)^3 - a_0$.

We remark that the number of cases increases, since Cardano always wants positive coefficients in the transformed equations. Let us consider for instance this case. If we perform the transformation $x = y - \frac{a_2}{3}$, we get to the equation

$$y^3 - \frac{a_2^2}{3} y + \frac{2}{27} a_2^3 - a_0 = 0,$$

where the sign of $\frac{2}{27} a_2^3 - a_0$ varies depending on the coefficients of the equation in x. Then, Cardano needs to distinguish the cases in order to be enabled to write explicitly the transformed equations.

Note moreover that there is only a loose correspondence between these cases and the sign of the discriminant $\Delta_3 = \frac{q^2}{4} + \frac{p^3}{27}$ (see above, the formula (2.5.10) on page 36). Indeed, we have $p = -a_1' \leq 0$

[60] Expressed as before, the cubic formula is

$$x = \sqrt[3]{\left(-\frac{a_2}{3}\right)^3 + \frac{a_0}{2} + \sqrt{\left(\left(-\frac{a_2}{3}\right)^3 + \frac{a_0}{2}\right)^2 - \left(\frac{a_2}{3}\right)^6}}$$
$$+ \sqrt[3]{\left(-\frac{a_2}{3}\right)^3 + \frac{a_0}{2} - \sqrt{\left(\left(-\frac{a_2}{3}\right)^3 + \frac{a_0}{2}\right)^2 - \left(\frac{a_2}{3}\right)^6}} - \frac{a_2}{3}.$$

Cardano does not provide the formula in this form.

and $q = a_0'$, which implies $q^2 \geq 0$. Anyway, this is not enough to completely determine the sign of Δ_3. We can only say that $\Delta_3 > 0$ in the first case $a_0' > 0$ and that $\Delta_3 \geq 0$ in the third case $a_0' = 0$.

Cardano firstly remarks that "[i]n this case we convert by the second method. The difference is that the first method showed the addition of one-third the coefficient of x^2, while the second shows its subtraction".[61] The "second method" of which Cardano is speaking is "from the part to the whole", according to the corresponding remark in Chapter XIV.

Cardano gives a proof only for the case $2\left(\frac{a_2}{3}\right)^3 < a_0$, that is $a_0' < 0$.

AM XV — *Proof.* Let the equation $\overline{AB}^3 + 6\overline{AB}^2 = 100$. [Cardano[62] considers the same diagram as in Chapter XII, see above, the figure 3.8 on page 111.] He takes $\overline{BC} = \frac{6}{3} = 2$.

[Then, Cardano shows that the equation $\overline{AB}^3 + 6\overline{AB}^2 = 100$ can be changed into $\overline{AC}^3 = 12\overline{AC} + 84$ *via* the substitution $\overline{AB} = \overline{AC} - 2$.]

By (AM VI.6) and by $\overline{AB}^3 + 6\overline{AB}^2 = 100$, Cardano gets

$$\begin{aligned}\overline{AC}^3 = (\overline{AB} + \overline{BC})^3 &= \overline{AB}^3 + 3\overline{AB}^2\,\overline{BC} + 3\overline{AB}\,\overline{BC}^2 + \overline{BC}^3 \\ &= \overline{AB}^3 + 6\overline{AB}^2 + 12\overline{AB} + \overline{BC}^3 \\ &= 100 + 12\overline{AB} + \overline{BC}^3 \\ [&= 100 + 12(\overline{AC} - \overline{BC}) + \overline{BC}^3].\end{aligned}$$

Since $\overline{BC} = 2$, we get $\overline{AC}^3 = 12\overline{AC} + 84$.

Now, Cardano knows how to solve the transformed equation in $\overline{AC}$ by (AM XII). The relation $\overline{AB} = \overline{AC} - \overline{BC}$ yields the solution of the equation in AB. □

[61] See [22, page 113] or "*[h]oc capitulum convertitur secundo modo, differentia autem est, quod primus modus ostendit addendam tertiam partem numeri quadratorum, et secundus minuendam*", see [4, Chapter XV, page 33v].

[62] See [4, Chapter XV, pages 33v-33r].

In the 1545 edition, Cardano adds a corollary at the end of the proof. Since it is about the solving method in Chapter XVI, we will comment on it in the next chapter.

Afterwards, Cardano gives another solution method for this equation. He says that it is due to Ludovico Ferrari.

AM XV bis. *Given the equation* $x^3 + a_2x^2 = a_0$, *a solution is*

$$x = y^2 - a_2,$$

with y *solution of* $y^3 = a_2y + \sqrt{a_0}$.

We remark that this is the same substitution in (AM VII.6-7).

We observe that the main advantage of this method is that Cardano and Ferrari do no longer need to explicitly distinguish the cases in the cubic formula. Indeed, performing the substitution, we get

$$y^6 - 2a_2y^4 + a_2y^2 = a_0,$$

that is

$$(y^3 - a_2y)^2 = a_0,$$

where the coefficients are as usual all positive. Then, solving $x^3 + a_2x^2 = a_0$ depends only on solving $y^3 = a_2y + \sqrt{a_0}$ in (AM XII). Note that, considering the cases in (AM XV), it sounds very reasonable that the equation $x^3 + a_2x^2 = a_0$ is brought back only to one case. Indeed, the transformed equation in y in the second case in (AM XV) has $\Delta_3 \lesseqgtr 0$, so that no restriction is put on the discriminant of the equation in x.

We moreover observe that this is no alternative method to (AM XV). Indeed, one can show with calculations that both (AM XV) and (AM XV bis) lead to the same cubic formula (see Footnote 60 on page 126).

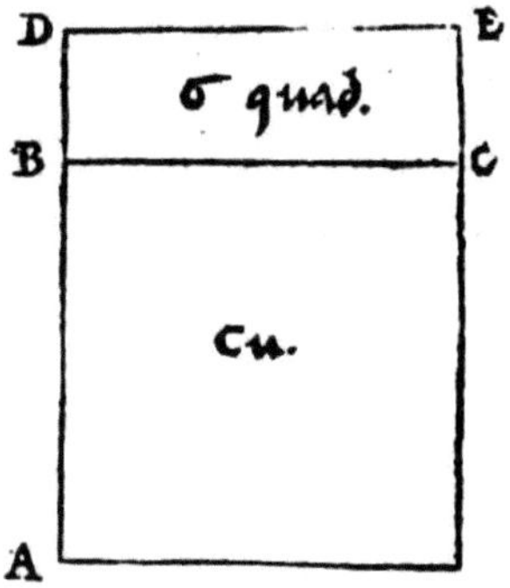

Figure 3.11. – Diagram from *Ars Magna* XV. It should be completed by drawing the parallelogram of height $\overline{AB}$ over the basis formed by the parallelepiped on AD, AB. Kind courtesy of Biblioteca Virtual del Patrimonio Bibliográfico and Universidad de Sevilla.

AM XV bis — *Proof*. Cardano and Ferrari[63] consider the diagram in Figure 3.11 on page 129 and the following equation $\overline{AB}^3+\overline{BD}\,\overline{AB}^2 = 100$, where $\overline{BD} = 6$. [They want to show that this equation can be changed into $\overline{AD}^3 = 6\overline{AD}+10$, where $\overline{AD}$ is a square quantity. Then $\overline{AB} = \overline{AD} - \overline{BD}$.]

They assume $\overline{AD} = y^2$.

Then, $\overline{AB} = \overline{AD} - \overline{DB} = y^2 - 6$ and $\overline{AB}^2 = y^4 - 12y^2 + 36$.

Therefore, $\overline{AD}\,\overline{AB}^2 = y^6 - 12y^4 + 36y^2$. But $\overline{AD}\,\overline{AB}^2[= (\overline{AB} + \overline{BD})\overline{AB}^2] = \overline{AB}^3 + \overline{BD}\,\overline{AB}^2 = 100$. Then,

$$y^6 - 12y^4 + 36y^2 = 100.$$

Taking the square root from both sides, they get

$$y^3 = 6y + 10.$$

Now, they know how to solve the transformed equation in y by

[63] See [4, Chapter XV, page 34v].

(AM XII). The relation $\overline{AB} = \overline{AD} - \overline{BC}$ yields then a solution of $\overline{AB}^3 + \overline{BD}\,\overline{AB}^2 = 100$. □

Cardano says that the proof is "similar to our general [demonstration] in Chapter VII". Indeed, the second transformation in (AM VII.6-7) implies as a particular case the following transformation

$$x^3 = a_1 x + a_0 \xrightarrow{y=x^2-a_1} y^3 + a_1 y^2 = a_0^2,$$

which is the same as the one in this chapter (note that x and y were exchanged in Chapter VII).

3.3.3. "On the cube and a number equal to some squares"

In **Chapter XVI** Cardano deals with the equation $x^3 + a_0 = a_2 x^2$.

AM XVI. *Given*[64] *the equation* $x^3 + a_0 = a_2 x^2$*, a solution is*

$$x = \frac{\left(\sqrt[3]{a_0}\right)^2}{y},$$

with y *solution of* $y^3 + a_0 = a_2 \sqrt[3]{a_0}\, y$.

Cardano gives no proof of this rule. He just affirms that "this rule is clearly evident from the demonstration of the seventh chapter [*[h]oc capitulum per se patet, ex demonstratione 7ⁱ capituli*]". Indeed, the

[64]Expressed as before, the cubic formula is

$$x = \sqrt[3]{\left(\frac{a_2}{3}\right)^3 - \frac{a_0}{2} + \sqrt{\left(\left(-\frac{a_2}{3}\right)^3 + \frac{a_0}{2}\right)^2 - \left(\frac{a_2}{3}\right)^6}} + \sqrt[3]{\left(\frac{a_2}{3}\right)^3 - \frac{a_0}{2} - \sqrt{\left(\left(-\frac{a_2}{3}\right)^3 + \frac{a_0}{2}\right)^2 - \left(\frac{a_2}{3}\right)^6}} + \frac{a_2}{3}.$$

Cardano does not provide the formula in this form.

We observe that it can be showed with calculations that this is the same cubic formula that one obtains using the substitution $x = y + \frac{a_2}{3}$.

first transformation in (AM VII.2-3) implies as a particular case the following one

$$x^3 + a_0 = a_2 x^2 \xrightarrow{y=\frac{(\sqrt[3]{a_0})^2}{x}} y^3 + a_0 = (\sqrt{a_0} a_2)\, y,$$

which is the one that we find in this chapter. The proof that Cardano gave in (AM VII.5) concerned just this particular case. We remind that in (AM VII.5) Cardano drawed a geometrical model for the equation $x^3 + a_0 = a_2 x^2$ taking $x = \overline{AD}$, $a_0 = \overline{AD}^2\, \overline{DM}$.

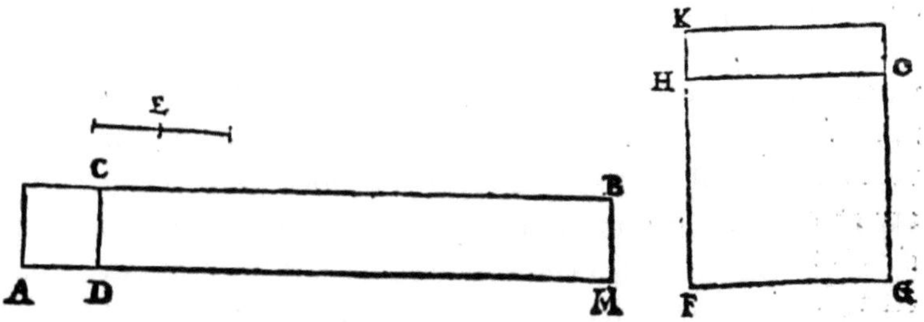

Diagram from *Ars Magna* VII.5. Kind courtesy of Biblioteca Virtual del Patrimonio Bibliográfico and Universidad de Sevilla.

Then, $a_2 = \overline{AM}$. Cardano showed that, assuming in addition that $a_1 = \sqrt[3]{a_0}\, a_2$ and $y = \frac{(\sqrt[3]{a_0})^2}{x}$, he managed to draw another geometrical model for the equation $y^3 + a_0 = a_1 y$.

I cannot figure out why Cardano does not use here the substitution $x = y + \frac{a_2}{3}$ in (AM XIV) to make disappear the term of second degree. I can only say that if he did it so, he would bring back the equation $x^3 + a_0 = a_2 x^2$ to the same three cases as in (AM XIV), whereas here he has just one case. Another possibility could have been to use one of the substitutions in Chapter VII, which bring back the equation $x^3 + a_0 = a_2 x^2$ to (AM XIII).

In the 1545 edition in Chapter XV, Cardano observes that

> [f]rom this is clear why the case of the cube and the constant equal to the square has not been demonstrated

> from the case of the cube and square equal to the constant. How, then, can the case of the cube and the number equal to the first power be demonstrated from the case of the cube equal to the first power and number? Since this case leads to the case of the cube and the number equal to the first power, it is better to proceed from the case of the cube and the number equal to the square directly to that of the cube and the number equal to the first power than to the same by the way of the case of the cube and square equal to the number. This avoids a far longer and more confusing demonstration.[65]

This quotation is quite tricky. We have seen that Cardano shows that the solving methods in (AM XIII) for $x^3 + a_0 = a_1 x$, in (AM XIV) for $x^3 = a_2 x^2 + a_0$, and partially in (AM XV) for $x^3 + a_2 x^2 = a_0$ depend directly on the solving method in (AM XII) for $x^3 = a_1 x + a_0$ (see below, on page 164). Also the solving method in (AM XV) depends for another part and directly on the solving method in (AM XIII), so that it depends indirectly on the solving method in (AM XII). Cardano explains why he does not want the solving method in (AM XVI) for $x^3 + a_0 = a_2 x^2$ to depend on the solving method in (AM XV) for $x^3 + a_2 x^2 = a_0$. Indeed, since (at least partially) the solving method in (AM XV) depends on the solving method in (AM XIII), which depends in turn on the solving method in (AM XII), he can skip a passage making the solving method in (AM XVI) depend directly on the solving method in (AM XIII).

[65]See [22, page 115] or "*[e]x hoc est manifestum, cur capitulum, cubi et numeri æqualium quadratis, non demonstratur ex capitulo cubi et quadratorum æqualium numero. Quemadmodum capitulum cubi et numeri, æqualium rebus, demonstratum est ex capitulo cubi æqualis rebus et numero. Nam cum capitulum hoc perveniat aliquando ad capitulum cubi et numeri æqualium rebus, melius est igitur ducere capitulum cubi et numeri æqualium quadratis, immediate ad capitulum cubi et numeri æqualium rebus, quam ad idem capitulum, medio capituli cubi et quadratorum æqualium numero, nam et operatio longior, et demonstratio magis confusa evaderet*", see [4, Chapter XV, pages 34r,34v].

But, in the 1570 and 1663 editions, the corollary disappears. I have no hypothesis on why. But I observe that the above quotation is not completely clear. Moreover, thanks to the alternative proof of (AM XV bis) in Chapter XV due to Ludovico Ferrari, the whole solving method in Chapter XV can be brought back to the solving method in (AM XII). In this way, one of Cardano's arguments in the above quotation seems to lose in strength.

Afterwards, Cardano gives a rule to find, given a solution of $x^3 + a_0 = a_2x^2$, the other (positive) solution.[66] Again, this would enable Cardano to avoid to use the cubic formula (AM XII) in Chapter XII (on which the cubic formula (AM XIII) in Chapter XIII depends, and in turn the solving method in Chapter XVI), provided that one can find a solution of $x^3 + a_0 = a_2x^2$.

AM XVI bis. *Let us consider the equation $x^3 + a_0 = a_2x^2$ and assume x_1 to be a solution. Then,*

$$x_2 = \sqrt{(a_2 - x_1)\left(x_1 + \frac{a_2 - x_1}{4}\right)} + \frac{a_2 - x_1}{2}$$

is another solution.

We observe that, as in (AM XIII bis), one needs to ask that $\Delta_3 < 0$ to guarantee the existence of the second real solution. Again, with a reasoning similar to the one on page 116, we could infer that Cardano got to this formula using (AM I.5-6), (AM VIII.2), and (AM XXV.14) (instead of (AM XXV.1)).

I will skip the proof in details, just sketching without calculations the idea behind it.

AM XVI bis (idea) — *Proof.* Considering the diagram in Figure 3.12 on page 134, Cardano[67] takes $x_1 = \overline{AB}$, $a_2 - \overline{AB} = \overline{AD}$ and

[66] Cardano knows by (AM I.8 ii) that in the case $\Delta_3 < 0$ the equation $x^3+a_0 = a_2x^2$ has two positive solutions.

[67] See [4, Chapter XVI, pages 35r-35v].

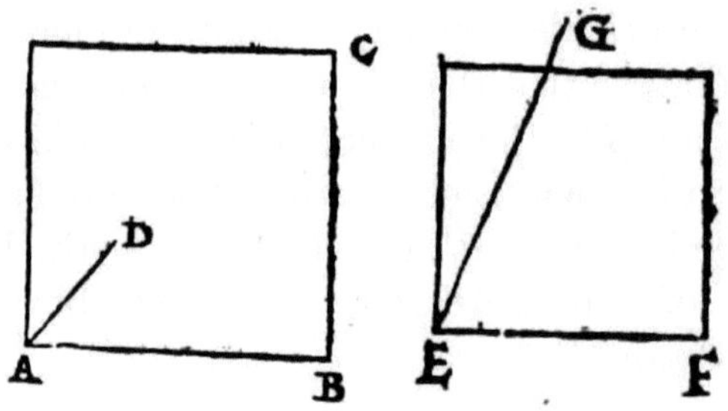

Figure 3.12. – Diagram from *Ars Magna* XVI bis. Kind courtesy of Biblioteca Virtual del Patrimonio Bibliográfico and Universidad de Sevilla.

$x_2 = \overline{EF}[= g]$. He wants to show that such a $\overline{EF}$ is a solution of $x^3 + a_0 = a_2 x^2$.

He knows that it exists $\overline{EG} = a_2 - \overline{EF}[= f]$. Now, it remains only to show that $\overline{EG}\,\overline{EF}^2 = a_0$ in order to infer by (AM VIII.2) that $\overline{EF}$ is a solution. For that, Cardano considers $\overline{AB}$ which is a solution of $x^3 + a_0 = a_2 x^2$ by hypothesis. Then, $a_0 = a_2\overline{AB}^2 - \overline{AB}^3 = (a_2 - \overline{AB})\overline{AB}^2 = \overline{AD}\,\overline{AB}^2$. Then, he finally shows that $\overline{EG}\,\overline{EF}^2 = \overline{AD}\,\overline{AB}^2$. This ends the proof. □

We remark that the proof is very similar to (AM XIII bis (idea)).

Again, this proposition does provide a formula to find a second solution for $x^3 + a_0 = a_2 x^2$ which bears no conditions to be satisfied. While this can help a bit in bypassing the *casus irreducibilis* for the second solution, it does not help at all for the first solution.

3.3.4. Summing up

Transformation of equations is the way of proceeding in these chapters. In Chapter VII we already came across some of the transformations used here – and, in fact, one of the proofs in Chapter XVI is skipped, explicitly referring to those in Chapter VII. The other proofs plainly consist in verifying in the usual way that the given substitution leads to the desired equation, once that the hypotheses are appropriately interpreted. The proofs employ (AM VI.6) and they make reference to the proofs in Chapters XI and XII.

Cardano's treatment is not homogeneous, even if one substitution (up to changing signs) could have fitted for all these equations, namely $y = x \pm \frac{a_2}{3}$ (and, by the way, Cardano knows this fact by (AM IV.2), see above, on page 81). He employs different kinds of substitutions, maybe for the sake of not increasing the number of sub-cases. Indeed, he uses $x = y + \frac{a_2}{3}$ in (AM XIV), $x = y - \frac{a_2}{3}$ or alternatively $x = y^2 - a_2$ in (AM XV), and $x = \frac{(\sqrt[3]{a_0})^2}{y}$ in (AM XVI). In Table 3.2 on page 136 are resumed all the substitutions that Cardano knows concerning cubic equations lacking in first degree term.

Note that, in order to apply the methods in (AM XV) and (AM XVI) to respectively solve $x^3 + a_2x^2 = a_0$ and $x^3 + a_0 = a_2x^2$, one should at least be able to find a solution of the family of equations in (AM XII), which is not always the case because of the *casus irreducibilis*. Then, in (AM XVI), as in (AM XIII), Cardano tries, given a solution of the considered equation, to provide a formula to find another solution. Both these methods could stand, on the one hand, as signs of Cardano's (unfulfilled) wish to find a way to skip the use of the cubic formula. On the other hand, they could have originated from parallel researches on transformations and "particular" solutions.

3.4. Solving complete equations

3.4.1. "On the cube, some squares, and some things equal to a number"

In **Chapter XVII** Cardano deals with the equation $x^3 + a_2x^2 + a_1x = a_0$.

Table 3.2. – Substitutions in cubic equations lacking in second degree term in *Ars Magna*, Chapters VII, XIV, XV, and XVI.

Equation	Substitution	Leads to	In chapter
$x^3 = a_2x^2 + a_0$	$x = y + \frac{a_2}{3}$	$y^3 = a_1y + \frac{2}{27}a_2^3 + a_0$	XIV
	$x = \frac{(\sqrt[3]{a_0})^2}{y}$	$y^3 + (\sqrt[3]{a_0}\, a_2)\, y = a_0$	VII
	$x = y^2 + a_2$	$y^3 + a_2y = \sqrt{a_0}$	VII
$x^3 + a_2x^2 = a_0$	$x = y - \frac{a_2}{3}$	$\left\{\begin{array}{l} y^3 + 2\left(\frac{a_2}{3}\right)^3 - a_o = 3\left(\frac{a_2}{3}\right)^2 y \\ y^3 = a_0 - 2\left(\frac{a_2}{3}\right)^3 + 3\left(\frac{a_2}{3}\right)^2 y \\ y^3 = 3\left(\frac{a_2}{3}\right)^2 y \end{array}\right\}$	XV
	$x = y^2 - a_2$	$y^3 = a_2y + \sqrt{a_0}$	VII, XV
	$x = \frac{\sqrt[3]{a_0}}{y}$	$y^3 = \left(\sqrt[3]{a_0}\right)^2 \frac{1}{a_2}y + a_0$	VII
$x^3 + a_0 = a_2x^2$	$x = \frac{(\sqrt[3]{a_0})^2}{y}$	$y^3 + a_0 = \left(a_2\sqrt[3]{a_0}\right) y$	VII, XVI
	$x = \pm(a_2 - y)$	$y^3 + \sqrt{a_0} = a_2y$	VII

AM XVII. *Given*[68] *the equation* $x^3 + a_2x^2 + a_1x = a_0$, *a solution is*

$$x = y - \frac{a_2}{3},$$

where y *satisfies the following conditions*

i. *if* $a'_1 = 0$, *then* $y^3 = -a'_0$,

ii. *if* $a'_1 > 0$, *then* $y^3 + a'_1y = -a'_0$,

iii. *if* $a'_1 < 0$, *then*

- ***a.*** *if* $a'_0 = 0$, *then* $y^3 = (-a'_1)y$,
- ***b.*** *if* $a'_0 < 0$, *then* $y^3 = (-a'_1)y + (-a'_0)$,
- ***c.*** *if* $a'_0 > 0$, *then* $y^3 + a'_0 = (-a'_1)y$,

with $a'_1 = -\frac{a_2^2}{3} + a_1$ *and* $a'_0 = -a'_1\frac{a_2}{3} - \left(\frac{a_2}{3}\right)^3 - a_0 [= -\frac{1}{3}a_1a_2 + \frac{2}{27}a_2^3 - a_0]$.

Note that Cardano uses '*Tpq̃d*' (which means '*Tertia pars numeri quadratorum*') to refer to $\frac{a_2}{3}$.

We remark that the number of cases increases more and more. Indeed, since Cardano always wants positive coefficients in the transformed equations, he has now to check the sign of a'_1 and a'_0 that can both vary. In particular, we observe that, in the case $a'_1 = 0$, we get $a'_0 = -\left(\frac{a_2}{3}\right)^3 - a_0 < 0$. The same $a'_0 = -a'_1\frac{a_2}{3} - \left(\frac{a_2}{3}\right)^3 - a_0 < 0$ also holds in the case $a'_1 > 0$. But, if $a'_1 < 0$, then the sign of a'_0 can vary further on. Similar observations can be made for the other chapters concerning the complete equations, and I will mostly omit them. Note moreover that there is only a loose correspondence between these cases and the sign of the discriminant $\Delta_3 = \frac{q^2}{4} + \frac{p^3}{27}$ (see above, the formula (2.5.10) on page 36). Indeed, we have[69] $p = a'_1$ and $q = a'_0$.

[68] From now on, I will not explicitly write down the cubic formula since it is a variation in signs of (2.5.9), see here page 36.

[69] It is not surprising that Cardano adds (in the 1545 edition) the following corollary. He remarks that the equation $x^3 + a_2x^2 + a_1x = a_0$ is brought back to $x^3 + a_1x = a_0$, $x^3 = a_0$, $x^3 = a_1x$, $x^3 = a_1x + a_0$, $x^3 + a_0 = a_1x$. Then, Cardano gives a clear insight of the method for complete equations.

Anyway, this is not enough to completely determine the sign of Δ_3. We can *a priori* only say that $\Delta_3 \geq 0$ in the first case $a'_1 = 0$ and that $\Delta_3 > 0$ in the second case $a'_1 > 0$.

Finally, we remark that the transformed equation in y can have more than one solution. In this case, the above rule preserves the number of (real) solutions (but, obviously, not necessarily their sign).

Cardano gives a proof for each one of the cases $a'_1 \lesseqgtr 0$. Concerning the third case, he only proves the sub-case $a'_0 < 0$.

AM XVII — *Proof*. Consider the diagram in Figure 3.13 on page 139 and let the equation $\overline{AB}^3 + 6\overline{AB}^2 + 20\overline{AB} = 100$. [We then have $a'_1 > 0$.] Cardano[70] takes $\overline{BC} = \frac{6}{3} = 2$. By (AM VI.6), Cardano

> In all four-term cases, however, it is a common [characteristic] that one-third the coefficient of x^2 is always added to or subtracted from the value arrived at after they have been resolved into two or three terms, just as in this rule it is always subtracted, and it is also common in every case that the coefficient of y and the [second] constant are built up in the same way, just as in this instance the coefficient of y is the difference between the coefficient of x given in the four-term equation and the product of the coefficient of x^2 and one-third itself; and the constant of the equation into which it [i.e., the original equation] resolves itself is the difference between the product of the coefficient of y and one-third the coefficient of x^2 and the sum of the cube of one-third the coefficient of x^2 and the constant of the first equation

or

> *in omnibus autem capitulis quatuor denominationum, c omune est, cum fuerint resoluta in capitulum trium vel duarum denominationum, ut æstimationi inventæ addatur aut minuatur Tpq̃d: ut in hoc capitulo semper minuitur, et commune est etiam omni capitulo, ut rerum numerus et numerus ipse constituantur eodem modo, velut hic numerus rerum, est differentia numeri rerum assumptarum in capitulo quatuor denominationum, et producti ex numero quadratorum in sui tertiam partem, et numeru capituli in quod resolvitur, est differentia producti ex numero rerum iam inventarum, in Tpq̃d: et aggregati ex cubo Tpq̃d: et numero æquationis primo,*

see [22, page 124] or [4, Chapter XVII, Corollary, page 36v].

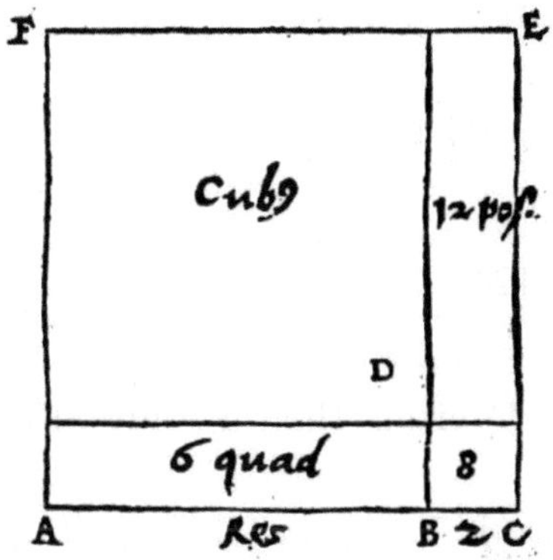

Figure 3.13. – Diagram from *Ars Magna* XVII. Kind courtesy of Biblioteca Virtual del Patrimonio Bibliográfico and Universidad de Sevilla.

knows that

$$\overline{AC}^3 = (\overline{AB} + \overline{BC})^3 = \overline{AB}^3 + 3\overline{AB}^2\,\overline{BC} + 3\overline{AB}\,\overline{BC}^2 + \overline{BC}^3,$$

that is, in this case,

$$\overline{AC}^3 = \overline{AB}^3 + 6\overline{AB}^2 + 12\overline{AB} + 8.$$

Then, Cardano has to add $8\overline{AB}$ in order to recover the relation given by the equation he wants to solve

$$\begin{aligned}\overline{AC}^3 + 8\overline{AB} &= (\overline{AB}^3 + 6\overline{AB}^2 + 12\overline{AB} + 8) + 8\overline{AB}\\ &= (\overline{AB}^3 + 6\overline{AB}^2 + 20\overline{AB}) + 8\\ &= 108.\end{aligned}$$

But Cardano wants a transformed equation where $\overline{AC}$ is the unknown. He knows that $8\overline{AC} = 8(\overline{AB} + \overline{BC}) = 8\overline{AB} + 8\overline{BC}$, so he just has to

[70] See [4, Chapter XVII, page 36r].

add $8BC = 16$ to both sides to get

$$\overline{AC^3} + 8\overline{AB} + 8\overline{BC} = 108 + 16,$$

that is

$$\overline{AC^3} + 8\overline{AC} = 124.$$

Now, he knows how to solve the transformed equation in $\overline{AC}$ by (AM XI). The relation $\overline{AB} = \overline{AC} - \overline{BC}$ yields then a solution of $\overline{AB}^3 + 6\overline{AB}^2 + 20\overline{AB} = 100$.

Again, always referring to the same diagram, let the equation $\overline{AB}^3 + 6\overline{AB}^2 + 12\overline{AB} = 100$. [We then have $a_1' = 0$.] In this case, Cardano has to add $\overline{BC}^3 = 8$ to the left side of the equation he wants to solve in order to recover $\overline{AC}^3$

$$(\overline{AB}^3 + 6\overline{AB^2} + 12\overline{AB}) + 8 = 108,$$

that is

$$\overline{AC}^3 = 108.$$

Then, $\overline{AC} = \sqrt[3]{108}$ and $\overline{AB} = \overline{AC} - \overline{BC} = \sqrt[3]{108} - 2$.

Again, always referring to the same diagram, let the equation $\overline{AB}^3 + 6\overline{AB}^2 + 2\overline{AB} = 100$. [We then have $a_1' < 0$.] In this case, Cardano has to add $10\overline{AB} + \overline{BC}^3 = 10\overline{AB} + 8$ to the left side of the equation he wants to solve in order to recover $\overline{AC}^3$

$$(\overline{AB}^3 + 6\overline{AB}^2 + 2\overline{AB}) + 10\overline{AB} + \overline{BC}^3 = 10\overline{AB} + 108,$$

that is

$$\overline{AC}^3 = 10\overline{AB} + 108.$$

But Cardano wants a transformed equation where $\overline{AC}$ is the unknown, so he adds again $10\overline{BC} = 20$ to both sides and he gets

$$\begin{aligned}\overline{AC}^3 + 20 &= 10\overline{AB} + 10\overline{BC} + 108\\ &= 10\overline{AC} + 108,\end{aligned}$$

that is

$$\overline{AC}^3 = 10\overline{AC} + 88.$$

Now, he knows how to solve the transformed equation in $\overline{AC}$ by (AM XII). The relation $\overline{AB} = \overline{AC} - \overline{BC}$ yields then a solution of $\overline{AB}^3 + 6\overline{AB}^2 + 20\overline{AB} = 100$. □

We observe that, in the three cases, the proof consists in the construction of the transformed equation completely relying on the substitution $x = y - \frac{a_2}{3}$, on the formula for the expansion of the cube of a binomial given in (AM VI.6), and obviously on the relation given by the original equation. Then, Cardano has only to adjust the coefficients in order to recover the original equation and he has to take care to obtain a transformed equation in $\overline{AC}$ with positive coefficients. In this way, he is able to bring back the complete equation $x^3 + a_2x^2 + a_1x = a_0$ to an equation that he knows how to solve thanks to (AM XI), (AM XII), or taking the cubic root of a number. We remark that the rule given above is a simple translation into an algorithmic language of the steps taken in the proof. Here again, similar observations can be made for all proofs in the following chapters, and I will mostly omit them.

3.4.2. "On the cube and some things equal to some squares and a number"

In **Chapter XVIII** Cardano deals with the equation $x^3 + a_1x = a_2x^2 + a_0$.

AM XVIII. *Given the equation* $x^3 + a_1 x = a_2 x^2 + a_0$, *a solution is*[71]

$$x = y + \frac{a_2}{3},$$

where y *satisfies the following conditions*

- ***i.*** *if* $a_1' = 0$, *then* $y^3 = -a_0'$,
- ***ii.*** *if* $a_1' > 0$, *then* $y^3 + a_1' y = -a_0'$,
- ***iii.*** *if* $a_1' < 0$, *then*
  - ***a.*** *if* $a_0' = 0$, *then* $y^3 = (-a_1')y$,
  - ***b.*** *if* $a_0' < 0$, *then* $y^3 = (-a_1')y + (-a_0')$,
  - ***c.*** *if* $a_0' > 0$, *then* $y^3 + a_0' = (-a_1')y$,

with $a_1' = -\frac{a_2^2}{3} + a_1$ *and*[72] $a_0' = a_1 \frac{a_2}{3} - 2\left(\frac{a_2}{3}\right)^3 - a_0$.

Note the loose correspondence between these cases and the sign of the discriminant $\Delta_3 = \frac{q^2}{4} + \frac{p^3}{27}$ (see above, the formula (2.5.10) on page 36). We have $p = a_1'$ and $q = a_0'$. Then, we can *a priori* only say that $\Delta_3 \geq 0$ in the first case $a_1' = 0$ and that $\Delta_3 > 0$ in the second case $a_1' > 0$.

Cardano gives a proof only for the case $a_1' < 0$.

[71] Cardano adds that "having derived the solution [for y], add or subtract one-third the coefficient of x^2 as you will learn from the examples." Indeed, in the first example, a solution of the transformed equation is $y = \sqrt[3]{17}$, which yields the solution of the original equation $x = \sqrt[3]{17} + 2$, whereas in the second example a solution of the transformed equation is $y = -\sqrt[3]{16}$, which yields the solution of the original equation $x = -\sqrt[3]{16} + 2$.

Cardano makes this remark only in this chapter, even if it holds also for other chapters. Indeed, as said, from a negative solution of the transformed equation we can have a positive solution of the original equation, adding $\frac{a_2}{3}$. He mentions quickly this fact again in Chapter XXI, see below, page 155, and at the end of Chapter XXIII, see below, page 160.

[72] We observe that $a_0' = a_1' \frac{a_2}{3} + \left(\frac{a_2}{3}\right)^3 - a_0$.

AM XVIII — *Proof.* Let the equation $\overline{AC}^3+33\overline{AC}=6\overline{AC}^2+100$. [From now on, the diagrams appear no longer in the proofs. This is because Cardano[73] makes always reference to the diagram in Chapter XVII, see above, the figure 3.13 on page 139.]

Cardano takes $\overline{BC}=\frac{6}{3}=2$. [By (AM VI.6), he knows that

$$\overline{AC}^3=(\overline{AB}+\overline{BC})^3=\overline{AB}^3+3\overline{AB}^2\,\overline{BC}+3\overline{AB}\,\overline{BC}^2+\overline{BC}^3,]$$

that is, in this case,

$$\overline{AC}^3=\overline{AB}^3+6\overline{AB}^2+12\overline{AB}+8.$$

Then, he has to add $33\overline{AC}=33\overline{AB}+33\overline{BC}=33\overline{AB}+66$ to $\overline{AC}^3$ in order to recover the relation given by the equation that he wants to solve

$$\begin{aligned}\overline{AC}^3+33\overline{AC}&=(\overline{AB}^3+6\overline{AB}^2+12\overline{AB}+8)+(33\overline{AB}+66)\\&=\overline{AB}^3+6\overline{AB}^2+45\overline{AB}+74\\&=6\overline{AC}^2+100.\end{aligned}$$

But Cardano wants a transformed equation where $\overline{AB}$ is the unknown. So, using $6\overline{AC}^2=6(\overline{AB}+\overline{BC})^2=6\overline{AB}^2+12\overline{AB}\,\overline{BC}+6\overline{BC}^2=6\overline{AB}^2+24\overline{AB}+24$, by *Elements* II,4 he gets

$$\overline{AB}^3+6\overline{AB}^2+45\overline{AB}+74=(6\overline{AB}^2+24\overline{AB}+24)+100,$$

that is

$$\overline{AB}^3+21\overline{AB}=50.$$

Now, Cardano knows how to solve the transformed equation in $\overline{AB}$ by (AM XI). The relation $\overline{AC}=\overline{AB}+\overline{BC}$ yields then a solution of $\overline{AC}^3+33\overline{AC}=6\overline{AC}^2+100$. □

[73] See [4, Chapter XVIII, page 37v].

At the end of this proof, Cardano adds a quite opaque passage.[74] Fortunately, thanks to the explicit setting out of the rule, one is able to complete and interpret it as describing the other cases which Cardano does not consider in the proof. He says that their proofs are similar to the ones in Chapter XVII.

Then, Cardano makes some examples. As usual, passing from the transformed equation to the original one, the rule preserves the number of (real) solutions. We observe however that, when $y = 0$, we have a sort of limit case. At the end of his fourth example $x^3+15x = 6x^2+10$, Cardano himself needs to explain better this fact.

> Hence it is evident in this case, where the cube and first power are equal to the number, that if the difference between the numbers were nothing – as if we put 14 in the place of 10 – the solution would be one-third the coefficient of x^2, or 2, since the derived equation would give us nothing to add or subtract, for $y^3 + 3y = 0$.[75]

[74]"It also clear that if the [number of] x's which were on the same side as the cube [in the original equation] had been equal to the product [of the coefficient of x^2 and one-third the same], we would have had a cube equal to a number alone [in the reduced equation]; and if [the number of] x's which were on the same side as the cube had been less [than that product], we would have had a first power on one side and a cube on the other. In this case, if the constant had been equal to a [certain] other [number], the first power would equal the cube, but if it [i.e., the new first power] were smaller [than the new cube] we would have had the first power and number equal to the cube and, if greater, the first power equal to the cube and a number, according to the same demonstration as in the preceding chapter", see [22, pages 128-9], or "*[m]anifestum est autem, quod ubi positione, quæ cum cubo erant, essent, æquales productis, haberemus cubum æqualem numero tantum, et ubi positiones quæ cum cubo erant, essent pauciores, haberemus rex ex una parte, et cubum ex alia, et tunc si numerus qui est cum cubo, foret æqualis alteri, essent positiones æquales cubo, et si maior, haberemus res æquales cubo et numero, ex eadem demonstratione, velut in præcedente capitulo*", see [4, Chapter XVIII, Proof, page 38r].

[75]See [22, page 132] or "*Ideo patet quod in hoc casu, ubi cubus et res, æquantur numero, si differentia numerorum nulla foret, vel si loco 10 posuissemus 14, æstimatio rei esset Tpq̃d, scilicet 2, quia in æquatione inventa, nihil haberemus addendum vel minuendum, quia cubus et 3 res, æquerentur nihil*", see [4, Chapter XVIII, fourth example, page 39r].

The equation $x^3 + 15x = 6x^2 + 14$ falls within the case $a'_1 > 0$, $a'_0 = 0$ (which is not completely detailed in the above rule). We get then the transformed equation $y^3 + 3y = 0$, a solution of which is 0. So, one of the solutions of the original equation is $x = y + \frac{a_2}{3} = 0 + 2 = 2$. Again, in his fifth example $x^3 + 10x = 6x^2 + 4$, the transformed equation is $y^3 = 2y$, the solutions of which are $\sqrt{2}$ [and $-\sqrt{2}$ and 0, both not mentioned by Cardano]. These yield the solutions of the original equation $2 + \sqrt{2}$, $2 - \sqrt{2}$, and 2, which nevertheless are all mentioned by Cardano. He will again hint at this fact in Chapter XXII, referring to both Chapters XXI and XXII, see here, page 155.

Afterwards, Cardano observes that

> [f]rom this it is evident that the coefficient of x^2, in the three examples in which there are three solutions for x, is always the sum of the three solutions. [...] Hence knowing two such solutions, the third always emerges. The reason for this appears at the beginning of this book.[76]

Cardano explicitly affirms that, in the case $x^3 + a_1 x = a_2 x^2 + a_0$ [when $\Delta_3 < 0$], he knows that $a_2 = x_1 + x_2 + x_3$, where x_1, x_2, x_3 are the three (real) solutions. To justify his statements, Cardano makes reference to Chapter I, where he spoke of "true" and "fictitious" solutions. Unfortunately, this reference is empty. Indeed, in Chapter I the equation $x^3 + a_1 x = a_2 x^2 + a_0$ is hardly mentioned. Anyway, in that same chapter, we find two statements (not justified) of similar rules for depressed cubic equations (see (AM I.5-6) with $a_2 = 0$) and for cubic equations lacking in the first degree term (see (AM I.8 ii)). Then, we can suppose that Cardano knows in general the relation between the coefficient of the term of second degree and the roots.

[76] See [22, page 134] or "*[e]x hoc patet, qod numerus quadratorum, in his tribus exemplis, in quibus æstimatio rei triplicatur, semper componitur ex tribus æstimationibus iunctis simul [...], ideo duabus cognitis, tertia semper emergit, et causa est cognita in initio huius libri*", see [4, Chapter XVIII, page 39v].

Again, Cardano observes that

> [i]t is also evident that whenever we arrive [at a case in which] the ys are separated from the cube, whether the number is joined to the y's or to y^3, three solutions always appear. The reason for this is likewise given in the same above place where we spoke of true and fictitious solutions. And it is also evident that all these methods can always be carried back to addition, for a minus that is added acts the same way as a plus that is subtracted.[77]

Cardano knows that, when $a_1' > 0$, the equation $x^3 + a_1x = a_2x^2 + a_0$ always has three solutions. Here again, Cardano refers back vainly to some justification in Chapter I. The only place where we can find it is an obscure addition to the 1570 and 1663 editions.[78]

Finally, Cardano adds another rule discovered by Ferrari.

---

[77] See [22, page 134] or "*[e]t manifestum est, quod cum pervenimus ad res, quæ a cubo separantur, seu numero rebus, seu cubo iungatur, semper emergunt tres æstimationes, et causa dicta est superius ibidem, ubi de vera et ficta æstimatione locuti sumus. Et patet etiam, quod omnes modi hi, ad additionem semper possunt referri, quamvis minus cum additur, vicem gerat plus cum detrahitur*", see [4, Chapter XVIII, page 39v].

[78] Here follows the quotation.

> When a constant and the highest power are equal to one or more middle powers, there are two solutions for x since, whatever their coefficients, the middle terms can exceed the extreme term, as $100x^2$ can be greater than x^3. Now let $\overline{AB}$ be the value of x.
>
> 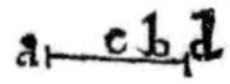
>
>
> The cube can then be made to equal $100x^2$ either by decreasing the estimated value of x and allowing the given number to remain the same (in which case [the value of x becomes] $\overline{AC}$) or by decreasing the cube and thus increasing the value of x (in which case it becomes $\overline{AD}$). Therefore $100x^2 = x^3 + 1$ has two solutions.
>
> Similarly, if there are many middle terms, even a hundred of them, changes all the middle powers likewise increase or decrease [*sic*]. But if the extreme powers and the middle powers, taken alternately, are equated to each other (as if the cube and first power are equated to the square and a constant), I say that there may be three solutions.

AM XVIII bis. *Given the equation* $x^3 + a_1 x = a_2 x^2 + a_0$, *a solution is*

- *if* $a'_1 = 0$, *then*
  - *if* $a'_0 < 0$, *then* $x = \frac{a_2}{3} + \sqrt[3]{a'_0}$,
  - *if* $a'_0 > 0$, *then* $x = \frac{a_2}{3} - \sqrt[3]{a'_0}$,
- *if* $a'_1 > 0$, *then*
  - *if* $a'_0 = 0$, *then* $x = \frac{a_2}{3}$,
  - *if* $a'_0 > 0$, *then* $x = \frac{a_2}{3} - y$, *where* y *is a solution of the equation* $y^3 + a'_1 y = a'_0$
  - *if* $a'_0 < 0$, *then* $x = \frac{a_2}{3} + y$, *where* y *is a solution of the equation* $y^3 + a'_1 y = -a'_0$,

---

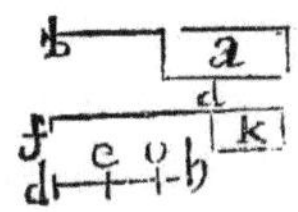

> Thus, let A, the constant, and B, the number of squares, equal K, the cube, and F, the number of x's, and let x be $\overline{DE}$. If F is assumed to be big and $\overline{DE}$ small, an equation can be made, since the squares and the cube become smaller because of the smallness of $\overline{DE}$. But if the squares exceed the cube and if the x's are placed beside the latter, as said, there will be two solutions – either one that is increased because of the magnitude of x^3 or one that is diminished by the increase in the x's. Therefore, there will be three solutions [in all],

see [22, pages 21-22].

Pietro Cossali explains the first part of the quotation as follows. Consider $x^3 + a_0 = a_2 x^2$ and suppose that it exists a real N such that $N^3 + a_0 = a_2 N^2$. Then, one can bring the inequality back to an equality in two ways, either increasing $a_2 N^2$ or decreasing N^3. This means decreasing N up to a certain N_1 which works, or increasing it up to N_2. Then, N_1 and N_2 are two solutions. Then, Cossali explains why this holds also for an equation of this kind with more than three terms, using the mathematics of his own time. After that, he continues dealing with the kind of equation in the second part of the quotation, but continuing in modern terms, see [57, Volume II, Chapter VII, item III, number 12, pages 354-7].

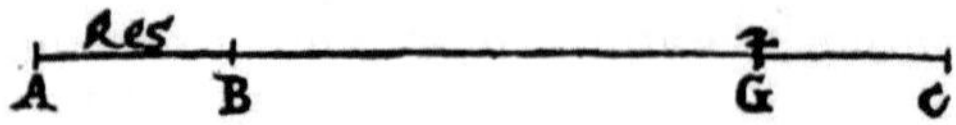

Figure 3.14. – Diagram from *Ars Magna* XVIII bis i. Kind courtesy of Biblioteca Virtual del Patrimonio Bibliográfico and Universidad de Sevilla.

- *if* $a_1' < 0$, *then* $x = y + \frac{a_2}{3}$, *where* y *satisfies the following conditions*
  - *if* $a_0' = 0$, *then* $y^3 = a_1' y$,
  - *if* $a_0' < 0$, *then* $y^3 = a_1' y + (-a_0')$,
  - *if* $a_0' > 0$, *then* $y^3 + a_0' = a_1' y$,

with $a_1' = -\frac{a_2^2}{3} + a_1$ *and* $a_0' = a_1' \frac{a_2}{3} + \left(\frac{a_2}{3}\right)^3 - a_0$.

We observe that this is substantially the same rule as before. Indeed, the transformed coefficients a_1' and a_0' are the same (see above, Footnote 72 on page 142) and this rule details all the cases depending on the sign of the solution y of the transformed equation, which were only sketched in the preceding rule (see above, Footnote 71 on page 142).

We firstly find a proof for the case $a_1' < 0$, $a_0' > 0$.

AM XVIII bis i — *Proof.* Consider the diagram in Figure 3.14 on page 148 and the equation $\overline{AB}^3 + 100\overline{AB} = 6\overline{AB}^2 + 10$. [We then have $a_1' < 0$, $a_0' > 0$.] Cardano and Ferrari[79] take $\overline{BC} = \frac{6}{3} = 2$ and mark G on $\overline{AC}$ such that $\overline{AG} = \overline{BC}$. Then, [$\overline{AB} = \overline{GC}$ and] $\overline{BG} = \overline{BC} - \overline{AB}$. By (AM VI.6), they know that

$$\begin{aligned}\overline{BG}^3 = (\overline{BC} - \overline{AB})^3 &= \overline{BC}^3 - 3\overline{BC}^2\,\overline{AB} + 3\overline{BC}\,\overline{AB}^2 - \overline{AB}^3 \\ &= \overline{BC}^3 + 3\overline{BC}\,\overline{AB}^2 - (3\overline{BC}^2\,\overline{AB} + \overline{AB}^3) \\ &= 8 + 6\overline{AB}^2 - (12\overline{AB} + \overline{AB}^3).\end{aligned}$$

[79] See [4, Chapter XVIII, pages 39v-40r].

Figure 3.15. – Diagram from *Ars Magna* XVIII bis ii. Kind courtesy of Biblioteca Virtual del Patrimonio Bibliográfico and Universidad de Sevilla.

They are looking for a transformed equation in $\overline{BG}$. By the given equation, they know that $6\overline{AB}^2 + 8 = (\overline{AB}^3 + 100\overline{AB}) - 2 = (\overline{AB}^3 + 12\overline{AB}) + 88\overline{AB} - 2$. Then, they get

$$\overline{BG}^3 = 88\overline{AB} - 2.$$

Since $\overline{BG} = \overline{BC} - \overline{AB} = 2 - \overline{AB}$, then $\overline{AB} = 2 - \overline{BG}$ and $\overline{BG}^3 + 2 = 88(2 - \overline{BG})$, that is

$$\overline{BG}^3 + 88\overline{BG} = 174.$$

Now, Cardano and Ferrari know how to solve the transformed equation in $\overline{BG}$ by (AM XI). The relation $\overline{AB} = 2 - \overline{BG}$ yields then a solution of $\overline{AB}^3 + 100x = 6x^2 + 10$. □

Afterwards, we find a proof for the case $a_1' > 0$, $a_0' < 0$.

AM XVIII bis ii — *Proof*. Again, consider the diagram in Figure 3.15 on page 149 and let the equation $\overline{EF}^3 + 5\overline{EF} = 6\overline{EF}^2 + 10$. [We then have $a_1' > 0$, $a_0' < 0$.] Cardano and Ferrari[80] take $\overline{DE} = \frac{6}{3} = 2$ [and mark H on $\overline{DF}$ such that $\overline{DH} = \overline{EF}$. Then $\overline{DE} = \overline{HF}$ and] $\overline{EH} = \overline{EF} - \overline{DE}$. Then, "from a demonstration similar to the foregoing" they get

$$\overline{EH}^3 = 7\overline{EH} + 16.$$

[80] See [4, Chapter XVIII, pages 39v-40r].

Now, Cardano and Ferrari know how to solve the transformed equation in $\overline{EH}$ by (AM XII). The relation $\overline{EF} = \overline{EH} + \overline{DE}$ yields then a solution of $\overline{AB}^3 + 100x = 6x^2 + 10$. □

No proof is given for the other cases.

3.4.3. "On the cube and some squares equal to some things and a number"

In **Chapter XIX** Cardano deals with the equation $x^3 + a_2x^2 = a_1x + a_0$.

AM XIX. *Given the equation $x^3 + a_2x^2 = a_1x + a_0$, a solution is*

$$x = y - \frac{a_2}{3},$$

where y satisfies the following conditions

i. if $a_0' = 0$, then $y^3 = (-a_1')y$,

ii. if $a_0' < 0$, then $y^3 = (-a_1')y + (-a_0')$,

iii. if $a_0' > 0$, then $y^3 + a_0' = (-a_1')y$,

with $a_1' = -\frac{a_2^2}{3} - a_1$ and $a_0' = -a_1'\frac{a_2}{3} - \left(\frac{a_2}{3}\right)^3 - a_0 [= \frac{1}{3}a_1a_2 + \left(\frac{a_2}{3}\right)^3 - a_0]$.

We remark that it is always $a_1' < 0$.

Moreover, note the loose correspondence between these cases and the sign of the discriminant $\Delta_3 = \frac{q^2}{4} + \frac{p^3}{27}$ (see above, the formula (2.5.10) on page 36). We have $p = a_1'$ and $q = a_0'$. Then, we can *a priori* only say that $\Delta_3 < 0$ in the first case $a_0' = 0$.

Cardano gives only a proof for the case $a_0' < 0$.

AM XIX — *Proof*. Let the equation $\overline{AB}^3 + 6\overline{AB}^2 = 20\overline{AB} + 200$. [Cardano[81] considers the same diagram as in Chapter XVII, see above, the figure 3.13 on page 139.]

[81]See [4, Chapter XIX, page 41r].

He takes $\overline{BC} = \frac{6}{3} = 2$. By (AM VI.6) Cardano knows that

$$\overline{AC}^3 = (\overline{AB} + \overline{BC})^3 = \overline{AB}^3 + 3\overline{AB}^2\,\overline{BC} + 3\overline{AB}\,\overline{BC}^2 + \overline{BC}^3,$$

that is, in this case,

$$\overline{AC}^3 = \overline{AB}^3 + 6\overline{AB}^2 + 12\overline{AB} + 8.$$

Then, using the relation given by the equation, Cardano gets

$$\begin{aligned}\overline{AC}^3 &= (20\overline{AB} + 200) + 12\overline{AB} + 8\overline{AB} \\ &= 32\overline{AB} + 208.\end{aligned}$$

But Cardano wants a transformed equation where $\overline{AC}$ is the unknown. He knows that $32\overline{AC} = 32\overline{AB} + 32\overline{BC}$, so he just has to add again $32\overline{BC} = 64$ to both sides to get

$$\overline{AC}^3 + 64 = 32\overline{AC} + 208,$$

that is

$$\overline{AC}^3 = 32\overline{AC} + 144.$$

Now, he knows how to solve the transformed equation in $\overline{AC}$ by (AM XII). The relation $\overline{AB} = \overline{AC} - \overline{BC}$ yields then a solution of $\overline{AB}^3 + 6\overline{AB}^2 = 20\overline{AB} + 200$. □

3.4.4. "On the cube equal to some squares, some things, and a number"

In **Chapter XX** Cardano deals with the equation $x^3 = a_2x^2 + a_1x + a_0$.

AM XX. *Given the equation* $x^3 = a_2x^2 + a_1x + a_0$, *a solution is*

$$x = y + \frac{a_2}{3},$$

where y is a solution of $y^3 = (-a_1')y + (-a_0')$, with $a_1' = -\frac{a_2^2}{3} - a_1$ and[82] $a_0' = a_1'\frac{a_2}{3} + (\frac{a_2}{3})^3 - a_0[= -\frac{1}{3}a_1a_2 - \frac{2}{27}a_2^3 - a_0]$.

We remark that it is always $a_1' < 0$. Moreover, we also always have $a_0' < 0$. Indeed,

$$\begin{aligned} a_0' &= (-\frac{a_2^2}{3} - a_1)\frac{a_2}{3} + \left(\frac{a_2}{3}\right)^3 - a_0 \\ &= -2\left(\frac{a_2}{3}\right)^3 - a_1\frac{a_2}{3} - a_0 < 0. \end{aligned}$$

Finally, note the loose correspondence between these cases and the sign of the discriminant $\Delta_3 = \frac{q^2}{4} + \frac{p^3}{27}$ (see above, the formula (2.5.10) on page 36). We have $p = a_1'$ and $q = a_0'$. Then, we can *a priori* only say that $\Delta_3 \neq 0$.

AM XX — *Proof.* Let the equation $\overline{AC}^3 = 6\overline{AC}^2 + 5\overline{AC} + 88$. [Cardano[83] considers the same diagram as in Chapter XVII, see above, the figure 3.13 on page 139.]

He takes $\overline{BC} = \frac{6}{3} = 2$. By (AM VI.6) Cardano knows that

$$\overline{AC}^3 = (\overline{AB} + \overline{BC})^3 = \overline{AB}^3 + 3\overline{AB}^2\,\overline{BC} + 3\overline{AB}\,\overline{BC}^2 + \overline{BC}^3,$$

[82]In Cardano's formulation of the rule there is a mistake concerning a_0'. Indeed, he says that "To find the number, moreover, multiply the coefficient of y by one-third the coefficient of x^2 and add the product to the constant of the equation. From this subtract the cube of one-third the coefficient of x^2 and the remainder is the number", see [22, page 144], or "*pro numero autem, duc numerum rerum secundum in Tpq̃d et productum adde numero æquationis a quo minue cubum Tpq̃d. Residuum est numerus*", see [4, Chapter XX, Rule, page 42r], which means $a_0' = a_1'\frac{a_2}{3} - (\frac{a_2}{3})^3 - a_0$.

But, in the equation $x^3 = 6x^2 + 5x + 88$ considered in the demonstration and in the only example $x^3 = 6x^2 + 72x + 729$ of this chapter, Cardano makes the right calculations (with the '+') to obtain the transformed equations $y^3 = 17y + 114$ and $y^3 = 84y + 889$.

[83]See [4, Chapter XX, pages 41v-42r].

that is, in this case,

$$\overline{AC}^3 = \overline{AB}^3 + 6\overline{AB}^2 + 12\overline{AB} + 8.$$

Then, using the relation given by the equation, Cardano gets

$$\overline{AB}^3 + 6\overline{AB}^2 + 12\overline{AB} + 8 = 6\overline{AC}^2 + 5\overline{AC} + 88,$$

that is

$$\overline{AB}^3 + 6\overline{AB}^2 + 12\overline{AB} = 6\overline{AC}^2 + 5\overline{AC} + 80,$$

that is

$$\begin{aligned}\overline{AB}^3 + 6\overline{AB}^2 + 12\overline{AB} &= 6(\overline{AB}^2 + 2\overline{AB}\,\overline{BC} + \overline{BC}^3) + 5(\overline{AB} + \overline{BC}) + 80\\ &= 6(\overline{AB}^2 + 4\overline{AB} + 4) + 5(\overline{AB} + 2) + 80\\ &= 6\overline{AB}^2 + 29\overline{AB} + 114,\end{aligned}$$

that is

$$\overline{AB}^3 = 17\overline{AB} + 114.$$

Now, Cardano knows how to solve the transformed equation in $\overline{AB}$ by (AM XII). The relation $\overline{AC} = \overline{AB} + \overline{BC}$ yields then a solution of $\overline{AC}^3 = 6\overline{AC}^2 + 5\overline{AC} + 88$. □

3.4.5. "On the cube and a number equal to some squares and some things"

In **Chapter XXI** Cardano deals with the equation $x^3 + a_0 = a_2x^2 + a_1x$.

AM XXI. *Given the equation* $x^3 + a_0 = a_2x^2 + a_1x$, *a solution is*

$$x = y + \frac{a_2}{3},$$

where y *satisfies the following conditions*

i. *if* $a'_0 = 0$, *then* $y^3 = (-a'_1)y$,

ii. *if* $a'_0 < 0$, *then* $y^3 = (-a'_1)y + (-a'_0)$,

iii. *if* $a'_0 > 0$, *then* $y^3 + a'_0 = (-a'_1)y$,

with $a'_1 = -\frac{a_2^2}{3} - a_1$ *and* $a'_0 = a'_1\frac{a_2}{3} + \left(\frac{a_2}{3}\right)^3 + a_0$.

We remark that it is always $a'_1 < 0$.

Moreover, note the loose correspondence between these cases and the sign of the discriminant $\Delta_3 = \frac{q^2}{4} + \frac{p^3}{27}$ (see above, the formula (2.5.10) on page 36). We have $p = a'_1$ and $q = a'_0$. Then, we can *a priori* only say that $\Delta_3 < 0$ in the first case $a'_0 = 0$.

Cardano gives only a proof for the case $a'_0 > 0$.

AM XXI — *Proof.* Let the equation $\overline{AC}^3 + 100 = 6\overline{AC}^2 + 24\overline{AC}$. [Cardano[84] considers the same diagram as in Chapter XVII, see above, the figure 3.13 on page 139.]

He takes $\overline{BC} = \frac{6}{3} = 2$. By (AM VI.6) Cardano knows that

$$\overline{AC}^3 = (\overline{AB} + \overline{BC})^3 = \overline{AB}^3 + 3\overline{AB}^2\,\overline{BC} + 3\overline{AB}\,\overline{BC}^2 + \overline{BC}^3,$$

that is, in this case,

$$\overline{AC}^3 = \overline{AB}^3 + 6\overline{AB}^2 + 12\overline{AB} + 8.$$

Then, using the relation given by the equation, Cardano gets

$$(\overline{AB}^3 + 6\overline{AB}^2 + 12\overline{AB} + 8) + 100 = 6\overline{AC}^2 + 24\overline{AC},$$

that is, since $\overline{BC} = 2$,

$$\begin{aligned}\overline{AB}^3 + 6\overline{AB}^2 + 12\overline{AB} + 108 &= 6(\overline{AB}^2 + 4\overline{AB} + 4) + 24(\overline{AB} + 2)\\ &= 6\overline{AB}^2 + 48\overline{AB} + 72,\end{aligned}$$

[84] See [4, Chapter XXI, page 42v].

that is

$$\overline{AB}^3 + 36 = 36\overline{AB}.$$

Now, Cardano knows how to solve the transformed equation in $\overline{AB}$ by (AM XIII). The relation $\overline{AC} = \overline{AB} + \overline{BC}$ yields then a solution of $\overline{AC}^3 + 100 = 6\overline{AC}^2 + 24\overline{AC}$. □

At the end of the proof, Cardano adds that, in the case $a'_0 < 0$, the transformed equation $y^3 = (-a'_1)y + (-a'_0)$ has three solutions, one positive and two negative ones. Since then we add $\frac{a_2}{3}$ to these solutions, $x^3 + a_0 = a_2x^2 + a_1x$ has two positive solutions and a negative one.[85]

Moreover, at the end of the next chapter while speaking of $x^3 + a_1x + a_0 = a_2x^2$, Cardano makes a remark that – he says – also holds for $x^3 + a_0 = a_2x^2 + a_1x$. In the case $a'_0 = 0$, the transformed equation is $y^3 = (-a'_1)y$ and "the fictitious solution does not differ from the true one in number. Therefore, for a second solution [for the principal equation], inasmuch as nothing is added to or subtracted from one-third the coefficient of x^2, this same one-third the coefficient of x^2 will be a true solution."[86] Indeed, if $a'_0 = 0$, we get $y_1 = \sqrt{-a'_1}$, $y_2 = \sqrt{-a'_1}$, and $y_3 = 0$. For the last solution, the corresponding solution of the equation in x is then $x_3 = \frac{a_2}{3}$.

[85]"Remember, moreover, that when the original equation turns into an equation of the cube equal to the first power and a number, the true solution of the latter will be added to one-third the coefficient of x^2 and the smaller of the fictitious solutions will [likewise] be added negatively. Thus you will have both [true] solutions for the equation of the cube and the number equal to the first power and square, although that of the cube equal to the first power and constant has only one true solution", see [22, page 147], or "*[m]emineris tamen, quod quando capitulum hoc pervenerit ad capitulum cubi æquali rebus et numero, addenda erit vera æstimatio eius, et ex his quæ fictæ sunt minor per m: Tpq̃d ut habeas utramque æstimationem capituli cubi et numeri æqualis rebus et quadratis, cum capitulum cubi æqualis rebus et numero, unam tantum veram æstimationem habeat*", see [4, Chapter XXI, Rule, page 42v].

[86]See [22, pages 151-2].

3.4.6. "On the cube, some things, and a number equal to some squares"

In **Chapter XXII** Cardano deals with the equation $x^3 + a_1x + a_0 = a_2x^2$.

AM XXII. *Given the equation $x^3 + a_1x + a_0 = a_2x^2$, a solution is*

$$x = y + \frac{a_2}{3},$$

where y satisfies the following conditions

i. *if $a_0' = 0$, then $y^3 = (-a_1')y$,*

ii. *if $a_0' < 0$, then $y^3 = (-a_1')y + (-a_0')$,*

iii. *if $a_0' > 0$, then $y^3 + a_0' = (-a_1')y$,*

with $a_1' = -\frac{a_2^2}{3} + a_1$ and[87] $a_0' = a_1\frac{a_2}{3} - 2\left(\frac{a_2}{3}\right)^3 + a_0$.

Cardano remarks that, if $a_1' > 0$, then "the case is impossible of a true solution", that is[88] $x < 0$. So, we assume $a_1' < 0$.

Note then the loose correspondence between these cases and the sign of the discriminant $\Delta_3 = \frac{q^2}{4} + \frac{p^3}{27}$ (see above, the formula (2.5.10) on page 36). We have $p = a_1'$ and $q = a_0'$. Then, we can *a priori* only say that $\Delta_3 < 0$ in the first case $a_0' = 0$.

[87] We observe that $a_0' = a_1'\frac{a_2}{3} + \left(\frac{a_2}{3}\right)^3 + a_0$.

[88] If $a_1' > 0$, then $a_0' = a_1'\frac{a_2}{3} + \left(\frac{a_2}{3}\right)^3 + a_0 > 0$. We would then have the transformed equation $y^3 + a_1'y + a_0' = 0$, the real solution of which is negative.

Indeed, by the cubic formula

$$y < 0 \iff \sqrt[3]{-\frac{a_0}{2} + \sqrt{\frac{a_0^2}{4} + \frac{a_1^3}{27}}} + \sqrt[3]{-\frac{a_0}{2} - \sqrt{\frac{a_0^2}{4} + \frac{a_1^3}{27}}} \iff -\frac{a_0}{2} < \frac{a_0}{2},$$

which is always true since we took $a_0 > 0$.

Anyway, Cardano gives no hints on how he discovered these results, whether by a similar calculation or generalising experience.

We remind[89] that, in the case $a'_0 = 0$, $\frac{a_2}{3}$ is also a solution of $x^3 + a_1x + a_0 = a_2x^2$.

Cardano gives a proof for both cases $a'_0 > 0$ and $a'_0 < 0$.

AM XXII — *Proof*. Let the equation $\overline{AC}^3 + 4\overline{AC} + 16 = 6\overline{AC}^2$. [We then have $a'_0 > 0$.] [Cardano[90] considers the same diagram as in Chapter XVII, see above, the figure 3.13 on page 139.]

He takes $\overline{BC} = \frac{6}{3} = 2$. By (AM VI.6) Cardano knows that

$$\overline{AC}^3 = (\overline{AB} + \overline{BC})^3 = \overline{AB}^3 + 3\overline{AB}^2\,\overline{BC} + 3\overline{AB}\,\overline{BC}^2 + \overline{BC}^3,$$

that is, in this case,

$$\overline{AC}^3 = \overline{AB}^3 + 6\overline{AB^2} + 12\overline{AB} + 8.$$

Then, using the relation given by the equation, Cardano gets

$$(\overline{AB}^3 + 6\overline{AB}^2 + 12\overline{AB} + 8) + 4\overline{AC} + 16 = 6\overline{AC}^2,$$

that is

$$\overline{AB}^3 + 6\overline{AB}^2 + 12\overline{AB} + 8 + 4(\overline{AB} + \overline{BC}) + 16 = 6(\overline{AB}^2 + 2\overline{AB}\,\overline{BC} + \overline{BC}^2),$$

that is

$$\overline{AB}^3 + 6\overline{AB}^2 + 12\overline{AB} + 8 + 4(\overline{AB} + 2) + 16 = 6(\overline{AB}^2 + 4\overline{AB} + 4),$$

that is

$$\overline{AB}^3 + 8 = 8\overline{AB}.$$

Now, Cardano knows how to solve the transformed equation in $\overline{AB}$ by (AM XIII). The relation $\overline{AC} = \overline{AB} + \overline{BC}$ yields then a solution of $\overline{AC}^3 + 4\overline{AC} + 16 = 6\overline{AC}^2$.

89 See above, on page 155.

90 See [4, Chapter XXII, page 43r].

Again, always referring to the same diagram, let the equation $\overline{AC}^3 + 4\overline{AC} + 1 = 6\overline{AC}^2$. [We then have $a'_0 < 0$.] Cardano takes $\overline{BC} = \frac{6}{3} = 2$. As before, using the relation given by the equation, Cardano gets

$$(\overline{AB}^3 + 6\overline{AB}^2 + 12\overline{AB} + 8) + 4\overline{AC} + 1 = 6\overline{AC}^2,$$

that is

$$\overline{AB}^3 = 8\overline{AB} + 7.$$

Now, Cardano knows how to solve the transformed equation in $\overline{AB}$ by (AM XII). The relation $\overline{AC} = \overline{AB} + \overline{BC}$ yields then a solution of $\overline{AC}^3 + 4\overline{AC} + 16 = 6\overline{AC}^2$. □

The same remark on page 155 also holds here. Then, if $a'_0 = 0$, we get to $y_1 = \sqrt{-a'_1}$, $y_2 = \sqrt{-a'_1}$, and $y_3 = 0$. For the last solution, the corresponding solution of the equation in x is then $x_3 = \frac{a_2}{3}$.

3.4.7. "On the cube, some squares, and a number equal to some things"

In **Chapter XXIII** Cardano deals with the equation $x^3 + a_2x^2 + a_0 = a_1x$.

AM XXIII. *Given the equation $x^3 + a_2x^2 + a_0 = a_1x$, a solution is*

$$x = y - \frac{a_2}{3},$$

where y is a solution of $y^3 + a'_0 = (-a'_1)y$, with $a'_1 = -\frac{a_2^2}{3} - a_1$ and $a'_0 = -a'_1\frac{a_2}{3} - \left(\frac{a_2}{3}\right)^3 + a_0[= \frac{1}{3}a_1a_2 + \frac{2}{27}a_2^3 + a_0]$.

We remark that $a'_1 < 0$. Moreover, we have $a'_0 > 0$. Indeed,

$$\begin{aligned} a'_0 &= (\frac{a_2^2}{3} + a_1)\frac{a_2}{3} - \left(\frac{a_2}{3}\right)^3 + a_0 \\ &= 2\left(\frac{a_2}{3}\right)^3 + a_1\frac{a_2}{3} + a_0 > 0. \end{aligned}$$

Finally, note the loose correspondence between these cases and the sign of the discriminant $\Delta_3 = \frac{q^2}{4} + \frac{p^3}{27}$ (see above, the formula (2.5.10) on page 36). We have $p = a'_1$ and $q = a'_0$. Then, we can *a priori* only say that $\Delta_3 \neq 0$.

AM XXIII — *Proof*. Let the equation $\overline{AB}^3 + 6\overline{AB}^2 + 4 = 41\overline{AB}$. [Cardano[91] considers the same diagram as in Chapter XVII, see above, the figure 3.13 on page 139.]

He takes $\overline{BC} = \frac{6}{3} = 2$. By (AM VI.6) Cardano knows that

$$\overline{AC}^3 = (\overline{AB} + \overline{BC})^3 = \overline{AB}^3 + 3\overline{AB}^2\,\overline{BC} + 3\overline{AB}\,\overline{BC}^2 + \overline{BC}^3,$$

that is, in this case,

$$\overline{AC}^3 = \overline{AB}^3 + 6\overline{AB}^2 + 12\overline{AB} + 8.$$

Then, using the relation given by the equation, Cardano gets

$$\begin{aligned} \overline{AC}^3 &= (\overline{AB}^3 + 6\overline{AB}^2 + 4) + 12\overline{AB} + 4 \\ &= 41\overline{AB} + 12\overline{AB} + 4 \\ &= 53\overline{AB} + 4. \end{aligned}$$

[91] See [4, Chapter XXIII, page 44r].

But Cardano wants a transformed equation where $\overline{AC}$ is the unknown. He knows that $53\overline{AC} = 53\overline{AB} + 53\overline{BC}$, that is $53\overline{AB} = 53\overline{AC} - 53\overline{BC} = 53\overline{AC} - 106$, so he gets

$$\overline{AC}^3 = (53\overline{AC} - 106) + 4,$$

that is

$$\overline{AC}^3 + 102 = 53\overline{AC}.$$

Now, Cardano knows how to solve the transformed equation in $\overline{AC}$ by (AM XIII). The relation $\overline{AB} = \overline{AC} - \overline{BC}$ yields then a solution of $\overline{AB}^3 + 6\overline{AB}^2 + 4 = 41\overline{AB}$. □

At the end of the chapter, Cardano adds that

> [y]ou must remember, however, that all the solutions for these cases are obtained always by adding from the true and the fictitious solutions of the equations into which they are resolved one-third of the coefficient of x^2. Then, provided that a number remains, even if that which is added is a pure negative, this remainder is a true solution for x. They can also be resolved into other cases of four powers when this is convenient.[92]

[92]"*Memineris igitur quod homnes horum capitulorum æstimationes, habentur, addendo semper veras et fictas æstimationes capitulorum in quo resolvuntur Tpq̃d, et dummodo numerus relinquatur, etiam id quod additur sit m: purum, illus relictum est rei vera æstimatio. Possunt etiam resolvi in capitula alia quatuor denominationum, ut liquet*", see [4, Chapter XXIII, page 44r]. The translation is mine. Witmer changes the text, substituting "subtracting" and "subtracted" in the place of "adding" and "added", as he judges that this comment refers only to this chapter, see [22, page 154]. But in my opinion, this comment refers to all (AM XVII)-(AM XXIII). Accordingly, it seems to me more substantive that, adding a pure negative number to another number, Cardano has to take care that it makes no zero. Moreover, Witmer's translation of "*capitulorum*" as "cases" seems to refer to the sub-cases in the immediately preceding examples (all concerning the equation $x^3 + a_2x^2 + a_0 = a_1x$ as they are in Chapter XXIII), whereas "*capitula*" always refers to the whole cubic equation.

3.4.8. Summing up

We observe that Chapters XVII-XXIII have a similar structure to the preceding ones. They all contain a rule, a complete or partial proof of the rule, and some examples.

Moreover, the rules and the proofs themselves have all a similar structure. The general strategy is to bring back a complete cubic equation either to an equation lacking in second degree term (which Cardano already knows how to solve) or to take the cubic root of a number. As said about Chapter XVII, once that the given hypotheses are interpreted in a geometrical environment, the proofs consist in deriving the transformed equation completely relying on the substitution $x = y \pm \frac{a_2}{3}$, on the formula for the expansion of the cube of a binomial given in (AM VI.6), and obviously on the relation given by the given complete equation. The link between the rules and the corresponding proof is translating from an algorithmic language to a geometric language, and backwards.

All the proofs of (AM XVII)–(AM XXIII) follow a very similar pattern. As a consequence of the repetition of so many similar proofs, the diagrams disappear, or better the proofs of (AM XVIII)–(AM XXIII) make reference to the same diagram in (AM XVII) (even if only (AM XVIII) explicitly mentions it), which is nevertheless no longer reproduced in the text. These repetitions give the impression that the proceeding is somehow going to become automatic and that the possible geometrical arguments that could justify (AM VI.6) get buried deeper and deeper.

3.5. Interdependencies between the cubic equations in *Ars Magna*, Chapters XI-XXIII and the *casus irreducibilis*

Now, we would like to summarise the links that bind the cubic equations in Chapters XI-XXIII.

Note in particular that, even if the rules and the proofs in Chapters XI and XII follow a similar pattern, they never make reference to each other. Anyway, it exists a link between them, which is indeed twisted. Indeed, in Chapter VII, Cardano provides some transformations of equations (see pages 69 and 71 above). In particular, he can pass from $x^3+a_1x = a_0$ in Chapter XI to $x^3 = a_2x^2+a_0$ in Chapter XIV and *vice versa*, which means that, if he knows a solution of $x^3 + a_1x = a_0$, then he would also know a solution of $x^3 = a_2x^2 + a_0$, and *vice versa.* In turn, the solving method for the equation in Chapter XIV (allegedly) refers back to Chapter XII. In the end, these remarks link the solving methods in Chapters XI and XII.

Concerning the solving methods in all the other chapters, there are one-way connections between them, as shown by the respective rules and proofs in Chapters XIII-XXIII. A further (two-way) connection is sometimes given by the transformations in Chapter VII. So, when the given equation is connected to another one and when it happens that this last one falls into the *casus irreducibilis*, then also the given equation has to fall into it.

We sum up all the connections between the equations in Chapters XI-XXIII in Figure 3.16 on page 164. I write '(AM ⟨*chapter number*⟩.⟨*case*⟩.⟨*sub-case*⟩)' to refer to the case '⟨*case*⟩' (or sub-case of a case '⟨*case*⟩.⟨*sub-case*⟩') of the solving method for the equation with which the proposition '⟨*Chapter number*⟩' deals. I use three kinds of arrows. I write '(AM y) $\longrightarrow$ (AM x)' to mean that (AM x) gives rise to (AM y). This one-way connection is usually found in the solving method of the equation (AM y). When there moreover exists a transformation which enables us to pass between the equation in Chapter x and the equation in Chapter y, and *vice versa*, I will write a double arrow. A dashed arrow means that the transformation is described neither in Chapter x nor in Chapter y, but it is anyway present in Cardano's text (namely, in Chapter VII). Note that when a dashed arrow and solid arrow overlap, only the solid arrow is visible. Actually, this does not happen but in two cases, namely between Chapters XI and XIV and between Chapters XVI and XVIII. Finally, when the arrow is dotted, the connection is not explicitly described in Cardano's text. This only happens once, namely between Chapter XIV and XII. There, with regard to Cardano's overall flow of argumentation, I made a guess,[93] which is in my opinion a very reasonable one, since it is consistent with the rest of Cardano's text.

Note that I do not represent the trivial sub-cases where the transformed equation can be solved by taking the third or second root. Note moreover that I pointed out the sign of the discriminant whenever it is always $\Delta_3 > 0$.

Let us take a closer look at the diagram in Figure 3.16. We observe that (directly or not) there is a link between the solution of the equation $x^3 = a_1x + a_0$ in Chapter XII and all the others. Since that equation sometimes runs into the *casus irreducibilis*, it can *a priori* affect all the other equations, the solving method of which directly refers to $x^3 = a_1x + a_0$ (or to $x^3 + a_0 = a_1x$). The left side of the diagram, where (AM XI), (AM XIV), (AM XVII.ii), and (AM XVIII.ii) are, and (AM XV.i) is the exception, since these families of equations always have $\Delta_3 > 0$.

We can then appreciate at best the extent of the problem set by the *casus irreducibilis*. Firstly, it undermines the generality of the cubic formula for $x^3 = a_1x + a_0$ in Chapter XII, since Cardano has to add a condition on the coefficients, which is equivalent to ask $\Delta_3 \geq 0$. But also all the other families of equations that can have (at least, a sub-case with) a negative discriminant (that is, all the other families of equations, except the ones in (AM XI) and (AM XIV)) are affected by this major drawback. If one wants to reliably anchor the solving methods for cubic equations, the problem set by the *casus irreducibilis* must be cleared. In the *Ars Magna*, we find only one passage where Cardano actively tries to do it. It is in Chapter XXV. We have already remarked[94] that Cardano's propositions in that chapter could help to guess a solution of a cubic equation, so that one can lower the degree by polynomial division. But his strategy does not settle at all the problem, even if in some cases it enables one to at least avoid having

[93] See page 125 above.

[94] See above, Section 3.1.4, on page 97.

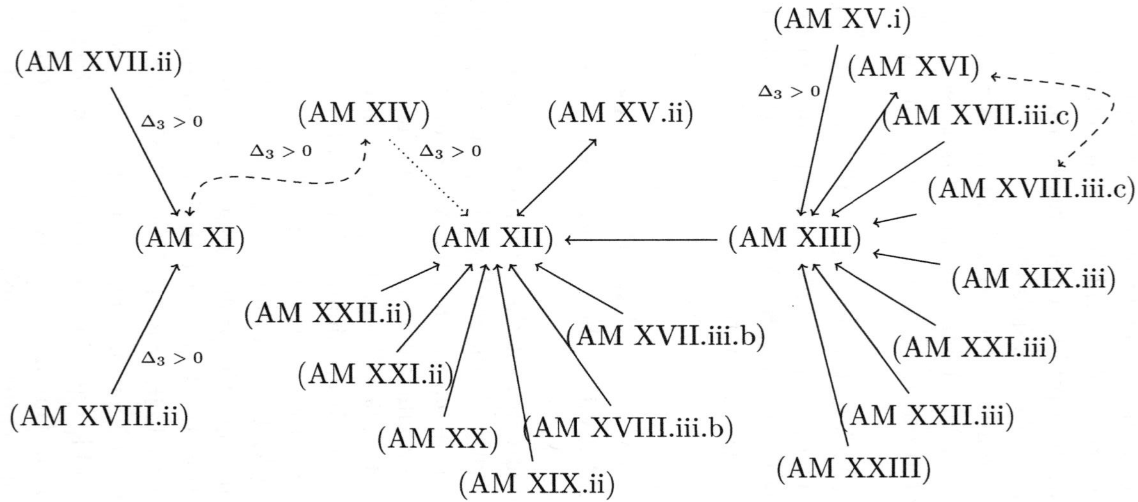

Figure 3.16. – Inter dependencies between the solving methods of the cubic equations in the *Ars Magna*.

to use the cubic formula. Obviously, Cardano is looking for a solution that works at all turns. Indeed, in Chapter XII of the the 1570 and 1663 editions of the *Ars Magna* the reference to Chapter XXV is given up for the benefit of the *De Regula Aliza.* Therefore, we expect to find there a method that will settle once and for all (and possibly in an algorithmic way) the *casus irreducibilis.*

3.6. Solving quartic equations

The next-to-last chapter, Chapter XXXIX, is entitled "On the rule by which we find an unknown quantity in several stages". It comes as an annex after Chapters XXVI-XXXVIII, where Cardano solves miscellaneous problems thanks to a mix of his cubic formulae and traditional algebraic rules, such as the Golden rule or the Rule of equal position. All the problems considered there lead to quadratic or cubic equations, or to "derivative cases".

3.6.1. Cardano's list of the "most general" quartic equations

Before starting, a brief quotation from **Chapter I** follows.

> Although a long series of rules might be added and a long discourse given about them, we conclude our detailed consideration with the cubic, others being merely mentioned, even if generally, in passing. For, as *positio* [the first power] refers to a line, *quadratum* [the square] to a surface, and *cubum* [the cube] to a solid body, it would be very foolish for us to go beyond this point. Nature does not permit it. Thus, it will be seen, all those matters up to and including the cubic are fully demonstrated, but the others which we will add, either by necessity or out of curiosity, we do not go beyond barely setting out.[95]

[95] See [22, page 9] or "*[e]t quanque longus sermo de his haberi posset, ac innumera capitulorum series subiungi, finem tunc exquisitæ considerationi in cubo*

Despite these words, Cardano not only devotes Chapter XXXIX to solve quartic equations, but also gives some proofs.

In **Chapter XXXIX** Cardano tells how, thanks to a method discovered by Ferrari, they found "all the solutions for equations of the fourth power, square, first power, and number, or of the fourth power, cube, square, and number".[96] Let us try to see in which sense one has to interpret this statement. Cardano makes a list of twenty quartic equations as in Figure 3.17 on page 167. We can reasonably expect to find a list of the quartic equations that one can solve by Ferrari's method. We can also reasonably expect not to find the quartic equations that lower in degree, that is, that can be solved using the methods for degrees up to the third in an easy way (for instance, biquadratic equations). But we remark that two (actually, three) kinds of quartic equations which are truly quartic equations do not appear in the classification. They are the complete quartic equations, the three-term quartic equations where only the second degree term is lacking (that is, the equations where the terms x^4, a_3x^3, a_1x, a_0 appear), and the two-term quartic equation $x^4 = a_0$ (but one only needs to take the fourth root to solve it, so it is sensible that it does not appear in Cardano's classification). At a closer view, we remark that in this list there are only quartic equations where x^3 and x are never coupled.

Moreover, Cardano observes that

> [i]n all these cases, therefore, which are indeed [only] the most general as there are 67 others, it is convenient to

faciemus, cætera, etiam si generaliter, quasi tamen per transennam tractantes. Nanque cum positio lineam, quadrato superficiem, cubus corpus solidum referat, nec utique stultum fuerit, nos ultra progredi, quo naturæ non licet. Itaque satis perfecte docuisse videbitur, qui omnia quæ usque ad cubum sunt, tradiderit, relique, quæ adijcimus, quasi coacti aut incitati, non ultra tradimus", see [4, Chapter I, Paragraph 1, pages 3r-v].

[96] See [22, page 237] or "*per eam habemus omnes æstimationes ferme capitulorum q̃d quadrati et quadrati rerum, et numeri, vel q̃d quadrati cubi, quadrati et numeri*", see [4, Chapter XXXIX, Paragraph 2, page 72v].

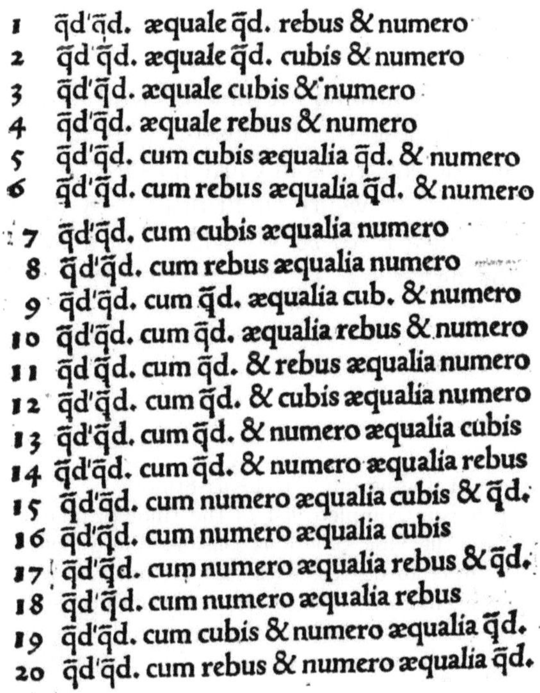

1 q̄d'q̄d. æquale q̄d. rebus & numero
2 q̄d'q̄d. æquale q̄d. cubis & numero
3 q̄d'q̄d. æquale cubis & numero
4 q̄d'q̄d. æquale rebus & numero
5 q̄d'q̄d. cum cubis æqualia q̄d. & numero
6 q̄d'q̄d. cum rebus æqualia q̄d. & numero
7 q̄d'q̄d. cum cubis æqualia numero
8 q̄d'q̄d. cum rebus æqualia numero
9 q̄d'q̄d. cum q̄d. æqualia cub. & numero
10 q̄d'q̄d. cum q̄d. æqualia rebus & numero
11 q̄d'q̄d. cum q̄d. & rebus æqualia numero
12 q̄d'q̄d. cum q̄d. & cubis æqualia numero
13 q̄d'q̄d. cum q̄d. & numero æqualia cubis
14 q̄d'q̄d. cum q̄d. & numero æqualia rebus
15 q̄d'q̄d. cum numero æqualia cubis & q̄d.
16 q̄d'q̄d. cum numero æqualia cubis
17 q̄d'q̄d. cum numero æqualia rebus & q̄d.
18 q̄d'q̄d. cum numero æqualia rebus
19 q̄d'q̄d. cum cubis & numero æqualia q̄d.
20 q̄d'q̄d. cum rebus & numero æqualia q̄d.

Figure 3.17. – List of the quartic equations considered in *Ars Magna*, Chapter XXXIX. Kind courtesy of Biblioteca Virtual del Patrimonio Bibliográfico and Universidad de Sevilla.

> reduce those [equations] involving the cube to equations in which x is present,[97]

such as – he goes on – reducing $x^4 + a_3x^3 = a_0$ to $x^4 = a_1x + a_0$ or $x^4 = a_2x^2 + a_3x^3 + a_0$ to $x^4 = a_2x^2 + a_1x + a_0$. I cannot figure out how Cardano counts the others 67 equations, to the amount of

[97] See [22, pages 237-8] or "*[i]n his igitur omnibus capitulis, quæ quidem sunt generalissima, ut reliqua omnia sexaginta septem superiora, oportet reducere capitula, in quibus ingreditur cubus, ad capitula, in quibus ingreditur res*", see [4, Chapter XXXIX, second rule].

87 equations. Maybe he includes some of the derivative cases, as he already did for cubic equations.[98]

These introductory remarks by Cardano are not as clear as one might expect. The following questions can be raised. Why are x^3 and x never coupled in the quartic equations in Cardano's list? Why does Cardano call 'general' his list of twenty equations? Why does Cardano judge convenient to reduce x^3 to x? We will try to answer to these questions in the following.

[98]See [22, pages 24-6] or [4, Chapter II, Paragraph 2, pages 7r-8r].

If one wants nevertheless to have a complete list of the (monic) quartic equations with the same features of the cubic equations considered in Chapters XI-XXIII, namely that do not lower in degree, with at least one real root (which means, since all the coefficients are taken positive, never to consider an equation where all the terms are on one side of the equal and 0 on the other side), and that cannot be solved by simply taking the fourth root (that is, $x^4 = a_0$), one finds

- 15 cases for complete equations, where the terms $x^4, a_3x^3, a_2x^2, a_1x, a_0$ appear,
- 7 cases for each of the following 3 families of four-term equations, the one lacking in the third degree term (where the terms x^4, a_2x^2, a_1x, a_0 appear), in the second degree term (where the terms x^4, a_3x^3, a_1x, a_0 appear), and in the first degree term (where the terms x^4, a_3x^3, a_2x^2, a_0 appear),
- 3 cases for each of the following 2 families of three-term equations, the ones where the terms x^4, a_3x^3, a_0 or x^4, a_1x, a_0 appear.

That makes 42 equations, out of which Cardano only lists twenty equations. If one adds the equations that lower in degree (7 cases for the family of four-term equations where $x^4, a_3x^3, a_2x^2, a_1x$ appear, 3 cases for each of the following 4 families of three-term equations where the terms x^4, a_3x^3, a_2x^2, x^4, a_3x^3, a_1x, x^4, a_2x^2, a_1x appear, and 1 case for each of the following 3 families of two-term equations where the terms x^4, a_3x^3, x^4, a_2x^2, and x^4, a_1x appear), the 3 biquadratic equations where the terms x^4, a_2x^2, a_0 appear, and the one that can be solved by taking the fourth root one finds 68 equations. They are in no way 87. Anyway, it is more likely that Cardano includes some arbitrarily chosen derivative cases such as biquadratic equations than the other equations which lower in degree or which can be solved by taking the fourth root.

3.6.2. Solving "the square square, some squares, and a number equal to some things"

I will not provide a general quartic formula, since it is so complicated to it becomes pointless in practice. In any case, neither Cardano does. He rather gives a step-by-step, explicit procedure.

Firstly, Cardano[99] shows the formula for the square of a trinomial, as he already did in (AM VI.6) for the binomial. He does not explicitly formulate the statement, so I put it into square brackets.

AM XXXIX.3. *[The formula for the square of a trinomial is*

$$(AB + BC + CG)^2 = AB^2+BC^2+CG^2+2AB\,BC+2BC\,CG+2CG\,AB,$$

or

$$\left(\overline{AB} + \overline{BC} + \overline{CG}\right)^2 = \overline{AB}^2+\overline{BC}^2+\overline{CG}^2+2\overline{AB}\,\overline{BC}+2\overline{BC}\,\overline{CG}+2\overline{CG}\,\overline{AB}]$$

I do not transcribe the proof, but only present the diagram that Cardano uses as a hint (see Figure 3.18 on page 170). Indeed, there are no substantial differences compared to the proof in (AM VI.6). We remark that Cardano will need to apply this formula to quartic equations taking $\overline{AB} = x^2$ and $\overline{BC}, \overline{CG}$ depending on the coefficients.

In the next step, Cardano[100] considers the equation $x^4+a_2x^2+a_0 = a_1x$. He remarks that, if $a_2 = 2\sqrt{a_0}$ holds, then it is easy to reduce the left side $x^4 + a_2x^2 + a_0$ of the equation to the perfect square $(x^2 + \sqrt{a_0})^2$. This is achieved by adding $(2\sqrt{a_0} - a_2)x^2$ to both sides of the equation.

The last step[101] will be to add "as many squares and such a number" to both sides of the equation in order to have the square of a trinomial on the left side and the square of a binomial on the right side in accordance with (AM XXXIX.3). Then, taking the square root, the solution will be easily derived by (AM V.4-6).

---

[99] See [22, page 238] or [4, Chapter XXXIX, Paragraph 3, pages 73r-73v].
[100] See [22, pages 238-9] or [4, Chapter XXXIX, Paragraph 4, page 73v].
[101] See [22, page 239] or [4, Chapter XXXIX, Paragraph 5, page 73v].

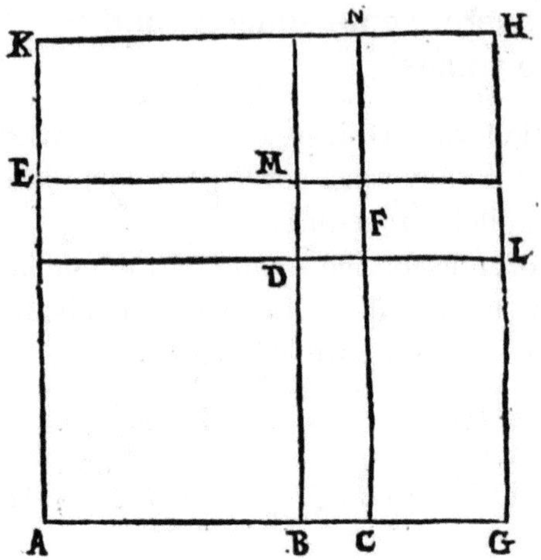

Figure 3.18. – Diagram from *Ars Magna* XXXIX.3. Kind courtesy of Biblioteca Virtual del Patrimonio Bibliográfico and Universidad de Sevilla.

In the end, we get a very loose explanation of a general course of actions, especially concerning the last step. Cardano needs to specify it better with the example to "divide 10 into three proportional parts, the product of the first and the second of which is 6".[102] This is one of the problems with which Zuanne de Tonini da Coi used to challenge his adversaries, believing that it could not be solved. It yields to the following quartic equation.

AM XXXIX Problem V. *To solve the equation* $x^4+6x^2+36 = 60x$.

AM XXXIX Problem V — *Solution.* Cardano and Ferrari[103] add $6x^2$ to both sides of the equation in order to get the square of a

[102]See [22, page 239] or "*[f]ac ex 10 tres partes proportionales, ex quarum ductu primæ in secundam producantur 6*", see [4, Chapter XXXIX, problem V, page 73v].

[103]See [4, Chapter XXXIX, pages 73v-74r].

binomial on the left side

$$(x^2+6)^2 = 6x^2 + 60x.$$

In order to obtain a square also on the right side, they want to add to both sides "enough squares and a number [*tot quadrati et numerus*]", say $b_2, b_0 > 0$,

$$x^4 + 12x^2 + 36 + (b_2x^2 + b_0) = 6x^2 + 60x + (b_2x^2 + b_0)$$

[such that the equation becomes

$$(x^2 + 6 + \gamma)^2 = (\delta x + \epsilon)^2,$$

where the real γ, δ, ϵ are to be further specified depending on the coefficients.] [Let $\overline{CG}$ be a real number (Cardano calls it in such a way to make reference to the diagram for (AM XXXIX.3)).] If they take $b_2 = 2\overline{CG}$ and $b_0 = \overline{CG}^2 + 12\overline{CG}$, then they know by (AM XXXIX.3) that it holds that

$$x^4+(12+2\overline{CG})x^2+(36+\overline{CG}^2+12\overline{CG})=(6+2\overline{CG})x^2+60x+(\overline{CG}^2+12\overline{CG}),$$

that is

$$x^4+(12+2\overline{CG})x^2+(6+\overline{CG})^2 = (6+2\overline{CG})x^2+60x+(\overline{CG}^2+12\overline{CG})$$

where the left side is the square $\left(x^2 + (6+\overline{CG})\right)^2$.

[This means that $\left(x^2 + (6+\overline{CG})\right)^2$, considered as a polynomial in x, is a square for each $\overline{CG}$. We look for a $\overline{CG}$ such that $(6+2\overline{CG})x^2 + 60x + (\overline{CG}^2 + 12\overline{CG})$ is also a square, considered as a polynomial in x. Take α a real such that $\alpha^2 = 6 + 2\overline{CG}$ and take β a real such that $\beta^2 = \overline{CG}^2 + 12\overline{CG}$. Then $\alpha^2x^2 + 60x + \beta^2 = (\alpha x + \beta)^2$ if and only if $60x = 2\alpha x\beta$, that is, if and only if $(60x)^2 = (2\alpha x\beta)^2$, that is if and only if $\alpha^2\beta^2 = 900$. We get $(6 + 2\overline{CG})(\overline{CG}^2 + 12\overline{CG}) = 900$.] Then,

$$\overline{CG^3} + 15\overline{CG}^2 + 36\overline{CG} = 450.$$

By the cubic formula in (AM XVII), Cardano and Ferrari find

$$\overline{CG} = \sqrt[3]{190 + \sqrt{33903}} + \sqrt[3]{190 - \sqrt{33903}} - 5.$$

This means that Cardano and Ferrari are able to transform the given quartic equation into another quartic where there are two squares on both sides of the equal, provided that they manage to solve a certain cubic equation. This can be done by the cubic formulae in (AM XI)-(AM XXIII). Then – as said – they take the square root on both sides of the equal and solve the quadratic equation that comes out by (AM V.4-6). In this way, both the quadratic and cubic formulae have to be used in order to solve quartic equations.

We moreover remark that Cardano and Ferrari need as usual a formula for the square of the sum of certain quantities – a trinomial, in this case. Once hidden the possibly geometrical argument in the proof of (AM XXXIX.3), no other geometrical argument appears in the solution of (AM XXXIX Problem V). Besides, any geometrical argument cannot appear at all. Indeed, Cardano and Ferrari consider $\overline{CG}^2 + 12\overline{CG}$. For homogeneity reasons, this forces the interpretation that all the quantities there have to be considered as numbers, and then the diagram becomes pointless.

Cardano also infers the following formula, to directly draw the cubic equation

$$\overline{CG}^3 + (1 + \frac{1}{4})2\sqrt{a_0}\overline{CG}^2 + a_0\overline{CG} = \frac{1}{2}\left(\frac{a_1}{2}\right)^2.$$

3.6.3. What about the other quartic equations?

Using the above method, Cardano is able to solve the equations of the family $x^4 + a_2x^2 + a_0 = a_1x$. Then, some examples follow. We will check them in order to try to find out how Cardano solves the other 19 families of quartic equations that he listed at the beginning. Except (AM XXXIX Problem V), there are eight examples more – called "problems [*quæstiones*]". Problem VIII on $x^4 + 32x^2 + 16 = 48x$ uses the same Ferrari's methods as in (AM XXXIX Problem V).

Problem X on $x^4 = x^2 + 1$ does not need Ferrari's method to be solved, since it is a derivative case.[104]

Problem VI on $x^4 = x + 2$ is a simplified version[105] of Ferrari's method, since x^4 on the left side of the equation is already a perfect square. Then, one step can be skipped and the remaining part of Ferrari's method is unchanged. Namely, Cardano takes $b_2 = 2n$ and $b_0 = n^2$ where n is a (real) number. Then, $\alpha^2 = 2n$ and $\beta^2 = 2 + n^2$, and he has to solve the cubic equation $n^3 + 2n = \frac{1}{8}$.[106]

Problem IX on $x^4+4x+8 = 10x^2$ and Problem XII on $x^4+3 = 12x$ need a little handling before applying Ferrari's method. Indeed, in order to find on the left side of the equation a perfect square where x^2 appears, Cardano[107] needs basically to arrange the second and first degree terms on the left side of the equation and the constant term on the right side. In this way, one can solve all the quartic equations where the third degree term does not appear.

Then, it remains to show how the equations where the third degree term appears – which involve half of the twenty equations that Cardano listed above – can be solved by Ferrari's method. Problem VII is on $x^4 + 6x^3 = 64$. Cardano[108] uses (AM VII.6-7) (but also (AM VII.2-3) works) to transform $x^4 + 6x^3 = 64$ into $y^4 = 6y + 4$. Then,

---

[104]See [22, pages 250-1] or [4, Chapter XXXIX, problem X, pages 77v-78r].

[105]See [22, pages 243-6] or [4, Chapter XXXIX, problem VI, pages 75r-75v].

[106]Concerning again Problem VI, Cardano provides another way to solve the equation. He manages to lower the degree of the equation dividing both sides by an appropriate polynomial. Anyway, this is a particular method, since the possibility to find a polynomial which divides the equation depends on one's ability to guess a solution of the equation.

[107]See [22, pages 246-8] or [4, Chapter XXXIX, problem VIII, pages 76r-76v], see [22, pages 248-50] or [4, Chapter XXXIX, problem IX, pages 76v-77v], and see [22, page 252] or [4, Chapter XXXIX, problem XII, page 78v].

[108]See [22, page 76r] or [4, Chapter XXXIX, problem VII, page 76r].

he can apply Ferrari's method in his simplified version, since the left side of the equation x^4 is already a perfect square. With a similar justification, one can bring back also the equations $x^4 = a_3x^3 + a_0$ and $x^4 + a_0 = a_3x^3$ respectively to $y^4 + b_1y = b_0$ and $y^4 + b_0 = b_1y$. Moreover, we remark that (AM VII.12) enables us to pass from $x^4 + a_2x + a_0 = a_3x^3$ to $y^4 + a_2y^2 + a_0 = a_1y$.

Also Problem XI on $x^4 - 3x^3 = 64$ could be solved in this way – as Cardano says, but he rather prefers to deal with it without transformations. Indeed, using the transformations in Chapter VII, one can only deal with the few above cases. Cardano states then a slight variation of Ferrari's method (the main idea to get two squares on both sides of the equal remaining unchanged) to solve all the quartic equations that he listed, where the third degree term appears.

> You ought to know two things: First, just as x should always remain on the side on which the number and the second power are and should not be on the side of x^4, so the cube, whether positive or negative, should remain with x^4. Secondly, just as the coefficient of x should not change, so also the coefficient of x^3 should not vary. To these we can add a third, namely that [just as] where there are x's we will arrive at the fourth power plus the square plus the number equal to the second power plus or minus the x's plus a number, so in this case we will arrive at the fourth and second powers plus the number equal to the fourth power plus or minus the cube plus the square.[109]

Let us now check how Cardano applies these prescriptions to Problem XI.

[109]See [22, page 251] or "*debes igitur scire duo. Primum, quod ut res debent semper manere ab alia parte, a qua est numerus cum quadratis, et non a arte q̃dq̃drati, sic cubi, seu p: seu m: debent manere cum q̃dq̃drato. Secundum, quod ut numerus rerum nunque debet variari, sic nec numerus cuborum. Et possumus addere tertius his, scilicet, quod ubi sunt res, pervenimus ad 1 q̀dq̀dratu p: quad.p: numero, æqualia quad.rebus p: vel m: et numero p: sic hic pervenimus ad q̃dq̃dratu p: quad.p: numero, æqualia q̃dq̃d. cubis p: vel m: et quad.p:*", see [4, Chapter XXXIX, Problem XI, page].

AM XXXIX Problem XI. *To solve the equation* $x^4 - 3x^3 = 64$.

AM XXXIX Problem XI — *Solution.* [We remark that 64 in the right side of the equation is already a perfect square.] Cardano[110] adds $\frac{n^2}{64}x^4 + 2nx^2$ (where n is a real number) to both sides in order to find, on the right side of the equation, a perfect square in which x^2 appears. Then,

$$x^4 - 3x^3 + \left(\frac{n^2}{64}x^4 + 2nx^2\right) = 64 + \left(\frac{n^2}{64}x^4 + 2nx^2\right),$$

that is

$$\left(\frac{n^2}{64} + 1\right)x^4 - 3x^3 + 2nx^2 = \left(8 + \frac{n}{8}x^2\right)^2.$$

[As before, this means that $\left(\frac{n^2}{64} + 1\right)x^4 - 3x^3 + 2nx^2$ is also a square, considered as a polynomial in x. We look for such a n. Take a real α such that $\alpha^2 = \frac{n^2}{64} + 1$ and a real β such that $\beta^2 = 2n$. Then $\alpha^2x^4 - 3x^3 + \beta^2x^2 = (\alpha x^2 + \beta x)^2$ if and only if $\alpha^2\beta^2 = \frac{9}{4}$. We get $(\frac{n^2}{64} + 1)2n = \frac{9}{4}$.] Then,

$$n^3 + 64n = 72.$$

[By (AM XI), Cardano finds a value for n. Then, since on both sides of the equal two squares appear, he gets a quadratic equation by taking the square root. Thanks to (AM V.4-6), Cardano knows how to solve it.]

And finally, what about the cases where a_3x^3 and a_1x appear together in the same equation? There are 22 (monic) quartic equations more that do not lower in degree, with at least one real solution, and that cannot be solved by simply taking the fourth root (that is, equations that should have had the right to appear in Cardano's list, but that he does not mention), where both a_3x^3 and a_1x appear. They are

110 See [22, pages 251-2] or [4, Chapter XXXIX, problem XI, pages 78r-78v].

the 15 complete equations (which are not in Cardano's list) and the 7 cases for the family of equations where the terms x^4, a_3x^3, a_1x, a_0 appear (which instead are there).

Cardano solves an equation of this last family. Problem XIII is indeed on $x^4 + 2x^3 = x + 1$. Unfortunately, Cardano[111] deals very shortly and confusingly with this example. Anyway, at the end, he gives the right (positive, real) solution and affirms that "by this you know the methods for these rules if you have paid careful attention to the examples and the operations".[112] Indeed, we can easily check that the methods in (AM XXXIX Problem V) – or, better, in (AM XXXIX Problem XI) – can be adapted to this case.[113]

[111]See [22, pages 252-3] or [4, Chapter XXXIX, problem XIII, pages 78v-79r].

[112]See [22, page 253] or "*per hæc intelligis modos harum regularum, si exempla hæc diligenter cum suis operationibus animadvertas*", see [4, Chapter XXXIX, problem XIII, page 79r].

[113]Following Cardano's suggestion, if we add x^2 to both sides of the equation $x^4 + 2x^3 = x + 1$, its left side becomes a perfect square $(x^2 + x)^2$. We want instead to add $b_2x^2 + b_1x + b_0$ to both sides

$$x^4 + 2x^3 + (b_2x^2 + b_1x + b_0) = x + 1 + (b_2x^2 + b_1x + b_0)$$

such that we get a perfect square of a trinomial, where the first two terms are x^2, x on the left side and another perfect square (*a priori* of a binomial) on the right side. Take the real γ to be the third term of the trinomial. Applying the formula for the square of a trinomial to $x^2 + x + \gamma$ and by comparison of the coefficients, we get $b_2 = 1 + 2\gamma$, $b_1 = 2\gamma$, and $b_0 = \gamma^2$. Then,

$$(x^2 + x + \gamma)^2 = (1 + 2\gamma)x^2 + (1 + 2\gamma)x + 1 + \gamma^2.$$

This means that $(1 + 2\gamma)x^2 + (1 + 2\gamma)x + 1 + \gamma^2$ is also a square considered as a polynomial in x. We look for one of such γs. Take a real α such that $\alpha^2 = 1 + 2\gamma$ and take a real β such that $\beta^2 = 1 + \gamma^2$. Then,

$$(1 + 2\gamma)(1 + \gamma^2)x^2 = \left(\frac{1 + 2\gamma}{2}x\right)^2,$$

that is

$$8\gamma^2 + 4\gamma + 3 = 0.$$

3.6.4. Summing up

Summing up, Cardano has Ferrari's method available to solve equations like $x^4 + a_2x^2 + a_0 = a_1x$. Thanks to this method, one performs a double reduction to two squares (*a priori* of a trinomial and of a binomial) on both sides of the considered equation. In order to do that, a formula analogous to (AM VI.6) for the square of a trinomial is needed. It is in Proposition (AM XXXIX.3). Moreover, Cardano needs to introduce a certain quantity to be further specified thanks to a technical condition imposed on one of the squares. The condition leads to a cubic equation, which he solves by (AM XI)–(AM XXIII). Then, taking the square root and using (AM V.4-6), Cardano finds (at least) a (real) solution of the quartic equations.

This method can be plainly adapted to solve all the quartic equations that Cardano is likely to consider (that is, monic quartic equations that do not lower in degree, with at least one real root, and that cannot be solved by taking the fourth root). If the quartic equation contains only x^4, a_2x^2, a_1x, a_0 or x^4, a_1x, a_0 (that is, equations number 1, 4, 6, 8, 10, 11, 14, 17, 18, and 20 in Cardano's list), then one needs to apply Ferrari's method in (AM XXXIX Problem V), or the simplified version in Problem VI, or again the slight modification consisting in rearranging the terms as in Problems IX and XII. If the quartic equation contains x^4, a_3x^3, a_0 (that is, equations number 3, 7, and 16 in Cardano's list), one first of all transforms the given equation in one that contains only x^4, a_1x, a_0 thanks to (AM VII.2-3) or (AM VII.6-7)

The real solution being $\gamma = -\frac{1}{2}$, we get $b_2 = 0$, $b_1 = -1$, and $b_0 = \frac{1}{4}$. The quartic equation becomes

$$(x^2 + x - \frac{1}{2})^2 = \frac{5}{4},$$

from which it follows that

$$x^2 + x - \frac{1}{2} - \frac{5}{4} = 0$$

whose solutions are $x = \frac{1}{2}\left(-1 - \sqrt{3 + 2\sqrt{5}}\right)$ and $x = \frac{1}{2}\left(\sqrt{3 + 2\sqrt{5}} - 1\right)$. The positive solution is the one mentioned by Cardano.

as in Problem VII, and then goes on as above. Or one can also adapt Ferrari's method as in Problem XI. Finally, if the quartic equation contains x^4, a_3x^3, a_1x, a_0 (that is, equations number 2, 5, 9, 12, 13, 15, and 19 in Cardano's list), it is also possible to adapt Ferrari's method (see Problem XIII). In this case, one has to add a more composite number on both sides of the equation in order to get two squares on both sides, so that in the end we have a slightly more complicated version.

For the remaining 15 families of quartic equations where a_3x^3 and a_1x appear together, no hints are given by Cardano himself. We can maybe suppose that Cardano knows the transformation to make the term of third degree vanish, so that he gets an equation in which only x^4, a_2x^2, a_1x, a_0 appear. Then, it is possible that Cardano calls 'general' the quartic equations in his list excluding the complete ones because their solving method is the one that can be widely applied to the others, once that the other families of equations are transformed to fit it.

We finally recall that, in this context, geometry possibly enters only in the proof for (AM XXXIX.3), as it was the case in (AM VI.6). But then, once having anchored this formula on a solid ground, no longer geometry is used in manipulating quartic equations.

4. Cardano's earlier encounters with some general treatments of equations

In the *Ars Magna*, we have seen what were the most advanced strategies and methods at Cardano's disposal to deal with equations. We have also remarked that his solving methods for cubic equations were supposedly strongly influenced by Tartaglia's poem, even if Cardano implemented it in a relevant way. Let us now take a step backwards in order to look for the foreshadows of these methods in the earlier algebraic treatises. We now focus on the *Practica Arithmeticæ* and on the *Ars Magna Arithmeticæ*, since these are the only two other works – beside the *Ars Magna* and the *De Regula Aliza* – in which Cardano deals with cubic equations.

4.1. Various encounters with equations in the *Practica arithmeticæ*

Cardano relates[1] that, between 1536 and 1537, he got back to one of his earlier mathematical works, called *Arithmetica rudimenta*, and revised it. In 1539 the outcome was published in Milan by Ioannis Antonins Castellioneus for Bernardini Calusci under the title *Hieronymi C. Cardani medici mediolanensis, Practica arithmetice, et mensurandi singularis, in quaque preter alias continentur, versa pagina demonstrabit* or *Practica Arithmeticæ*. We also have a posthumous edition in the fourth volume of the 1663 edition of Cardano's *Opera Omnia*,

[1]See [17, page 41].

printed in Lyon by Ioannis Antonii Huguetan and Marci Antonii Ravaud. No manuscript is known.[2]

The *Practica Arithmeticæ* can be divided into two parts, namely Chapters I-LXV, devoted to the theory, and Chapters LXVI-LXVII, which gather some examples and problems. Most of the theoretical part concerns numbers and operations with numbers in a traditional style. Nevertheless, we find in Chapters XLV-LI a sort of short appendix dealing with equations.

In the next section, we take a glance at the statements concerning equations. We focus in particular on the statements in Chapter LI concerning cubic equations, which are the only ones to present innovations comparing to the older *abaco* treatises.

4.1.1. The appendix devoted to equations

The *Practica Arithmeticæ*, Chapter XLVIII deals with two-term equations from first to fourth degree where, in order to find a solution, one has simply to take the nth-root, with $n = 1, 2, 3, 4$. Once and for all, Cardano observes that one has always to reduce to unity the coefficient of the higher degree term. Then, Chapter XLIX deals with the "minor composite chapters [*capitula minora composita*]",[3] that is, with quadratic equations and with equations that reduce to them (in

[2]See [44].

[3]The term 'chapter [*capitulum*]' is commonly used to refer to family of equations. Indeed, as Jeff Oaks argues in a private communication (thanks Jeff!), this term is very probably a mistranslation from the Arabic '*bāb*', which means 'door; gate; opening, gateway; entrance, chapter, section, column, rubric; group, class, category; field, domain' among other things. In most of the Arabic books '*bāb*' means 'chapter'. In the algebra books of Al-Khwārizmī, Abū Kāmil, and others it also takes the meaning of 'class, category, type, kind' when for instance they speak of the "six types" of equations. This is probably were the Latin translators such as Robert of Chester or Gerardo of Cremona were confused. Indeed, both of them translate Al-Khwārizmī's '*bāb*' as '*capitulum*' when they should have chosen a word which means 'kind' or 'type'. So they will often write the "*sex capitula*".

We moreover remark that the mistranslated term was part of the common mathematical terminology of the Italian algebraists of the time. Indeed, we find for instance a "*Capitolo di potenze, e Tanti eguali à numero*" in Bombelli's

the *Ars Magna* they are called 'derivative'). Chapter L deals with the "greater composite chapters [*capitula composita maiora*]", that is, with biquadratic equations and with equations that reduce to them. Here, Cardano applies only the usual rules.

Finally, **Chapter LI** "On the imperfect cases [*De modis omnibus imperfectis*]" is a miscellany. There, Cardano deals with surd numbers, continued proportions, taking n^{th}-roots, equations that can be reduced to quadratic equations, the rules *de modo*, *de duplici*, and *de medio*,[4] and with cubic equations. We only consider a choice of interesting paragraphs from this chapter. We remark that, unlike in the *Ars Magna*, all the results in the *Practica Arithmeticæ* are stated by the means of numerical examples, as it was usual for that time. Cardano does in no way try to resume the results in a general statement or formula – which I nevertheless do.

In the *Practica Arithmeticæ*, there are mainly two strategies to deal with cubic equations, namely polynomial division and transformation of equations. Concerning polynomial division, Cardano provides the following proposition.

PA LI.25. *The following equalities hold:*

$$\frac{x^3+n^3}{x+n} = x^2 - nx + n^2 \quad \textit{and} \quad \frac{x^3-n^3}{x-n} = x^2 + nx + n^2,$$

with n natural.

The two rules directly come from Chapter XXII, Paragraphs 11, 12, and 16 where they are simply stated without any justification.

Then, Cardano shows how to lower some families of cubic equations to the second degree using these two formulae.

Algebra and some "*Capitoli de Cosa, e Cubo equal à numero ed altri suoi aderenti*" in Tartaglia's *Quesiti et inventioni deverse.*

[4]See [62] and [64].

PA LI.26. *Consider the depressed cubic equations* $x^3 = a_1x + a_0$ *and* $x^3 + a_0 = a_1x$*. If the following condition is fulfilled*

$$na_1 = n^3 + a_0,$$

with n *natural, then one gets a quadratic equation by polynomial division.*

Indeed, consider firstly the equation $x^3 = a_1x + a_0$. If $na_1 = n^3 + a_0$, then $a_0 = na_1 - n^3$ and the equation becomes $x^3 + n^3 = a_1(x + n)$. Therefore $x + n$ divides the left side of the equation by (PA LI.25), and the right side too. In this way, we obtain the quadratic equation $x^2 + n^2 = nx + a_1$. Similarly, the second equation becomes $x^3 - n^3 = a_1(x - n)$ and, dividing both sides by $x - n$, we get the quadratic equation $x^2 + nx + n^2 = a_1$.

PA LI.27 i. *Consider the complete cubic equations* $x^3 + a_1x = a_2x^2 + a_0$, $x^3 + a_1x + a_0 = a_2x^2$, $x^3 + a_2x^2 = a_1x + a_0$, $x^3 + a_0 = a_2x^2 + a_1x$, *and* $x^3 = a_2x^2 + a_1x + a_0$. *If* $\pm n$, *with* n *natural, is a solution, then one gets a quadratic equation by polynomial division.*

This proposition works in a way similar to (PA LI.26). The two propositions are justified through some examples. Note that they partially correspond to (AM XXV.6) and (AM XXV.10), taking $n = f$.

We remark that, in all the examples by Cardano, n is natural, but we can with no extra difficulty assume that n is real, since the only requirement imposed on it is to know how to take its third power n^3 or, conversely, its third root $\sqrt[3]{n}$. This was common knowledge at that time, and Cardano taught in the preceding chapters how to perform this operation on real numbers.

At the very end of Paragraph 27, Cardano remarks the following.

PA LI.27 ii. *In the equations where the "extreme denominations [*denominationes extremæ*]" (that is, the highest degree term and the constant term) are on the same side of the equal compared to the other terms, if there is a solution, there are always two.*

Again, his only justifications here are the examples where, in the case of cubic equations, given a solution, Cardano usually employs polynomial division to lower the degree of the equation. Since he obtains a quadratic equation, he knows how to solve it. The examples mentioned by Cardano are equations of the families $x^3 + a_0 = a_1 x$ and[5] $x^3 + a_0 = a_2 x^2$.

Concerning transformation of equations, in Paragraph 28, Cardano makes one particularly remarkable statement (among many others) concerning continued proportions.

PA LI.28. *Consider a, b, c, d real such that $a : b = b : c = c : d$. Then*

$$(a+b)(a+d) = \left(\frac{a+d}{b} + \frac{a+d}{c}\right) b^2.$$

As usual, Cardano does not justify it but by some examples.[6] He observes that this rule is useful to solve $x^3 + a_2 x^2 = a_0$, $x^3 + a_0 = a_2 x^2$, and $x^3 = a_2 x^2 + a_0$, but we remark that only $x^3 + a_0 = a_2 x^2$ will be mentioned later on.

In Paragraph 32, Cardano shows how to pass from $x^3 + a_0 = a_2 x^2$ to $y^3 + a_0 = a_1 y$. He considers $x^3 + 64 = 18x^2$ and takes $x = a$, $18 = a + d$, and $\sqrt[3]{64} = b$, where $a : b = b : c = c : d$. (We write $y = c$.) By the continued proportion, it follows that $d = \frac{c^2}{b} = \frac{y^2}{\sqrt[3]{64}}$. Then, $a = 18 - d = 18 - \frac{y^2}{\sqrt[3]{64}}$. By Paragraph 28, Cardano knows that it

[5] We remind that Cardano makes the same consideration in the *Ars Magna* at the end of Chapter I, see here note 78, page 147. Anyway, here no further explanations are given.

[6] Anyway, it is a straight calculation to check it. Indeed,

$$(a+b)(a+d) = a^2 + ad + ba + bd.$$

By the continued proportions follows that $b^2 = ac$. Then

$$\left(\frac{a+d}{b} + \frac{a+d}{c}\right) b^2 = \frac{a+d}{b} b^2 + \frac{a+d}{c} ac = ab + db + a^2 + ad.$$

holds

$$\left(18-\frac{y^2}{\sqrt[3]{64}}+\sqrt[3]{64}\right)18=18\sqrt[3]{64}+\frac{18}{y}\sqrt[3]{64},$$

that is $\sqrt[3]{\frac{729}{8}x^3}+\sqrt[3]{23887872}=324x$, that is $x^3+64=72x$. Moreover, Cardano gets that the first of the four continued proportionals a is solution of $x^3+b^3=(a+d)x^2$ and that the third c is solution of $x^3+b^3=b(a+d)x$. We can resume the rule as follows.

PA LI.32 (resumed). *Let us consider the following transformation of equations:*

$$x^3+a_0=a_2x^2 \xleftrightarrow{x=a_2-\frac{y^2}{\sqrt[3]{a_0}}} y^3+a_0=\left(a_2\sqrt[3]{a_0}\right)y.$$

We remark that, by the continued proportion, it follows that $ac=b^2$, that is $xy=\left(\sqrt[3]{a_0}\right)^2$. Therefore, we can restate the rule as follows.

PA LI.32 (restated). *Let us consider the following transformation of equations:*

$$x^3+a_0=a_2x^2 \xleftrightarrow{y=\frac{\left(\sqrt[3]{a_0}\right)^2}{x}} y^3+a_0=\left(a_2\sqrt[3]{a_0}\right)y.$$

We already met this transformation in the *Ars Magna*. Indeed,[7] this is a particular case of the rule (AM VII.2-3). We are now able to observe that Cardano achieved this rule through a property of continued proportions.

4.1.2. Summing up

For the most part, the *Practica Arithmeticæ* reminds of an *abaco* treatise, insomuch as Cardano himself never considered it as a masterpiece.[8] Even if it does not contain much on cubic equations, it helps

[7]See here, on page 69.

[8]See [64, page 64].

to enlighten all the same on the directions undertaken by Cardano before 1539, when he finally received Tartaglia's poem.

In this treatise, we find two techniques to deal with cubic equations, namely polynomial division and transformations of equations. Using polynomial division, provided a (real) solution of a cubic equation, Cardano is able to find the remaining solutions (if they exist real). In certain cases, he is also able to discuss the number of solutions. Using transformation of equations, Cardano knows how to pass from $x^3 + a_0 = a_2x^2$ to $y^3 + a_0 = a_1y$, and *vice versa*. Both techniques come from surveys on continued proportions. Since these techniques do not ask for requirements on the coefficients of the equations, they are always effective. But neither of the two can provide by itself a way to get a solution of a cubic equation starting only from the coefficients, that is, a formula. Indeed, in order to fully apply these two techniques, one has to already have at hand a solution.

Therefore, in the *Practica Arithmeticæ*, we come across Cardano's earlier investigations on cubic equations. He approaches at first the families of equations $x^3 = a_1x + a_0$, $x^3 + a_0 = a_1x$, and $x^3 + a_0 = a_2x^2$, but submits only non-systematical methods. At this first stage, each result on cubic equations comes after a corresponding result on continued proportion, so that in the *Practica Arithmeticæ* cubic equations come to be side effects of proportions, meaning that they are not already a subject of interest by themselves. As mentioned above, traces of this approach will still be there six years later in the *Ars Magna*.

4.2. A more systematic approach to cubic equations in the *Ars magna arithmeticæ*

As said,[9] only recently the *Hieronymi Cardani Ars magna arithmeticæ, seu liber quadraginta capitulorum et quadraginta quæstionum*, or *Ars Magna Arithmeticæ*, has been restored to its fair chronological position. As such, its composition date follows the *Practica Arithmeticæ* and

[9] See here, Section 2.1 from page 9.

foreruns the *Ars Magna*, and this is why we deal with it now. The common belief that the *Ars Magna Arithmeticæ* was subsequent to the *Ars Magna* had been raised by the fact that the only printed version that we have available is in the 1663 *Opera Omnia*. There the publishers together with the editor Spon were supposed to follow a chronological sequence of the treatises, but in fact they inverted the order for these two. We also have available the Manuscript N 187 at the Biblioteca Trivulziana in Milan, which partially attests for the *Ars Magna Arithmeticæ*.

The contents of the *Ars Magna Arithmeticæ* are clearly three-folded. In Chapters I-XVII the so-called arithmetisation of *Elements*, Book X is developed. There, Cardano deals with rules concerning the irrational numbers called *binomia* and *recisa*.[10] In particular, in Chapters I-XII Cardano establishes the terminological correspondences between the different kinds of lines in Euclid's classification and the *binomia* and *recisa*. Moreover, he sets the rules to perform the basic operations on these numbers. In Chapters XIII, XIV, and XVII, he explains how to handle the powers of these quantities. Finally, Chapters XV and XVI are devoted to continued proportions.

On the other hand, Chapters XVIII-LX deal directly with equations – this is why we mainly focus on them. In particular, in Chapters XVIII-XIX and XXII-XXVI Cardano states which types of *binomia* and *recisa* can or cannot be solutions of some families of cubic and quartic equations. Chapter XX is a list of all the equations that Cardano considers. Chapter XXI is on transformations of equations. Chapter XXVII deals with the ways to get in touch with solving equations. In Chapters XXVIII-XXX, XXXII, XXXIV, and XXXVI, Cardano solves depressed cubic equations and cubic equations lacking in the first degree term, whereas in Chapters XXXI, XXXIII, XXXV, and XXXVII he gives some "particular" rules on them. Chapter XXXIX contains a partial treatment of complete cubic equations. Finally, in Chapter LX, Cardano deals with the derivative cases and then presents, in the same style of *Ars Magna*, Chapters XXIV-XL,

[10]See page 80 above or [32, Book X, Definitions, pages 133r-133v].

miscellaneous fringe problems, which are the "questions" recalled by the title of the *Ars Magna Arithmeticæ.* We remark that no chapter is specifically devoted to quadratic equations and that Cardano only glances over some families of quartic equations in Chapter XXXVIII (and, in any case, he never gives rules to solve them).

Finally, after Chapter XL, we find a further (non-numbered) chapter that is a long miscellany of solved problems allocated to over forty questions.

In the following, we consider a choice of some interesting chapters, namely the ones that will turn out to be useful to interpret the *De Regula Aliza.* First of all and very briefly, I show in Section 4.2.1 that some considerations on the square roots of negative quantities very similar to the ones in the *Ars Magna* are also echoed in the *Ars Magna Arithmeticæ.* Going on to recall the topics already highlighted while analysing the *Ars Magna,* I report in Section 4.2.2 on Cardano's heuristics and in Section 4.2.3 on some techniques to transform equations. In Section 4.2.5 I deal with the cubic formulae. On the contrary, the study of "general" shapes for irrational solutions is the peculiarity of the *Ars Magna Arithmeticæ.* I deal with this subject in Section 4.2.4. Together with Section 4.2.6 on "particular" solving methods, it is also be useful to clarify the opposition between "particular" and "general".

4.2.1. Shortly again on the square roots of negative numbers

To firstly mention a topic that reveals its full significance only in Section 5, I very briefly deal with the square roots of negative numbers. We find them in the last miscellaneous chapter, **Question 38**, almost at the very end of the *Ars Magna Arithmeticæ.*

There, among other problems, Cardano solves the systems

$$\begin{cases} 6 = x + y \\ 16 = xy \end{cases}$$

and

$$\begin{cases} 6 = x + y \\ 20 = xy \end{cases},$$

the solutions of which respectively are $x = 3 \pm \sqrt{-7}$, and $x = 3 + \sqrt{-13}$ and $y = 3 - \sqrt{-11}$. Cardano simply gets to them applying the quadratic formula. Then, he verifies that those are the demanded values. For instance, in the first case, we have

$$(3 - \sqrt{-7})(3 - \sqrt{-7}) = 9 - (-7) = 16,$$

since "minus minus 7 makes plus 7".[11]

Finally, Cardano remarks that, since $\sqrt{9} = \pm 3$, then $\sqrt{-9}$ can be neither 3 nor -3, but rather it must be of a "third hidden nature".[12]

We observe that we find here all the ingredients of *Ars Magna*, Chapters I and XXXVII (and, by the way, the examples provided are very similar when not identical).[13] Basically, there are the usual sign rules to multiply negative quantities and the idea that the square roots of negative numbers are to be kept separated from the other numbers. Again, we implicitly find that $(\sqrt{-a})^2 = -a$ must hold.

4.2.2. Methods that lead to the solution of new cases

Cardano places **Chapter XXVII** "On the ways for finding new chapters [*De modis inveniendi capitula nova*]" immediately before the part strictly devoted to solve cubic equations. We account for it in advance. As we expect from the title, it is analogous to *Ars Magna*, Chapter VI.

Cardano[14] suggests the following heuristic methods. Accordingly one could solve new cases:

---

[11] "*Quod etiam m: m: 7 fit p: 7*", see [13, Question 38, page 372].

[12] "*Et nota quod R p: 9 est 3 p: vel 3m: nam p: [in p:] et m: in m: faciunt p:. Igitur R m: 9 non est 3 p: nec m: sed quædam tertia natura abscondita*", see [13, Question 38, page 373].

[13] See above, Section 3.1.1.

[14] See [13, Chapter XXVII, pages 338-41].

AMA XXVII.1 by geometric demonstrations ("the nicest one [*pulcherrima*]"),

AMA XXVII.2 by undertaking a known issue in another way ("very nice [*valde pulcher*]"),

AMA XXVII.3 by assuming the solution,

AMA XXVII.4 by derivation,

AMA XXVII.5 by similitude:

i increasing the solutions,

ii transforming equations,

iii of the solutions,

iv of the equations.

In Cardano's opinion, solving new cases "by a geometrical demonstration [*per demonstrationem Geometricam*]" (AMA XXVII.1) is the best way. It is "universal [*universalis*]" and "general [*generalis*]". As an example of cases solved by geometrical demonstrations, Cardano mentions the quadratic equations[15] and $x^3 + a_1x = a_0$. We have already found a similar argument in Chapter VI, Paragraph 5 in the *Ars Magna*, see above on page 100.

Next comes the method of solving new cases "by undertaking a known issue in another way [*ut accipias quætionem notam per aliquam*

[15]Cardano speaks about "the first three composite chapters of the algebra [*prima tria capitula Algebræ composita*]". I interpret these as being quadratic equations, since they are the equations which appear in Chapters XLIX "On the smaller composite chapters [*De capitulis minoribus compositis*]" and L "On the greater composite chapters [*De capitulis compositis maioribus*]" of the *Practica Arithmeticæ* and Chapter V "On finding the solutions for equations composed of minors [*Ostendit æstimationem capitulorum compositorum minorum, quæ sunt quadratorum, numeri, et rerum*]" of the *Ars Magna*. In this interpretation, since quadratic equations are the topic which comes usually after having solved simple equations like $x^n = a_0$ with n natural, "composite" means 'to have more than one term'.

viam]" (AMA XXVII.2), which is also a good way in Cardano's opinion. The example given by Cardano is to solve

$$\begin{cases} x+y=y^2 \\ xy=6 \end{cases}.$$

Depending on the substitution that one chooses, one obtains a depressed cubic equation or a cubic equation lacking in the second degree term. Then, being able to solve one of the two equations, one will also know the solution of the other. We have already found the same argument (and almost the same example, with 8 instead of 6) in *Ars Magna*, Chapter VI, Paragraph 4, see above on page 66. There, this method was called "by transmutation".

(AMA XXVII.3) consists in assuming a solution and then drawing a cubic equation – that is, the coefficients, given a certain family to which the cubic equation belongs – which has that same solution. (AMA XXVII.4) is the usual way to derive biquadratic equations (or, in general, higher degree equations) from the quadratic ones. These two methods also appear in the *Ars Magna* (and with no major variations), even if we did not stop and pay attention to them, see [4, Chapter VI, Paragraph 2 and 1, pages 14v-15v].

Finally, Cardano states the four-folded method of "similitude [*similitudo*]". It is directly comparable to the quotation from *Ars Magna*, Chapter VI, Paragraph 3 (see above on page 66). The method of "similitude by increasing the solutions [*similitudo augmentorum in æquationibus*]" (AMA XXVII.5 i) is not clearly explained: it seems to be a mixing of (AMA XXVII.3) and (AMA XXVII.4). Indeed, Cardano refers back to Chapter XXII, where he showed how to construct some quartic equations that lower to second degree starting from a given solution. It corresponds to *Ars Magna*, Chapter VI, Paragraph 3, point [2]. The method of "similitude by transformations of equations [*similitudo conversionis æquationis*]" (AMA XXVII.5 ii) corresponds to *Ars Magna*, Chapter VI, Paragraph 3, point [3] and to *Ars Magna*, Chapter XIV (see above, on page 122). We provide a detailed account on transformations of equations in the next section.

The method (AMA XXVII.5 iii) is by "similitude of the solutions [*similitudo æquationum*]". This means that, knowing, for instance, that $\sqrt[3]{4} - \sqrt[3]{2}$ is a solution of $x^3 + 6x = 2$, one should pay attention to the behaviour of the cubic roots appearing in x^3 and $6x$, once that the solution is substituted and the calculation developed. Indeed, since they are equal but opposite in sign, they vanish. Then, one should look for a similar behaviour while forcing $\sqrt[3]{4} - \sqrt[3]{2}$ to be a solution of a certain equation. It turns out that the equation $x^3 = 6x + 6$ fits. This method corresponds to *Ars Magna*, Chapter VI, Paragraph 3, point [4], or at least to a part of it. The method (AMA XXVII.5 iv) is by similitude of the equations. Cardano remarks that, since the quadratic formulae for $x^2 = a_1x + a_0$ and $x^2 + a_1x = a_0$ are composed by the same quantities up to changing signs, the same should[16] hold for $x^3 = a_1x + a_0$ and $x^3 + a_1x = a_0$. If one stops at the mere quotation, this method could correspond to *Ars Magna*, Chapter VI, Paragraph 3, point [1]. But we have seen that, in the context of the *Ars Magna*, this quotation had a deeper meaning.[17]

We remark that, in this chapter, almost the whole content (with the exception of the formula for the cube of a binomial) of *Ars Magna*, Chapter VI is covered in a dispersed order. The heuristic conceptions of the *Ars Magna*, then, are very close to the ones in the *Ars Magna Arithmeticæ*. Even more noticeable is the fact that Cardano's questioning gets more and more grounded in the *Ars Magna Arithmeticæ* conceptions, abandoning the starting point of the properties of continued proportions as it was in the *Practica Arithmeticæ*. Indeed, even if Cardano still mentions his results on continued proportions in Chapters XV and XVI, they are in no way mentioned in the section dealing with equations.

[16] We observe nevertheless that the parallelism is not complete, since in the cubic case the quantities under the square root vary in sign.

[17] Namely, in the *Ars Magna* there is much more than a simple relation between the shapes of the cubic formulae for $x^3 = a_1x + a_0$ and $x^3 + a_1x = a_0$. Indeed, also their proofs are strictly related, see above, on pages 110 and 161.

4.2.3. Transformations of equations

As introduction to **Chapter XXI** "On the reciprocal exchange of chapters [*De permutatione capitulorum invicem*]", Cardano[18] begins stating that, among cubic equations, there are two principal families, namely $x^3 + a_1x = a_0$ and $x^3 = a_1x + a_0$, whereas the others, namely $x^3 + a_0 = a_1x$, $x^3 + a_2x^2 = a_0$, $x^3 = a_2x^2 + a_0$, and $x^3 + a_0 = a_2x^2$, derive from those. We remark that complete cubic equations are not considered at all. Anyway, in the transformations that follow we find one family of complete cubic equations (and also some quartic equations).

AMA XXI.1. *Let us consider the following transformation of equations:*[19]

$$x^3 = a_2x^2 + a_0 \underset{y=\frac{a_0}{a_2 x}}{\longleftrightarrow} y^3 + \frac{a_0}{a_2}y = \frac{a_0^2}{a_2^3}.$$

Cardano says that this rule is taken "from the nineteenth question below [*ex decima nona quæstione inferius*]". Since he does not specify in which chapter the question should be, it is not easy to recover it. One could argue that he refers to the forty questions at the end of the last chapter of the *Ars Magna Arithmeticæ*, but the system solved there does not agree with the one that Cardano adds immediately hereafter. It is the following

$$\begin{cases} a : b = b : c \\ a + b = \frac{a_0}{a_2^2} \\ bc = \frac{a_0}{a_2} \end{cases}.$$

Cardano then suggests that (AMA XXI.1) comes from the resolution of the above system. Indeed, he details through an example the two ways to solve the system. Taking $b = \frac{a_0}{a_2c}$, developing calculations, and then taking $x = c$, he gets the equation on the left side in (AMA XXI.1). On the other hand, taking $c = \frac{a_0}{a_2b}$, developing calculations,

[18]See [13, Chapter XXI, pages 326-30].

[19]As already remarked on page 72,

and then taking $y = b$, he gets the equation on the right side in (AMA XXI.1). Developing calculations, the relations between the coefficients of the two equations appear. We remark that this is the same method as in *Ars Magna*, Chapter VI, Paragraph 3 (see above, on page 66).

AMA XXI.2. *Let us consider the equation* $x^3 + a_2x^2 = a_0$ *and take* $a'_0 = a_0 - \frac{2}{27}a_2^3$. *If* $a'_0 > 0$, *then the following transformation of equations holds:*

$$x^3 + a_2x^2 = a_0 \xleftrightarrow[y=\frac{a_2}{3}-x]{} y^3 = \frac{a_2^2}{3}y + a'_0.$$

If $a'_0 < 0$, *then the following transformation of equations holds:*

$$x^3 + a_2x^2 = a_0 \xleftrightarrow[y=\frac{a_2}{3}+x]{} y^3 + \left(a_0 - \frac{2}{27}a_2^3\right) = \frac{a_2^2}{3}y.$$

Moreover, the following transformation of equations holds:

$$x^3 + a_2x^2 = a_0 \xleftrightarrow[y=\frac{\sqrt{a_0}}{x}]{} y^3 = a_2y + \sqrt{a_0}.$$

We remark that the first two folds of the above rule are the same as in (AM XV). Here, the limit case $a'_0 = 0$ is lacking. On the contrary, the third fold is original. No further justifications are given for this rule.

AMA XXI.3. *Let us consider the following transformation of equations:*

$$x^3 + a_0 = a_1x \xleftrightarrow[x=\frac{y}{2}\pm\sqrt{a_1-3\left(\frac{y}{2}\right)^2}]{} y^3 = a_1y + a_0.$$

We remark that the above rule is the same as in (AM XIII). No further justifications are given for this rule.

AMA XXI.4. *Let us consider the following transformation of equations:*

$$x^3 + a_0 = a_2x^2 \xleftrightarrow[y=\frac{(\sqrt[3]{a_0})^2}{x}]{} y^3 + a_0 = \left(a_2\sqrt[3]{a_0}\right)y.$$

Cardano says that this rule is taken "from the thirty-second rule of chapter fifty-one [*ex trigesima secunda regula quinquagesimi primi capituli*]". The *Ars Magna Arithmeticæ* only contains forty chapters, but it exists a "chapter fifty-one" in the *Practica Arithmeticæ*. If we go and check, we find that, in Paragraph 32, there is exactly the same rule (see above, on page 184).[20] We also remind that this rule is (AM XVI) and a particular case of (AM VII.2-3) (see above, on page 69).

AMA XXI.5. *If* $x^4 = a_1x + a_0$ *and* $x = \sqrt{a} \pm b$, *then* $y^4 + a_1y = a_0$ *and* $y = \sqrt{a} \mp b$.

No further justifications are given for this rule.

AMA XXI.6. *Let us consider the following transformations of equations:*

$$x^4 + a_0 = a_3x^3 \xleftrightarrow[y=\frac{\sqrt{a_0}}{x}]{} y^4 + a_0 = \left(a_3\sqrt{a_0}\right)y,$$

$$x^4 = a_3x^3 + a_0 \longleftrightarrow y^4 + \left(a_3\sqrt{a_0}\right)y = a_0,$$

$$x^4 + a_3x^3 = a_0 \xleftrightarrow{y=\frac{\sqrt{a_0}}{x}} y^4 = \left(a_3\sqrt{a_0}\right)y + a_0.$$

Cardano says that this rule is taken "from three questions [*ex tribus quæestionibus*]", without specifying where they can be found. They are

$$\begin{cases} a : b = b : c \\ a + c = a_3 \\ ab = \sqrt{a_0} \end{cases} \quad \text{or} \quad \begin{cases} a : b = b : c \\ a = c + a_3 \\ ab = \sqrt{a_0} \end{cases} \quad \text{or} \quad \begin{cases} a : b = b : c \\ a + a_3 = c \\ ab = \sqrt{a_0} \end{cases} .$$

With the same method as in *Ars Magna*, Chapter VI, Paragraph 3 or (AMA XXI.1), it is easily checked that they respectively correspond to the three folds of the above rule. We moreover remark that the first and the third fold of the rule are particular cases of (AM VII.2-3) (see above, on page 69).

[20] This is, by the way, an argument to support Gavagna's opinion, see here Section 2.1 from page 10.

AMA XXI.7. *Let us consider the following transformation of equations:*

$$x^4 + a_2x^2 + a_0 = a_3x^3 \xleftrightarrow[y=\frac{\sqrt{a_0}}{x}]{} y^4 + a_2y^2 + a_0 = (a_3\sqrt{a_0})\, y.$$

Cardano says that this rule follows from

$$\begin{cases} a : b = b : c \\ a + b + c = a_3 \\ ab = a_2 \end{cases}$$

(by the same method of double resolution as above). We remark that this is the same as (AM VII.12) (see above, on page 72).

AMA XXI.8. *Let us consider the following transformation of equations:*

$$x^3 + a_0 = a_2x^2 \longleftrightarrow x^3 + \left(\frac{a_2}{2}\right)^2 x = a_2x^2 + \frac{a_0}{8}.$$

We remark that this is the same as (AM VII.13) (see above, on page 72).

AMA XXI.9. *Whenever the term a_2x^2 appears, two of the four substitutions $x = y + \frac{a_2}{3}$ and $x = y + \frac{2}{3}a_2$, or $x = y - \frac{a_2}{3}$ and $x = y - \frac{2}{3}a_2$ lead to a depressed cubic equation.*

Whenever the term a_1x appears, both substitutions $x = y + \sqrt{\frac{a_1}{3}}$ and $x = y - \sqrt{\frac{a_1}{3}}$ lead to a cubic equation lacking in the first degree term.

We remark that part of the first fold is the same as (AM IV.2) (see here, on page 81). No further justifications are given for this rule.

Summing up, the transformations of equations presented in the *Ars Magna Arithmeticæ* are very close to the ones in the *Ars Magna*. Most of them (except (AMA XXI.1), the third fold in (AMA XXI.2), (AMA

XXI.5), the second fold in (AMA XXI.6), and the second fold in (AMA XXI.9)) are exactly the same or particular cases of the transformations in the *Ars Magna*. But, whereas in the *Ars Magna* we had only the little hint given by Chapter VI on how these transformations could have been discovered, here there are still traces of their origins. Indeed, in three cases Cardano details the system the solution of which leads to the transformations and in one case he also mentions the *Practica Arithmeticæ*.

As Cardano explicitly says at the beginning of Chapter XXI, he plans to use these transformations to derive the solutions of the families of cubic equations lacking in first degree term and of $x^3 + a_0 = a_1x$ from the two principal depressed equations. But, at this stage, the connection[21] between (AM XIV) and (AM XI) is still missing. It will later be given as a generalisation of (AMA XXI.4) and partially of (AMA XXI.6) in Chapter VII of the *Ars Magna*.

Complete cubic equations do not appear in Cardano's general planning. Anyway, he knows the transformations in (AMA XXI.9) and, as a matter of fact, we have seen that using in particular the first transformation reproduces the *Ars Magna*'s way to reduce complete cubic equations to others that Cardano already knows how to solve.

4.2.4. "General" shapes for irrational solutions

In **Chapter XVIII** "On some necessary premises [*De aliquibus præmittendis necessariis*]", Cardano argues that, if one wants to discover the "universal [*universalis*]" rules of algebra, one should better investigate the *binomia* and *recisa* than the (rational) numbers, since a (rational) number can be written as a *binomium* or *recisum* but not the contrary.[22] Then, Cardano lists[23] all the types of *binomia*

[21]See above, Figure 3.16 concerning the *Ars Magna* on page 164.

[22]"*[C]um volueris invenire regulas aliquas capitulorum algebræ universales, aut condere ipsa capitula debes facere hoc per viam binomiorum et non simplicium numerorum, dato quod velis etiam experiri in numeris rationalibus*", see [13, Chapter XVIII, Paragraph 3, page 322].

[23]"*[N]otandum, subiecta capitulorum sunt per se usque ad cubum duodecim, primum est binomium primum et quartum, ut 7 p: R 2. Secundum est recisum*

and *recisa* that – in his opinion – can be solution of a cubic equation. They are

- the "*binomia* and *recisa* of the 1st or 4th, 2nd or 5th type":

$$a \pm \sqrt{b},\ \sqrt{a} \pm b;$$

---

primum, et quartum, ut 7 m: R 2. Tertium est binomium secundum et quintum, ut R 7 p: 2. Quartum est recisum secundum et quintum, ut R 7 m: 2. Quintum est binomium cubicum simplex, ut RCU: 32 p: RCU: 2 vel universale ut RCU: RV: 5 p: R 17 p: RCU: RV: 5 m: R 17 et in hoc animadverte, quod semper componuntur ex binomio primo vel quarto cum suo reciso. Sextum est recisum cubicum simplex, ut RCU: 9 m: RCU: 3 vel est recisum universale cubi, ut RV: cu: R 33 p: 5 m: RV: cu: R 33 m: 5. Et in hoc animadverte quod semper componuntur ex binomio secundo vel quinto cum suo reciso. Septimum est trinomium ex binomium duarum RV: cubarum primi vel quarti binomij cum suo reciso, et tertia pars est R numeri producibilis ex multiplicatione dictarum partium invicem detracta ab illo binomio veluti RV: cu: $9\frac{1}{2}$ p: R $89\frac{1}{4}$ m: RV: $9\frac{1}{2}$ m: R $89\frac{1}{4}$ m: l 1. Octavum est trinomium quod contat ex duabus RV: cubicis vel simplicibus, et tertia parte quæ est R numeri qui producitur ex una parte R cubicæ, in aliam veluti RCU: $65\frac{67}{125}$ p: RCU: $\frac{32}{125}$ p: $1\frac{3}{5}$. Nunum est recisum secundum mixtu, cuius una pars, est RCU: alia numerus, ut RCU: 30 m: 2. Decimum est binomium secundum mixtum, cuius una pars est RCU: alia numerus, ut RCU: 30 p: 2. Undecimum est binomium primum mixtum, cuius una pars est numerus, alia RCU: ut 3 p: RCU: 2. Duodecimum est recisum primum mixtum cuius una pars est numerus, alia RCU: veluti 3 m: RCU: 2 ut in figura vides", see [13, Chapter XVIII, Paragraph 5, page 322]. This quotation makes reference to the following figure.

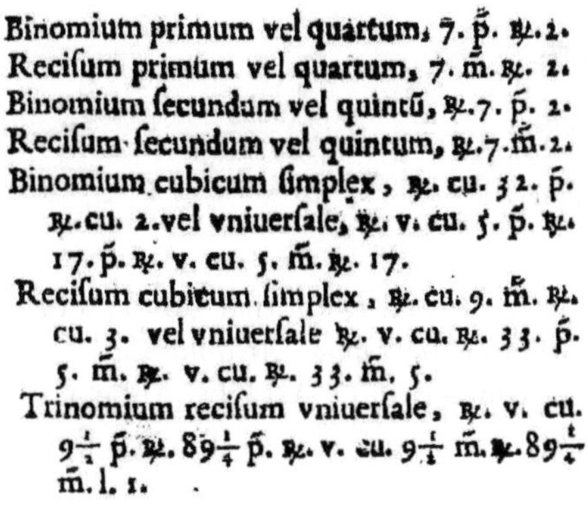
Binomium primum vel quartum, 7. p̄. ℞. 2.
Recifum primum vel quartum, 7. m̄. ℞. 2.
Binomium fecundum vel quintū, ℞. 7. p̄. 2.
Recifum fecundum vel quintum, ℞. 7. m̄. 2.
Binomium cubicum fimplex, ℞. cu. 32. p̄. ℞. cu. 2. vel vniuerfale, ℞. v. cu. 5. p̄. ℞. 17. p̄. ℞. v. cu. 5. m̄. ℞. 17.
Recifum cubicum fimplex, ℞. cu. 9. m̄. ℞. cu. 3. vel vniuerfale ℞. v. cu. ℞. 33. p̄. 5. m̄. ℞. v. cu. ℞. 33. m̄. 5.
Trinomium recifum vniuerfale, ℞. v. cu. 9½ p̄. ℞. 89¼ p̄. ℞. v. cu. 9½ m̄. ℞. 89¼ m̄. l. 1.

- the "simple cubic *binomia* and *recisa*":

$$\sqrt[3]{a} \pm \sqrt[3]{b},$$

  (where $\sqrt[3]{ab}$ is rational);

- the "universal cubic *binomia*":

$$\sqrt[3]{a+\sqrt{b}} + \sqrt[3]{a-\sqrt{b}},$$

  where $a \pm \sqrt{b}$ are a *binomium* and *recisum* of the 1st or 4th type, and the "universal cubic *recisa*":

$$\sqrt[3]{\sqrt{a}+b} - \sqrt[3]{\sqrt{a}-b},$$

  where $\sqrt{a} \pm b$ are a *binomium* and *recisum* of the 2nd or 5th type;

- the "universal *recisa trinomia*" such that

$$\sqrt[3]{a+\sqrt{b}} + \sqrt[3]{a-\sqrt{b}} - \sqrt{c},$$

  where $a \pm \sqrt{b}$ are a *binomium* and *recisum* of the 1st or 4th type and $c = (a+\sqrt{b})(a-\sqrt{b})$;

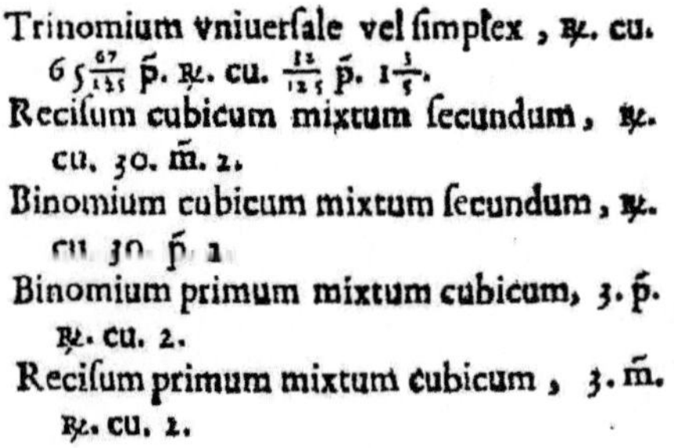
Trinomium vniuersale vel simplex, R. cu. 65 67/135 p. R. cu. 12/125 p. 1 1/5.
Recisum cubicum mixtum secundum, R. cu. 30. m. 2.
Binomium cubicum mixtum secundum, R. cu. 30. p. 2.
Binomium primum mixtum cubicum, 3. p. R. cu. 2.
Recisum primum mixtum cubicum, 3. m. R. cu. 2.

Figure 4.1. – List of the types of *binomia* and *recisa* that can be solution of a cubic equation in *Ars Magna Arithmeticæ*, Chapter XVIII. Kind courtesy of OPAL Libri Antichi (Università di Torino).

- the "simple cubic *trinomia*" such that

$$\sqrt[3]{a} + \sqrt[3]{b} \pm \sqrt{c},$$

where $c = \sqrt[3]{ab}$, and the "universal cubic *trinomia*" such that

$$\sqrt[3]{A} + \sqrt[3]{B} \pm \sqrt{c},$$

where A, B are a *binomium* and *recisum* (not specified by Cardano) and $c = \sqrt[3]{AB}$;

- the "mixed (cubic) *binomia* and *recisa* of the 1st, 2nd type" (later called "of the 1st or 4th, 2nd or 5th type"):

$$a \pm \sqrt[3]{b}, \ \sqrt[3]{a} \pm b,$$

taking $a > b$ rational, positive, non zero[24] such that the square and cubic roots do not respectively vanish (and paying attention to combine the signs one with the other in order not to fall down on negative quantities). Cardano's list is for now not completely clear in all its details.

In **Chapter XIX** "On the properties of 3rd and 6th *binomium* and *recisum*, and on their inability to [fit] the chapters [*De proprietatibus tertij et sexti binomij et recisi, et inhabilitate ad capitula*]", Cardano states the following proposition.

AMA XIX. *The* binomia *and* recisa *of the 3rd or 6th type* $\sqrt{a} \pm \sqrt{b}$ *cannot be solution of whatever degree equations [with rational coefficients].*[25]

[24] The assumption $a > b$ must be required, since it comes from the classification of the *binomia* and *recisa*, see above, on page 80

[25] "*[D]ico quod in tertio et sexto binomio et reciso non verificatur aliquod capitulum algebræ, nec de inventis nec de non inventis usque ad infinitum*", see [13, Chapter XIX, Paragraph 1, page 323].

First of all, he observes that, if $\sqrt{a}, \sqrt{b}$ are irrational, then $\sqrt{a} \pm \sqrt{b}$ is irrational.[26] Then, he justifies the main statement looking at some particular cases of equations of second, third, fourth, and fifth degrees. Concerning cubic equations, Cardano says that $\sqrt{a} \pm \sqrt{b}$ can be solution neither of $x^3 = a_1x + a_0$ nor of $x^3 = a_2x^2 + a_0$. He justifies the first part of the statement saying that, if one replaces the value $\sqrt{a} \pm \sqrt{b}$ in $x^3 = a_1x + a_0$, "all [the terms from x^3] are square roots, then one cannot makes the number neither by addition by the first hypothesis nor by subtraction".[27] Indeed, by substitution one gets

$$(\sqrt{a} \pm \sqrt{b})^3 = a_1(\sqrt{a} \pm \sqrt{b}) + a_0,$$

that is

$$(a + 3b)\sqrt{a} \pm (b + 3a)\sqrt{b} = a_1\sqrt{a} \pm a_1\sqrt{b} + a_0,$$

where, by comparison, he remarks that there is no rational term on the left side of the equation that can correspond to the rational term of the right side. Similar arguments based on the comparison of rational and irrational terms after substitution also hold for whatever cubic equation (matching Cardano's requirements on the coefficients), even though Cardano does not make it explicit. Then, Cardano ends up saying that $\sqrt{a} \pm \sqrt{b}$ cannot be a solution of any equation

[26] Cardano says that "*oportet scire quod ex duabus R simplicibus iunctis, numquam provenit numerus, cum ex duabus superficiebus simpliciter medialibus sive communicantes sint sive non, non possit fieri superficies rationalis*", see [13, Chapter XIX, page 323].

We remark that Cardano's justification is a paraphrase of what he wants to justify, since he replaces "*R*" by "*superficies medialis*". Cardano's statement is nevertheless true. Suppose by contradiction to have two rational, positive, non zero a, b such that $\sqrt{a}, \sqrt{b}$ is irrational and $\sqrt{a} \pm \sqrt{b}$ is rational. Then, $(\sqrt{a} \pm \sqrt{b})^2 = a \pm 2\sqrt{ab} + b$ is rational, which means that also $\sqrt{ab}$ is rational. This implies also that $\sqrt{\frac{a}{b}}$ is rational, since $\sqrt{\frac{a}{b}} = \sqrt{\frac{a}{b}\frac{a}{a}} = \sqrt{\frac{a^2}{ab}} = \frac{a}{\sqrt{ab}}$. Then, $\sqrt{a} \pm \sqrt{b} = \sqrt{b}\left(\sqrt{\frac{a}{b}} \pm 1\right)$ implies that $\sqrt{b}$ is rational, which is absurd. Then $\sqrt{a} \pm \sqrt{b}$ is irrational.

[27] "*[U]t vides omnes sunt R igitur non potest per additionem fieri numerus ex primo supposito, nec per diminutionem*", see [13, Chapter XIX, page 324].

(without justifying it).[28] Nevertheless, we remark that Cardano's statement is not true in full generality. Despite his words, $\sqrt{a} \pm \sqrt{b}$ can be solution of a cubic equation with rational coefficients, namely of $x^3 - \lambda x^2 - (a \pm 2\sqrt{ab} + b)x + \lambda(a \pm 2\sqrt{ab} + b) = 0$, where λ and $\sqrt{ab}$ are rational.[29] Remark that we always have either a complete equation (for $\lambda \neq 0$) or an equation the degree of which easily lowers to two (for $\lambda = 0$).

It is clear that Cardano implicitly assumes that the coefficients are rational. Indeed, at the end of the chapter, Cardano also says that it can happen that $\sqrt{a} \pm \sqrt{b}$ is a solution of some equations, if one admits that the coefficients are real.[30] Retrospectively, in order to give consistency to Cardano's words, we should assume that, in Chapter XVIII (and wherever else Cardano speaks of the shape of the solutions of cubic equations in terms of *binomia* and *recisa*), he is considering equations with rational coefficients. Note that in the vast majority of the examples Cardano's equations really have rational coefficients.

From Chapters XXII to XXVI, we find a clarification of Cardano's statements in Chapter XVIII. In **Chapter XXII** "On the examination of the first, second, fourth, and fifth *binomium* with its *recisum* [*De examine primi, secundi, quarti et quinti binomij cum suis recisis*]" Cardano firstly affirms that

---

[28]"*Relinquitur igitur quod tam tertium quam sextum binomium et recisum sunt inutilia ad capitula omnia algebræ usque ad infinitum*", see [13, Chapter XIX, page 324].

[29]Indeed, $\sqrt{a} \pm \sqrt{b}$ is a root of a (unique) monic quadratic polynomial with rational coefficients if and only if $\sqrt{ab}$ is rational. In this case, the polynomial is $x^2 - (a \pm 2\sqrt{ab} + b)$. In particular, $\sqrt{a} \pm \sqrt{b}$ is also a root of the cubic polynomials $(x - \lambda)(x^2 - (a \pm 2\sqrt{ab} + b))$ for any rational λ.

[30]"*Dico tamen quod per accidens fiunt æquatione in tertio et sexto binomio et suis recisi, et etiam in quantitatibus universalibus ligatis, et est quod cum in aliquo capitulo fuerit numerus rerum, aut cuborum in R alicuius numeri*", see [13, Chapter XIX, page 324]. The examples that Cardano makes are $x^2 = \sqrt{8}x + 10$ and $x^2 + \sqrt{8}x = 10$.

AMA XXII.i. *The* binomia *and* recisa *of the 1st or 4th type $a \pm \sqrt{b}$ and the* recisa *of the 2nd or 5th type $\sqrt{a} - b$ can be solutions of $x^3 + a_0 = a_1 x$.*

AMA XXII.i — *Proof.* Cardano[31] first considers a *binomium* or *recisum* of the 1st or 4th type $a \pm \sqrt{b}$. [He wants to show that it can be solution of a certain equation of the family $x^3 + a_0 = a_1 x$, that is, he wants to find the coefficients a_1, a_0 rational, positive, non zero depending on a, b such that $a \pm \sqrt{b}$ verifies the equation.]

By Chapter XVII, Paragraphs 1 and 2, Cardano[32] knows that, writing $(a \pm \sqrt{b})^3 = c \pm \sqrt{d}$, it holds that $\frac{\sqrt{d}}{\sqrt{b}} > \frac{c}{a}$. Then, $\frac{\sqrt{d}}{\sqrt{b}}(a \pm \sqrt{b}) - (c \pm \sqrt{d})$ is rational. Cardano only needs to take $a_1 = \frac{\sqrt{d}}{\sqrt{b}}$ and $a_0 = \frac{\sqrt{d}}{\sqrt{b}}(a \pm \sqrt{b}) - (c \pm \sqrt{d})$. In this way, $x = a \pm \sqrt{b}$ is a solution of $x^3 + a_0 = a_1 x$.

Cardano proves in an analogous way the statements on the *recisa* of the 1st or 4th type $a - \sqrt{b}$ and of the 2nd or 5th type $\sqrt{a} - b$. □

For instance, take $a = 3$ and $b = 2$. Then, $a + \sqrt{b} = 3 + \sqrt{2}$ and $(a + \sqrt{b})^3 = 45 + 29\sqrt{2}$. Then, $\frac{29\sqrt{2}}{\sqrt{2}} > \frac{45}{3}$ and it is enough to take $a_1 = 29$ and $a_0 = 42$.

We observe that, if we replace x by $a \pm \sqrt{b}$ in $x^3 + a_0 = a_1 x$ and develop calculations, we will act as a matter of fact in the same way as Cardano does. In all the following statements in Chapters XXIII-XXVI, Cardano uses similar ways of arguing (sometimes in shorter forms, sometimes more articulately), so that I will not go into more detail in this respect.

Cardano[33] shows the following statements.

[31] See [13, Chapter XXII, pages 331-332].

[32] Indeed, we remark that $c = a^3 \pm 3ab = (a^2 \pm 3b)a$ and $\sqrt{d} = (3a^2 \pm b)\sqrt{b}$. Then, $\frac{\sqrt{d}}{\sqrt{b}} = 3a^2 \pm b$ and $\frac{c}{a} = a^2 \pm 3b$. Moreover, $3a^2 \pm b > a^2 \pm 3b$, since $a^2 > b$ by the definition of *binomium* and *recisum*, see above, on page 80.

[33] See [13, Chapter XXII, pages 331-333].

AMA XXII.ii. *The* binomia *of the 2nd or 5th type $\sqrt{a}+b$ can be solution of $x^3 = a_1x + a_0$.*

He observes that $x^3 + a_1x = a_0$ cannot be solved by a *binomium* or *recisum* of the above types. His argument is that no longer possible *binomia* or *recisa* are left, since $\sqrt{a} \pm \sqrt{b}$ cannot be solutions and $a \pm \sqrt{b}$ and $\sqrt{a} \pm b$ have already been assigned.[34]

He also observes that, since a solution of $x^3 + a_0 = a_1x$ can go under three shapes, namely $a \pm \sqrt{b}$ and $\sqrt{a} - b$ where[35] respectively $a > \sqrt{b}$ and $\sqrt{a} > b$, one cannot assign to them a "title [*titulus*]" nor

[34] "*[N]am cum tertio et sextum binomium cum suis recisis non sint utilia ad capitula per hanc regulam, primum vero et quartum binomium cum suis recisis et recisum secundum et quintum inserviunt tantum capitulo de rebus æqualibus cubo et numero, binomium vero secundum et quintum inserviunt tantum capitulo de cubo æquali rebus et numero relinquitur a sufficienti divisione quod cubus et res non possunt æquari numero in aliquo genere binomiorum aut recisorum*", see [13, Chapter XXII, page 332].

Let us try to see why a *binomium* or *recisum* of the 1st-6th type cannot be solution of $x^3 + a_1x = a_0$. Suppose $x = a \pm \sqrt{b}$ *binomium* or *recisum* of the 1st or 4th type. Then, we want to determine the rational coefficients a_1, a_0. We have

$$(a \pm \sqrt{b})^3 + a_1(a \pm \sqrt{b}) = a_0,$$

that is

$$(a^2 + 3b + a_1)a + (3a^2 + b + a_1)\sqrt{b} = a_0.$$

Then,

$$\begin{cases} a_1 = -3a^2 - b \\ a_0 = a^2 + 3b + a_1 \end{cases}.$$

This means that $a_1, a_0 < 0$. So, $x = a \pm \sqrt{b}$ satisfies $x^3 + a_0 = a_1x$.

Similar observations hold supposing $x = \sqrt{a} \pm b$ *binomium* or *recisum* of the 2nd or 5th type.

On the contrary, supposing $x = \sqrt{a} \pm \sqrt{b}$ *binomium* or *recisum* of the 3rd or 6th type, we get to the contradiction

$$\begin{cases} a_1 = -3a - b \\ a_1 = -a - 3b. \\ a_0 = 0 \end{cases}$$

[35] See above, on page 80.

find a "general chapter [*capitulum generale*]".[36] It seems then that Cardano affirms here that no *binomium* and *recisum* of the 1st, 2nd, 4th, or 5th type is suited to the role of a more comprehensive shape to encompass the remaining ones.

AMA XXII.iii. *The* binomia *and* recisa *of the 1st or 4th type* $a \pm \sqrt{b}$ *and the* binomia *of the 2nd or 5th type* $\sqrt{a} + b$ *can be solutions of* $x^3 + a_0 = a_2 x^2$.

AMA XXII.iv. *The* recisa *of the 2nd or 5th type* $\sqrt{a} - b$ *can be solution of* $x^3 + a_2 x^2 = a_0$.

As before and by similar arguments, Cardano observes that $x^3 = a_2 x^2 + a_0$ cannot be solved by a *binomium* or *recisum* of the 1st-6th type. He also observes that there cannot be a more comprehensive shape to encompass the *binomia* and *recisa* that are solutions of $x^3 + a_0 = a_2 x^2$.

Chapter XXIII "On the examination of the cubic *binomium* and its *recisum* [*De examine binomij cubici et sui recisi*]" contains the following statements.

AMA XXIII.i. *The cubic* binomia $\sqrt[3]{a} + \sqrt[3]{b}$ *[such that* $\sqrt[3]{ab}$ *is rational] can be solution of* $x^3 = a_1 x + a_0$, *where* $a_1 = 3\sqrt[3]{ab}$ *and* $a_0 = a + b$.

[36] "*Notabile est quod cum res æqualis cubo et numero verificetur et de binomio primo et de suo reciso et de reciso secundo, quorum tria natura est diversa nec potest uno titulo assignari, quia in binomio primo numerus est maior R̶ et uterque p: in reciso primo R̶ est m: in binomio secundo numerus est minor R̶ igitur nullum poterit assignari capitulum generale, quia æquatio cadit super tres omnino diversas [...] manet firmum sed in capitulo de rebus æqualibus cubo et numero inveniuntur tres æquationes diversæ nihil continentes firmu ita quod illud quod est in una æquatione sit in duabus reliquis igitur sequitur quod tale capitulum non potest habere regulam generalem*", see [13, Chapter XXII, page 332].

AMA XXIII.ii. *The cubic* recisa $\sqrt[3]{a}-\sqrt[3]{b}$ *[such that* $\sqrt[3]{ab}$ *is rational] can be solution of* $x^3+a_1x=a_0$, *where* $a_1=3\sqrt[3]{ab}$ *and*[37] $a_0=a-b$ [>0].

Note that these rationality conditions put some structure on the *binomia*. Let us call the rational $\sqrt[3]{ab}=k$. Then, $\sqrt[3]{b}=\frac{k}{\sqrt[3]{a}}=\frac{k}{a}\sqrt[3]{a^2}$, and the *binomium* becomes

$$\sqrt[3]{a}\pm\frac{k}{a}\sqrt[3]{a^2}.$$

These statements[38] are proved in the same way as (AMA XXII.i). As a matter of fact, they are equivalent to replace the considered *binomium* or *recisum* in the equation and drawing the sign of the coefficients by developing calculations.

Concerning the first statement, we remark that, while detailing the example on $\sqrt[3]{6}+\sqrt[3]{2}$, Cardano comes across the equation $x^3=\sqrt[3]{324}x+8$, where the coefficient of the first degree term is not rational. This could contradict my previous guess on the fact that, in the context of discussing the shape of the solutions of the cubic equations in terms of *binomia* or *recisa*, we should consider (positive, non zero) rational coefficients. Anyway, immediately after the example, Cardano adds the condition that "to have the things in a number it is always necessary that the roots, one multiplied by the other, produce a number".[39] This is why I included this condition in square brackets in the Propositions (AMA XXIII.i) and (AMA XXIII.ii).

---

[37]We remark that here Cardano necessarily needs to assume that $a>b$ in order to avoid getting to the equation $x^3+a_1x+a_0=0$.

[38]We remark that one can prove by Galois theory that $\sqrt[3]{a}\pm\sqrt[3]{b}$ is a root of a cubic polynomial if and only if either $\sqrt[3]{ab}$ or $\frac{\sqrt[3]{a}}{\sqrt[3]{b}}$ is rational. Note that in the second case, assuming $\frac{\sqrt[3]{a}}{\sqrt[3]{b}}=k$, we can write the root (the *binomium*) as $(k\pm1)\sqrt[3]{b}$. Then the equation must have $a_1=0$ and $a_0=(k\pm1)^3b$. This is why Cardano does not consider this second condition.

[39]"*[U]t autem habeas res in numero semper oportet ut partes R invicem multiplicatæ producant numerum*", see [13, Chapter XXIII, page 335].

Cardano finally observes (without justification) that the cubic *binomium* or *recisum* $\sqrt[3]{a} \pm \sqrt[3]{b}$ cannot be solution of a cubic equation lacking in the first degree term (otherwise – I add – either a or b must be zero, since $a_1 = 3\sqrt[3]{ab}$).

In **Chapter XXIV** "On the examination of the mixed, cubic *recisa* and their *binomium* [*De examine recisorum cubicurum mixtorum, et suorum binomiorum*]", Cardano states the following propositions.[40]

AMA XXIV.i. *The mixed, cubic* binomia *of the 1st or 4th type* $a + \sqrt[3]{b}$ *and the cubic mixed* binomia *and* recisa *of the 2nd or 5th type* $\sqrt[3]{a} \pm b$ *can be solutions of the equation* $x^3 + a_1 x = a_2 x^2 + a_0$.

AMA XXIV.ii. *The mixed, cubic* recisum *of the 1st or 4th type* $a - \sqrt[3]{b}$ *can be solution of the equation* $x^3 + a_2 x^2 + a_1 x = a_0$.

Finally, Cardano observes (without justification) that the mixed, cubic *binomia* or *recisa* cannot be solution of a non-complete cubic equation.

In **Chapter XXV** "On the examination of useful *trinomia* and their *recisa* [*De examinatione trinomiorum utilium cum suis recisis*]",

[40] Cardano never specifies the classification of cubic *binomia* and *recisa*. Anyway, analogously to the square (see above, on page 80) we could say that a cubic *binomium* is an expression that can be written in the form $\alpha + \beta$, where $\alpha^3 > \beta^3$. Moreover, when $\sqrt[3]{\alpha^3 - \beta^3}$ is commensurable with α, we say that it is of the

1st type if $\alpha = a$ is rational and $\beta = \sqrt[3]{b}$ is irrational, that is $\alpha + \beta = a + \sqrt[3]{b}$,

2nd type if $\alpha = \sqrt[3]{a}$ is irrational and $\beta = b$ is rational, that is $\alpha + \beta = \sqrt[3]{a} + b$,

3rd type if $\alpha = \sqrt[3]{a}$ is irrational and $\beta = \sqrt[3]{b}$ is irrational, that is $\alpha + \beta = \sqrt[3]{a} + \sqrt[3]{b}$,

and, when $\sqrt[3]{\alpha^3 - \beta^3}$ is incommensurable with α, we say that it is of the

4th type if $\alpha = a$ is rational and $\beta = \sqrt[3]{b}$ is irrational, that is $\alpha + \beta = a + \sqrt[3]{b}$,

5th type if $\alpha = \sqrt[3]{a}$ is irrational and $\beta = b$ is rational, that is $\alpha + \beta = \sqrt[3]{a} + b$,

6th type if $\alpha = \sqrt[3]{a}$ is irrational and $\beta = \sqrt[3]{b}$ is irrational, that is $\alpha + \beta = \sqrt[3]{a} + \sqrt[3]{b}$.

Likewise for a *recisum*, which is written in the form $\alpha - \beta$. Moreover, Cardano calls 'mixed, cubic *binomia* and *recisa*' the ones of the 1st, 2nd, 4th, and 5th type. We could then soundly call 'of the 3rd and 6th type' the cubic *binomia* and *recisa* in Chapter XXIII.

Cardano lists six types of *trinomia* and *recisa*, where two out of the three terms are cubic roots.

$$\sqrt[3]{a}+\sqrt[3]{b}+c,$$
$$\sqrt[3]{a}+\sqrt[3]{b}-c,$$
$$\sqrt[3]{a}-\sqrt[3]{b}+c,$$
$$\sqrt[3]{a}-\sqrt[3]{b}-c,$$
$$-\sqrt[3]{a}+\sqrt[3]{b}+c,$$
$$-\sqrt[3]{a}-\sqrt[3]{b}+c,$$

with $\sqrt[3]{a}$, $\sqrt[3]{b}$ irrational, positive, non zero and c rational, positive, non zero. Note that the list is not complete, since $-\sqrt[3]{a}-\sqrt[3]{b}-c$ and $-\sqrt[3]{a}+\sqrt[3]{b}-c$ are missing. If we moreover assume that $a>b$ (as Cardano implicitly did in his list in Chapter XVIII, see on page 197), then the two above *trinomia* are negative for each rational, positive, non zero c, which justifies their absence.

Afterwards, Cardano states the following propositions.

AMA XXV.i. *The* trinomia $\sqrt[3]{a}+\sqrt[3]{b}+c$ *such that* $c^2=\sqrt[3]{ab}$ *can be solution of the equation* $x^3=a_2x^2+a_0$.

AMA XXV.ii. *The* recisa $\sqrt[3]{a}+\sqrt[3]{b}-c$ *such that* $c^2=\sqrt[3]{ab}$ *can be solution of the equation* $x^3+a_2x^2=a_0$.

The *trinomia* of the above type are called 'proportioned [*proportionata*]', since the continued proportion $\sqrt[3]{a}:c=c:\sqrt[3]{b}$ holds. Note that this proportion puts some structure on the *trinomium*. Let us call the ratio $k=\frac{\sqrt[3]{a}}{c}=\frac{c}{\sqrt[3]{b}}$. Then, we remark that $c=k\sqrt[3]{b}$ and $\sqrt[3]{a}=kc=k^2\sqrt[3]{b}$ and that the *trinomium* is $k^2\sqrt[3]{b}+k\sqrt[3]{b}+\sqrt[3]{b}$. Moreover, since c is rational, also $k\sqrt[3]{b}$ must be rational, then $k=\sqrt[3]{b^2}$. Finally, the *trinomium* becomes

$$b\sqrt[3]{b^2}+b+\sqrt[3]{b},$$

and it holds that $k^2\sqrt[3]{b^2} = k^3\sqrt[3]{b}\sqrt[3]{b}$, which was the original condition written by Cardano.

Cardano observes at first that all the listed *trinomia* and *recisa* cannot be solution of a depressed cubic equation. Then, he specifies (without justification) that the remaining ones $\sqrt[3]{a}-\sqrt[3]{b}+c$, $\sqrt[3]{a}-\sqrt[3]{b}-c$, $\sqrt[3]{a}+\sqrt[3]{b}+c$, $-\sqrt[3]{a}+\sqrt[3]{b}+c$, $-\sqrt[3]{a}-\sqrt[3]{b}+c$ cannot be solution of any cubic equation.

In **Chapter XXVI** "On the remaining types of *binomia*, *trinomia*, and *quadrinomia* not useful to the chapters of algebra, and on their *recisa* [*De reliquis speciebus binomiorum, et quadrinomiorum, inutilium ad capitula algebra, et de recisis eorum*]", Cardano says that the following quantities "cannot make a chapter general and not even be part of a general value".[41] They are

- the "simple roots"[42]:

$$\sqrt{a};$$

- the *binomia* or *recisa* of the 3rd or 6th type:

$$\sqrt{a} \pm \sqrt{b};$$

- the cubic roots:

$$\sqrt[3]{a};$$

- the "median [*mediales*]" roots:

$$\sqrt[4]{a};$$

[41] "*[Q]uantitates autem quæ sunt R simplices et binomia vel recisa ex tertio et sexto genere et R cubicæ et mediales, id est RR et binomiales ex R et R cubica et trinomiales non proportionatæ et RV: quadratæ triplicatæ sive proportionatæ sive non, et trinomia ex radicibus quadratis, et trinomia cubica in quibus prima vel tertia pars est m: vel est numerus, nullo modo faciunt capitulum generale, nec possunt esse pars æquationis generalis ita quod totum genus illius quantitatis inserviat uni capitulo*", see [13, Chapter XXVI, page 338].

[42] Note that here "simple" is opposed to "cubic", "median", ..., but not to "universal" as before.

- the *binomia* such that

$$\sqrt{a} \pm \sqrt[3]{b},\ \sqrt[3]{a} \pm \sqrt{b};$$

- the "non-proportioned" *trinomia*:

$$\sqrt[3]{a} + \sqrt[3]{b} \pm \sqrt{c},$$

where $c \neq \sqrt[3]{ab}$;

- the "proportioned or not *trinomia* from square and universal roots";
- the "*trinomia* from square roots":

$$\pm\sqrt{a} \pm \sqrt{b} \pm c;$$

- the cubic *trinomia* such that

$$-\sqrt[3]{a} + \sqrt[3]{b} + \sqrt[3]{c},\ \sqrt[3]{a} + \sqrt[3]{b} - \sqrt[3]{c},\ a + \sqrt[3]{b} - \sqrt[3]{c},\ \sqrt[3]{a} + \sqrt[3]{b} \pm c,$$

taking $a > b$ rational, positive, non zero such that the square and cubic roots do not respectively vanish (and paying attention to combine the signs one with the other in order not to fall down on negative quantities). This last classification is quite unclear. Obviously, some of the above listed numbers can be solution of a somewhat equation. Then, Cardano's statement needs to be detailed. Cardano goes on arguing that,[43] even if it is really possible that $\sqrt{a}$, or a *trinomium*

[43] "*Quia bene possibile est (ut dictum est in capitulo) quod æuatio census æqualis rebus et numero perveniat ad æquationem unius R tantum, et similiter possibile est invenire trinomia ex RV: et R et numero, vel binomia ex duabus RV: vel ex RV: et R simplici quæ æquivalent numero vel alteri R in quam cadat æquatio capituli, sed hoc non est per se generale. Quod autem ita sit patet, nam si maxime hoc esset, posset esse in trinomio quadrato proportionaliter, cuius pars media sit numerus; veluti R 8 p: 2 p: R 2. Quod autem hoc non ita sit patet, nam in cubo proveniunt plus quam tres partem non communicantes, quare non potest æquari rebus et numero, et ut videas quod eadem ratione*", see [13, Chapter XXVI, page 338].

composed by a universal root, a square root, and a number, or a *binomium* composed by two universal roots, or a *binomium* composed by a universal root and a square root, or a *trinomium* composed by square proportioned roots could be solutions of a family of quadratic equations ($x^2 = a_1x + a_0$, for instance), they would not be at the same time solutions of the corresponding families of higher degree equations (which is $x^3 = a'_1x + a'_0$), and so on. Cardano ends this chapter with a quite obscure remark, by which he dismisses any other irrational quantity to be the shape of a solution of a cubic equation.[44]

Summing up, Cardano inquires the shapes under which an irrational solution of a cubic equation can go, assuming that the equation has rational coefficients. The Table 4.1 on page 211 resumes Cardano's results in Chapters XVIII-XIX and XXII-XXVI. The usual way in which Cardano proves his statement is to replace the assumed value of the solution and then to develop calculations in order to find out the expected sign of the coefficients – if they exist rational.

Among the kind of irrational quantities that Cardano lists at the beginning as possible shapes,[45] there are $\sqrt[3]{a+\sqrt{b}}+\sqrt[3]{a-\sqrt{b}}$ and $\sqrt[3]{\sqrt{a}+b}-\sqrt[3]{\sqrt{a}-b}$ with some appropriate *binomia* and *recisa* under the cubic roots. They are no longer mentioned after Chapter XVIII, but they can be studied in a way[46] similar to $\sqrt[3]{a} \pm \sqrt[3]{b}$, as Cardano probably also does with the "universal cubic *trinomia*" mentioned in Chapter XVIII. In the list also appears $\sqrt[3]{a+\sqrt{b}}+\sqrt[3]{a-\sqrt{b}}-\sqrt{c}$, where $a \pm \sqrt{b}$ is a *binomium* or *recisum* of the 1st or 4th type and $c = (a+\sqrt{b})(a-\sqrt{b})$. Except for the subsequent example $x^3+3x^2 = 21$,

---

[44] "*Si quis igitur dicat ad quid sunt aliæ quantitates irrationales non bimediales quæ possunt producere æquationem. Dico quod etsi non producunt æquationem per se, producunt per accidens, ut in capitulo decimo nono*", see [13, Chapter XXVI, page 338].

[45] See above, on page 197.

[46] Let us rewrite the *binomia* and *recisa* at issue as $\sqrt[3]{A} \pm \sqrt[3]{B}$, where A, B are a *binomium* and *recisum* respectively of the 1st or 4th, 2nd or 5th type. In the list from Chapter XVIII, I wrote in a similar way the "universal cubic *trinomia*" $\sqrt[3]{A}+\sqrt[3]{B} \pm \sqrt{c}$, where $c = \sqrt[3]{AB}$.

Table 4.1. – Shapes of irrational solutions to cubic equations (with rational coefficients) in *Ars Magna Arithmeticæ*, Chapters XXII-XXV.

Equation	Shape of an irrational solution			
	binomia and/or *recisa*			*trinomia*
	(normal)	cubic	mixed cubic	
$x^3 + a_1x = a_0$		$\sqrt[3]{a} - \frac{k}{a}\sqrt[3]{a^2}$		
$x^3 = a_1x + a_0$	$\sqrt{a} + b$	$\sqrt[3]{a} + \frac{k}{a}\sqrt[3]{a^2}$		
$x^3 + a_0 = a_1x$	$a \pm \sqrt{b}$, $\sqrt{a} - b$			
$x^3 = a_2x^2 + a_0$				$a\sqrt[3]{a^2} + \sqrt[3]{a} + a$
$x^3 + a_2x^2 = a_0$	$\sqrt{a} - b$			$a\sqrt[3]{a^2} + \sqrt[3]{a} - a$
$x^3 + a_0 = a_2x^2$	$a \pm \sqrt{b}$, $\sqrt{a} + b$			
$x^3 + a_2x^2 + a_1x = a_0$			$a - \sqrt[3]{b}$	
$x^3 + a_1x = a_2x^2 + a_0$			$a + \sqrt[3]{b}$, $\sqrt[3]{a} \pm b$	
Condition		$k \in \mathbb{Q}$		

a solution of which is $\sqrt[3]{\frac{19}{2}+\sqrt{\frac{357}{4}}}+\sqrt[3]{\frac{19}{2}-\sqrt{\frac{357}{4}}}-1$, Cardano does no longer mention this shape in the following, so that at the end he does not justify this statement.[47]

In Chapters XXII and XXVI the term 'general' (referred to solutions) occurs. Similarly to the *Ars Magna*, Chapter IV, a "general solution" is a solution the shape of which can encompass other solutions as "particular" – that is, more specific – cases. At the end of Chapter XXVI, it even seems that a "general" shape for a solution should not be restricted to classes of equations depending on their degrees. We have now more elements to try to specify a weaker version of our interpretation of "general solutions" and, as a consequence, of "particular" solutions on page 82. I say that a "general solution" is a solution that goes under certain shapes that are usually expounded thanks to the *binomia* and *recisa*. We remind that Cardano establishes at the beginning of the *Ars Magna Arithmeticæ* a correspondence between the terms '*binomium*' and '*recisum*' and the traditional classification of the lines in *Elements*, Book X. In this context, 'universal' is used by Cardano as a synonym of 'general'. By opposition, a "particular" solution is instead a solution that is not "general". Then, a "general" solution has a shape which can encompass the one of a "particular" solution, but not in other way round.

[47] From a modern viewpoint, understanding for which a, b such a number $t = \sqrt[3]{a+\sqrt{b}}+\sqrt[3]{a-\sqrt{b}}-\sqrt{c}$, with $c=(a+\sqrt{b})(a-\sqrt{b})$, is a root of a polynomial of degree 3 is far from being a trivial problem.

Indeed, t is always a root of a polynomial of degree 18. For random a, b this is the lowest possible degree. A sufficient condition for the degree to be 3 is that $a^2-b=u^6$ for some rational u. In this case, the polynomial is

$$x^3+3u^3x^2+(3u^6-3u^2)x+u^9-3u^5-2a=0.$$

Note that for such coefficients $a=-\frac{q}{2}$ and $b=\frac{q^2}{4}+\frac{p^3}{27}=\Delta_3$ where p, q are defined in (2.5.5), see above on page 34. Cardano's example $x^3+3x^2=21$ is a particular case where $a=\frac{19}{2}$ and $b=\frac{357}{4}$, and then $u=1$.

But the converse is false. Indeed, if $a=1$ and $b=\frac{80}{81}$, then t is a root of $x^3+\frac{1}{3}x^2+\frac{1}{27}x-\frac{2186}{729}$. For such a polynomial we have $-\frac{q}{2}=\frac{3}{2}$ and $\Delta_3=\frac{9}{4}$.

We will see in the next section that in the *Ars Magna Arithmeticæ* there is a good number of cubic formulae, or at least they are *in nuce* in the procedure that Cardano describes for solving the equation. In the abstract, it is perfectly possible that Cardano got inspired by the method by which he solves certain equations to discuss the possible shapes for irrational solutions to certain cubic equations with real coefficients. But then one can only explain the presence of the chapters of which we spoke in this section by assuming Cardano's will to proceed in the arithmetisation of *Elements*, Book X that takes place in the first part of the *Ars Magna Arithmeticæ*. In practice, it seems to me by far more likely that he acted the other way round: the inquiries on the possible shapes existed before the solving methods. It is for sure easier to try to solve, or to guess a solution of, a cubic equation with rational coefficients rather than of an equation with real ones. Then, I interpret the study of *binomia* and *recisa* in the *Ars Magna Arithmeticæ* as one of Cardano's access points to the study of the structure of the solutions of an equation. But then, in the *Ars Magna* the situation radically changes. We find some similar statements on the shape of solutions in Chapter IV, namely (AM IV.5) which corresponds both to (AMA XXIII.i) and (AMA XXIII.ii), and (AM IV.6) which corresponds both to (AMA XXV.i) and (AMA XXV.ii). I am likely to interpret these as some late remnants rather than early results still to be developed. Indeed, they simply cannot point at the way to discover a formula. Moreover, in the *Ars Magna* the considerations about the shape of the solutions really become marginal, leaving full room to the cubic formulae, which come to play the main role. We may notice in particular that in the *Ars Magna* all the statements concerning the impossibility for a *binomium* or *recisum* to be solution disappear. This is maybe due to the fact that the results of this type can only be drawn for equations with (positive, non zero) rational coefficients, whereas in the *Ars Magna* they are completely arbitrary (that is, real, positive, non zero coefficients).

4.2.5. The cubic formulae, or what gets closer to them

In **Chapter XX** "On the number of the chapters to be enquired [*De numero capitulorum quærendorum*]" Cardano lists some equations (see Figure 4.2 on page 215) from second to tenth degrees. He puts a small cross next to the ones which are "known in a universal [that is, general] way [*univeraliter nota*]" and a star next to the ones which are "known in a particular way [*particulariter nota*]". Otherwise, no symbol is added.[48] Looking at the list, we observe that Cardano declares to know how to solve "in a universal way" the depressed cubic equations and the cubic equations lacking in the first degree term. He also knows how to solve "in a particular way" two families of complete cubic equations, but no hints are given on the remaining ones. We find Cardano's "universal" solving methods for these equations in Chapters XXVIII-XXX, XXXII, XXXIV, XXXVI, and XXXIX. Moreover, we remark that Cardano has no hints at all concerning the solution of quartic equations.

In **Chapter XXVIII** "On the general chapter of the cube and things equal to a number, by *Magister* Niccolò Tartaglia from Brescia [*De capitulo generali cubi, et rerum æqualium numero, Magistri Nicolai, Tartabliæ, Brixiensis*]" Cardano deals with the equation $x^3 + a_1 x = a_0$. He also mentions his "second book on Euclid".[49]

Cardano says that the solution goes under the shape $\sqrt[3]{\sqrt{b} + a} \pm \sqrt[3]{\sqrt{b} - a}$, where $\sqrt{b} \pm a$ is a *binomium* or *recisum* of the 2nd or 5th type. Sometimes – he continues – the solution occurs as cubic roots of numbers or even just as a number. Therefore, the above shape is "general".

Then, Cardano provides the cubic formula.

---

[48] "*Et nota quod ubi apponitur † talia capitula sunt universaliter nota, ubi ⋆ vero ponitur sunt particulariter nota tantum, ubi nihil, sunt ignota*", see [13, Chapter XX, page 326].

[49] See here, Footnote 31, page 95.

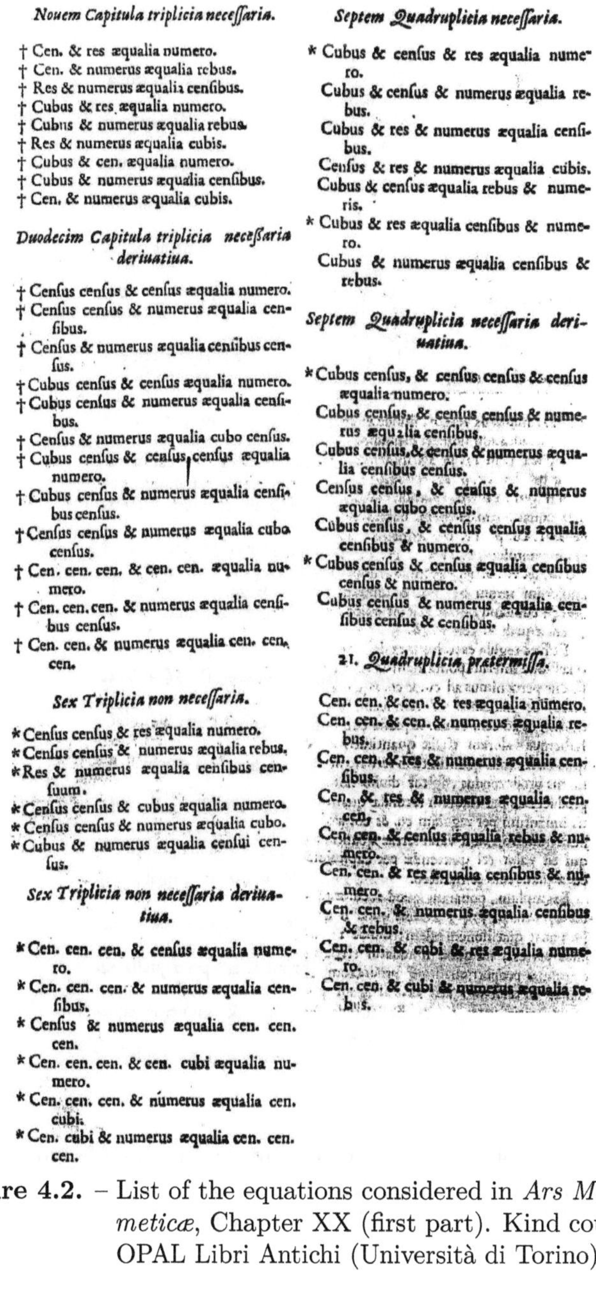

Nouem Capitula triplicia neceſſaria.

† Cen. & res æqualia numero.
† Cen. & numerus æqualia rebus.
† Res & numerus æqualia cenſibus.
† Cubus & res æqualia numero.
† Cubus & numerus æqualia rebus.
† Res & numerus æqualia cubis.
† Cubus & cen. æqualia numero.
† Cubus & numerus æqualia cenſibus.
† Cen. & numerus æqualia cubis.

Duodecim Capitula triplicia neceſſaria deriuatiua.

† Cenſus cenſus & cenſus æqualia numero.
† Cenſus cenſus & numerus æqualia cenſibus.
† Cenſus & numerus æqualia cenſibus cenſus.
† Cubus cenſus & cenſus æqualia numero.
† Cubus cenſus & numerus æqualia cenſibus.
† Cenſus & numerus æqualia cubo cenſus.
† Cubus cenſus & cenſus cenſus æqualia numero.
† Cubus cenſus & numerus æqualia cenſibus cenſus.
† Cenſus cenſus & numerus æqualia cubo cenſus.
† Cen. cen. cen. & cen. cen. æqualia numero.
† Cen. cen. cen. & numerus æqualia cenſibus cenſus.
† Cen. cen. & numerus æqualia cen. cen. cen.

Sex Triplicia non neceſſaria.

* Cenſus cenſus & res æqualia numero.
* Cenſus cenſus & numerus æqualia rebus.
* Res & numerus æqualia cenſibus cenſuum.
* Cenſus cenſus & cubus æqualia numero.
* Cenſus cenſus & numerus æqualia cubo.
* Cubus & numerus æqualia cenſui cenſus.

Sex Triplicia non neceſſaria deriuatiua.

* Cen. cen. cen. & cenſus æqualia numero.
* Cen. cen. cen. & numerus æqualia cenſibus.
* Cenſus & numerus æqualia cen. cen. cen.
* Cen. cen. cen. & cen. cubi æqualia numero.
* Cen. cen. cen. & numerus æqualia cen. cubi.
* Cen. cubi & numerus æqualia cen. cen. cen.

Septem Quadruplicia neceſſaria.

* Cubus & cenſus & res æqualia numero.
Cubus & cenſus & numerus æqualia rebus.
Cubus & res & numerus æqualia cenſibus.
Cenſus & res & numerus æqualia cubis.
Cubus & cenſus æqualia rebus & numeris.
* Cubus & res æqualia cenſibus & numero.
Cubus & numerus æqualia cenſibus & rebus.

Septem Quadruplicia neceſſaria deriuatiua.

* Cubus cenſus, & cenſus cenſus & cenſus æqualia numero.
Cubus cenſus, & cenſus cenſus & numerus æqualia cenſibus.
Cubus cenſus, & cenſus & numerus æqualia cenſibus cenſus.
Cenſus cenſus, & cenſus & numerus æqualia cubo cenſus.
Cubus cenſus, & cenſus cenſus æqualia cenſibus & numero.
* Cubus cenſus & cenſus æqualia cenſibus cenſus & numero.
Cubus cenſus & numerus æqualia cenſibus cenſus & cenſibus.

21. *Quadruplicia prætermiſſa.*

Cen. cen. & cen. & res æqualia numero.
Cen. cen. & cen. & numerus æqualia rebus.
Cen. cen. & res & numerus æqualia cenſibus.
Cen. & res & numerus æqualia cen. cen.
Cen. cen. & cenſus æqualia rebus & numero.
Cen. cen. & res æqualia cenſibus & numero.
Cen. cen. & numerus æqualia cenſibus & rebus.
Cen. cen. & cubi & res æqualia numero.
Cen. cen. & cubi & numerus æqualia rebus.

Figure 4.2. – List of the equations considered in *Ars Magna Arithmeticæ*, Chapter XX (first part). Kind courtesy of OPAL Libri Antichi (Università di Torino).

Cen. cen. & res & numerus æqualia cubis.

Cubus & res & numerus æqualia cenſibus cenſus.

Cen. cen. & cubi æqualia rebus & numero.

Cen. cen. & res æqualia cubis & numero.

Cen. cen. & numerus æqualia cubis & rebus.

Cen. cen. & cubi & cenſus æqualia numero.

Cen. cen. & cubi & numerus æqualia cenſibus.

Cen. cen. & cenſus & numerus æqualia cubis.

Cubus & cenſus & numerus æqualia cen. cen.

Cen. cen. & cubi æqualia cenſibus & numero.

Cen. cen. & cenſus æqualia cubo & numero.

Cen. cen. & numerus æqualia cubo & cenſibus.

21. Capitula quadruplicia prætermiſſa deriuatiua non poſui, quia ex præcedentibus faciliter cognoſcuntur, & difficile eſt peruenire ad æquationem, & rariſſimè cadunt in vſum.

Quinque Capitula quincuplicia prætermiſſa.

Cen. cen. & cubi & cenſus & res æqualia numero.

Cen. cen. & cubi & cenſus & numerus æqualia rebus.

Cen. cen. & cubi & res & numerus æqualia cenſibus.

Cen. cen. & cenſus & res & numerus æqualia cubis.

Cubus & cenſus & res & numerus æqualia cenſui cenſus.

Reliqua etiam ſua correſpondentia deriuatiua quinque capitula non poſui propter rationes ſuperius adductas.

Meminiſſe autem oportet, quòd cognito quocumque capitulo ſuum deriuatiuum ei correſpondens ſubitò cognoſcitur, ſiue ſit ex neceſſariis ſiue non, ſiue ex prætermiſſis ſiue ex triplicibus, ſiue ex quadruplicibus, ſiue quincuplicibus, nam æquatio deriuatiui ſemper eſt ℞. v. æquationis principalis, cuius exempla videbis in capitulo penultimo.

Eſt etiam aliud genus capitulorum quod vocatur aſſimilatorum & ſunt 9. & ſunt vniuerſalia, & aſſimilantur primis 9. & cognoſcuntur per illa ſine difficultate & ſunt hæc:

Nouem Capitula vniuerſalia aſſimilata.

† Cubus cenſus & cubus æqualia numero.

† Cubus cenſus & numerus æqualia cubis.

† Cubus & numerus æqualia cubo cenſus.

† Cubus cubi & cubus æqualia numero.

† Cubus cubi & numerus æqualia cubis.

† Cubus & numerus æqualia cubo cubi.

† Cenſus relati primi & ℞^{m} p^{m} æqualia numero.

† Cenſus relati primi & numerus æqualia ℞^{o} p^{o}.

† R^{m} p^{m} & numerus æqualia cenſui ℞^{i} primi.

Figure 4.3. – List of the equations considered in *Ars Magna Arithmeticæ*, Chapter XX (second part). Kind courtesy of OPAL Libri Antichi (Università di Torino).

AMA XXVIII. *Given the equation* $x^3 + a_1 x = a_0$*, a solution is*

$$x = \sqrt[3]{\sqrt{\left(\frac{a_0}{2}\right)^2 + \left(\frac{a_1}{3}\right)^3} + \frac{a_0}{2}} - \sqrt[3]{\sqrt{\left(\frac{a_0}{2}\right)^2 + \left(\frac{a_1}{3}\right)^3} - \frac{a_0}{2}}.$$

Only a few examples follow. Since the formula is abruptly stated with no justifications at all (not even a proof as in (AM XI)) and since Tartaglia is expressly recalled in the title of the chapter, it is very likely that Cardano received the formula from Tartaglia and did not spend too much time to work on it.

In **Chapter XXIX** "On the cube equal to the squares and a number in general [*De cubo æquali censibus, et numero generali*]" Cardano deals with the equation $x^3 = a_2x^2 + a_0$. He suggests two ways to get a solution. A few examples follow each way.

The first way is by transformations of equations. Cardano transforms $x^3 = a_2x^2 + a_0$ into $y^3 + a_1y = a_0'$, changing the coefficients according to (AMA XXI.1). Then, he draws a solution of $y^3 + a_1y = a_0'$ thanks to the formula in the preceding chapter and transforms back the solution of the equation in y into a solution of the equation in x again by (AMA XXI.1). See the diagram in the proposition below for a short version.

AMA XXIX.i. *Given the equation* $x^3 = a_2x^2 + a_0$*, one draws a solution in the following way.*

$$\begin{array}{ccc} x^3 = a_2x^2 + a_0 & \xrightarrow{\text{AMA XXI.1}} & y^3 + \frac{a_0}{a_2}y = \frac{a_0^2}{a_2^3} \\ & & \Big\downarrow {\scriptstyle \text{AMA XXVIII}} \\ x = \frac{a_0}{a_2 y} & \xleftarrow[\text{AMA XXI.1}]{} & y \end{array}$$

The second way is – in Cardano's opinion – "by far easier".

AMA XXIX.ii. *Given the equation $x^3 = a_2x^2 + a_0$, if f, g real exist such that*

$$\begin{cases} 2\left(\frac{a_2}{3}\right)^3 + a_0 = f + g \\ \left(\frac{a_2}{3}\right)^6 = fg \end{cases},$$

then $x = \sqrt[3]{f} + \sqrt[3]{g} + \frac{a_2}{3}$.

No hints are given on how Cardano discovered the propositions. In particular, we remark that (AMA XXIX.i) does not help in understanding why Cardano sets $2\left(\frac{a_2}{3}\right)^3 + a_0$ and $\left(\frac{a_2}{3}\right)^6$ apart. But, if we perform the substitution $x = y + \frac{a_2}{3}$ in (AMA XXI.9) (or in (AM IV.2)), we get to $y^3 = \frac{a_2^2}{3}y + (\frac{2}{27}a_2^3 + a_0)$, the coefficients of which appear in (AMA XXIX.ii).

We moreover remark that, in order to find such f, g, one only has to solve a quadratic equation. More precisely, it happens that

$$\begin{cases} q = f + g \\ \left(-\frac{p}{3}\right)^3 = fg \end{cases},$$

where p, q are defined in (2.5.5) (see above, on page 34), so that the quadratic equation $t^2 - qx + \left(-\frac{p}{3}\right)^3 = 0$ is the Lagrange's resolvent (up to the sign of the coefficients, see equation (2.5.8), on page 35). Then, (AMA XXIX.ii) is equivalent to the cubic formula[50] in (AMA

[50]See above (2.5.9), on page 36 (only the first line for real solutions) or the specific version for $x^3 = a_2x^2 + a_0$ in (AMA XIV), on page 123.

Indeed, assuming the hypothesis in the system in (AMA XXIX.ii) and the cubic formula in (AMA XIV), the formula boils down to

$$x = \sqrt[3]{\frac{f+g}{2} + \sqrt{\left(\frac{f+g}{2}\right)^2 - fg}} + \sqrt[3]{\frac{f+g}{2} - \sqrt{\left(\frac{f+g}{2}\right)^2 - fg}} + \frac{a_2}{3},$$

that is $x = \sqrt[3]{f} + \sqrt[3]{g} + \frac{a_2}{3}$.

Conversely, assuming (AMA XXIX.ii), we get

$$\begin{cases} f = 2\left(\frac{a_2}{3}\right)^3 + a_0 - g \\ \left(\frac{a_2}{3}\right)^6 = \left(2\left(\frac{a_2}{3}\right)^3 + a_0 - g\right)g \end{cases},$$

XIV), even if Cardano does not explicitly state it.

Cardano ends saying that "this solution has no obstacle, therefore the chapter is the most general".[51] We remark that, indeed, in this case we always have $\Delta_3 > 0$.

In **Chapter XXX** "On the general case which has only one exception when the cube is equal to some things and a number [*De caitulo generali habente tantum unam exceptionem, quando cubus æquatur rebus, et numero*]" Cardano deals with the equation $x^3 = a_1 x + a_0$. He says that this case is the converse of (AMA XXVIII).

In a very similar fashion, Cardano provides the cubic formula as in (AM XII). No justification is given, except one example to show that the formula works.

AMA XXX. *Given the equation $x^3 = a_1 x + a_0$, a solution is*

$$x = \sqrt[3]{\frac{a_0}{2} + \sqrt{\left(\frac{a_0}{2}\right)^2 - \left(\frac{a_1}{3}\right)^3}} + \sqrt[3]{\frac{a_0}{2} - \sqrt{\left(\frac{a_0}{2}\right)^2 - \left(\frac{a_1}{3}\right)^3}}.$$

Cardano observes that, if a condition equivalent to $\Delta_3 \geq 0$ does not hold, then "no solution can follow [*nulla potest sequi æquatio*]". By the way, we remark that, in the example $x^3 = 6x + 10$ provided by Cardano, it happens that $\Delta_3 > 0$.

Cardano ends saying that he found "this rule also by the rules of Chapter XXVII, and all the remaining [rules] except the ones of which [he] mentioned the authors in their places".[52]

In **Chapter XXXII** "On the chapter of the cube and squares equal to a number which is general with only one exception [*De capitulo cubi*

---

that is $g^2 - \left(2\left(\frac{a_2}{3}\right)^3 + a_0\right) g + \left(\frac{a_2}{3}\right)^6 = 0$, which is the resolvent. Its solutions are the ones expected (and the same for f) in order to get the cubic formula.

[51] "*[T]alis equatio non habet impedimentum, ideo capitulum est generalissimum*", see [13, Chapter XXIX, page 343].

[52] "*[H]oc etiam ergo inveni ex regulis capituli 27i sicut et omnia reliqua præter illa duo quibus assignavi auctores suis locis*", see [13, Chapter XXX, page 343].

et census æqualium numero, et est generale, habens unam exceptionem tantum]" Cardano deals with the equation $x^3 + a_2x^2 = a_0$. As in Chapter XXIX, he suggests three ways to get a solution.

The first way corresponds to (AMA XXIX.ii).

AMA XXXII.i. *Given the equation* $x^3 + a_2x^2 = a_0$*, if* f, g *real exist such that*

$$\begin{cases} a_0 - 2\left(\frac{a_2}{3}\right)^3 = f + g \\ \left(\frac{a_2}{3}\right)^6 = fg \end{cases},$$

then $x = \sqrt[3]{f} + \sqrt[3]{g} - \frac{a_2}{3}$.

Also in this case, the corresponding remarks hold. In particular, (AMA XXXII.i) is equivalent to the cubic formula (2.5.9) (only the first line for real solutions, see above, on page 36) or the specific version for $x^3 + a_2x^2 = a_0$ in the footnote 60 (see above, on page 126).

The second way is by transformations of equations and corresponds to (AMA XXIX.i).

AMA XXXII.ii. *Given the equation* $x^3 + a_2x^2 = a_0$*, one draws a solution in the following way.*

$$\begin{array}{ccc} x^3 + a_2x^2 = a_0 & \xrightarrow{\text{AMA XXI.2}} & y^3 = \frac{a_2^2}{3}y + \left(a_0 - \frac{2}{27}a_2^3\right) \\ & & \Big\downarrow * \\ x = y - \frac{a_2}{3} & \xleftarrow[\text{AMA XXI.2}]{} & y \end{array}$$

We expect $*$ to be (AMA XXX) in order to pursue the parallelism with (AMA XXIX.i). Instead, Cardano says that $*$ is

$$\begin{cases} a_0 - 2\left(\frac{a_2}{3}\right)^3 = f + g \\ \left(\frac{a_2}{3}\right)^6 = fg \end{cases}.$$

This means that, once applied the transformation in (AMA XXI.2), we fall again on (AMA XXXII.i), which was equivalent to the cubic formula for $x^3 + a_2x^2 = a_0$. In the end, $*$ is (AMA XXX), and $*$ together with the transformation in (AMA XXI.2) gives (AMA XXXII.i). In this case, we find in the text an insight on why Cardano sets $a_0 - 2\left(\frac{a_2}{3}\right)^3$ and $\left(\frac{a_2}{3}\right)^6$ apart, since the transformation in (AMA XXI.2) (or in (AMA XXI.9), or in (AM IV.2)) is explicitely recalled. We remark that this is basically the same idea as in (AM XV), even if less refined. Indeed, here Cardano considers only the case $a_0 - \frac{2}{27}a_2^3 > 0$, whereas there are also the cases < 0 and $= 0$.

The third way is, in short:

AMA XXXII.iii. *Given the equation $x^3 + a_2x^2 = a_0$, one draws a solution in the following way.*

$$\begin{array}{ccc} x^3 + a_2x^2 = a_0 & \xrightarrow{\textit{AMA XXI.2}} & y^3 = a_1y + \sqrt{a_0} \\ & & \Big\downarrow * \\ x = \dfrac{\sqrt{a_0}}{y} & \xleftarrow[\textit{AMA XXI.2}]{} & y \end{array}$$

I recall that (AMA XXI.2) was composed by three folds. In this proposition, Cardano uses the third one.[53] Concerning $*$, Cardano only suggests to "find the solution [*invenias æquationem*]" without further specifying (neither in the short example which follows). So, I suppose that here $*$ is again (AMA XXX).

In this chapter, Cardano never comes back on the "exception" mentioned in the title. Anyway, we remark that some "particular" rules are given for the equation $x^3 + a_2x^2 = a_0$ in Chapter XXXIII.

In **Chapter XXXIV** "On the general rule of the cube and number equal to the things [*De regula generali cubi et numeri æqualis rebus*]"

[53]Cardano speaks of "the transformation of the cube and squares equal to a number in the cube equal to things and a number by the second way of this transformation in Chapter XXI [*conversionem cubi et census æqualium numero in cubum æqualem rebus et numero per secundum modum talis conversionis 21. capituli*]", see [13, Chapter XXXII, page 346]. By the short example which follows it is clear that he is referring to the third fold of (AMA XXI.2).

Cardano deals with the equation $x^3 + a_0 = a_1 x$. He suggests a transformation to get a solution.

AMA XXXIV.i. *Given the equation $x^3 + a_0 = a_1 x$, one draws a solution in the following way.*

$$\begin{array}{ccc} x^3 + a_0 = a_1 x & \xrightarrow{\textit{AMA XXI.3}} & y^3 = a_1 y + a_0 \\ & & \big\downarrow \textit{AMA XXX} \\ x = \frac{y}{2} \pm \sqrt{a_1 - 3\left(\frac{y}{2}\right)^2} & \xleftarrow[\textit{AMA XXI.3}]{} & y \end{array}$$

The formula is the one in (AM XIII).

Then, Cardano provides the following proposition.

AMA XXXIV.ii. *The solutions of $x^3 + a_1 x = a_0$ and $x^3 = a_2 x^2 + a_0$ can be* "binomia *or* trinomia *by one or more* universalia *cubic roots".*[54]

The solutions of $x^3 = a_1 x + a_0$ and $x^3 + a_2 x^2 = a_0$ can be part "binomia *or* trinomia *by cubic root", part* "binomia *or* trinomia *by one or more square roots".*[55]

The solutions of $x^3 + a_0 = a_1 x$ and $x^3 + a_0 = a_2 x^2$ can be "binomia *or* trinomia *by one or more square roots".*[56]

We remark that one finds a *binomium* in the case of depressed equations and a *trinomium* in the case of cubic equations lacking in the first degree term. This proposition is simply obtained by observing the shape of the solutions in the previously mentioned chapters (and in Chapter XXXVI). Note that, now that Cardano has (explicitly or

[54] "*[Æ]quationes cubi et rerum æqualium numero, item cubi æqualis censibus et numero sunt RV: cubicæ bunimiales vel trinomiales*", see [13, Chapter XXXIV, page 348].

[55] "*[Æ]quationes cubi æqualis rebus et numero et cubi et censuum æqualium numero sunt partim binomiales aut trinomiales R cubicæ partim quadratæ*", see [13, Chapter XXXIV, page 348].

[56] "*[Æ]quationes cubi numeri æqualium censibus aut rebus sunt R binomiales aut trinomiales quadratæ*", see [13, Chapter XXXIV, page 348].

implicitly described by the solving method) some of the cubic formulae available, the qualitative description of the shape for the solutions radically changes. This could confirm my interpretation on page 212.

Then, Cardano ends saying that, since $x^3 + a_0 = a_1 x$ always has two (real, positive, non zero) solutions, then the following proposition holds.

AMA XXXIV.iii. *Given the equation $x^3 + a_0 = a_1 x$, let x_1 be a solution. Then*

$$x_2 = \sqrt{a_1 - x_1^2 + \frac{x_1^2}{4}} - \frac{x_1}{2}$$

is another solution.

This is (AM XIII bis). Cardano's words[57] remind of (AM VIII.2), but I did not find a similar statement in the *Ars Magna.*

In **Chapter XXXVI** "On the universal chapter of the cube and number equal to the squares [*De capitulo universali cubi et numeri æqualium censibus*]" Cardano deals with the equation $x^3 + a_0 = a_2 x^2$. Similarly to Chapter XXXIV, he suggests two ways to get a solution by transformation and a way to get another (real, positive, non zero) solution provided that one has a (real, positive, non zero) solution.

AMA XXXVI.i. *Given the equation $x^3 + a_0 = a_2 x^2$, one draws a solution in the following way.*

[57] "*Notum quod cum volueris invento uno valore æquationis cubi et numeri æqualium rebus habere reliquum valorem tu scis quod unus semper est R unius partis numeri radicum quæ multiplicata in reliquam partem producit numerum æquationis reliquus valor est etiam eiusmodi videlicet R unius partis quæ ducta in reliquam producit numerum æquationis*", see [13, Chapter XXXIV, page 348].

$$\begin{array}{ccc} x^3 + a_0 = a_2x^2 & \xrightarrow{\text{AMA XXI.4}} & y^3 + a_0 = a_2\sqrt[3]{a_2}y \\ & & \Big\downarrow \text{AMA XXXIV} \\ x = \frac{(\sqrt[3]{a_0})^2}{y} & \xleftarrow[\text{AMA XXI.4}]{} & y \end{array}$$

The formula is the one in (AM XVI).

AMA XXXVI.ii. *Given the equation $x^3 + a_0 = a_2x^2$, , one draws a solution in the following way.*

$$\begin{array}{ccc} x^3 + a_0 = a_2x^2 & \xrightarrow{?} & y^3 + a_2y^2 = a_0 \\ & & \Big\downarrow \text{AMA XXXII} \\ x & \xleftarrow[?]{} & y \end{array}$$

I have no hints on which transformation Cardano uses in *?*, since he never mentions it elsewhere.

Then, Cardano says that "the general solution of this chapter is the mixed solution".[58] This could have been seen from (AMA XXII.iii) or from the following proposition.

AMA XXXVI.iii. *Given the equation $x^3 + a_0 = a_2x^2$, let x_1 be a solution. Then*

$$x_2 = \sqrt{(a_2 - x_1)\left(x_1 + \frac{a_2 - x_1}{4}\right)} + \frac{a_2 - x_1}{2}$$

is another solution.

[58] "*[Æ]quatio generalis huius capituli est æquatio mixta*", see [13, Chapter XXXVI, page 350].

Again, this is (AM XVI bis). And again, Cardano's words[59] remind of (AM VIII.2).

In **Chapter XXXIX** "On the solutions of the chapters of *quadrinomia* and *quinomia* [*De æquationibus capitulorum quadrinomiorum et quinomiorum*]" Cardano deals with the complete cubic equations and some derivative cases. We follow him only on complete cubic equations.

Cardano begins with two rules, which, according to the list in Chapter XX (see above, on page 215), should be the only "universal" (that is, "general") ones in this chapter.

AMA XXXIX.1. *Consider* $x^3 + a_2x^2 + a_1x = a_0$ *with* $a_1 = \frac{a_2^2}{3}$. *Then* $x = \sqrt[3]{\left(\frac{a_2}{3}\right)^3 + a_0} - \frac{a_2}{3}$.

Cardano mentions Ferrari, who should have found the "geometrical demonstration" of this rule.[60] Actually, no proof follows.[61] We observe that Cardano made a similar, even if less detailed, remark in (AMA XXIV.ii). According to that, the above solution is a mixed, cubic *recisum* of the 1st or 4th type

AMA XXXIX.2. *Consider* $x^3 + a_1x = a_2x^2 + a_0$ *with* $a_1 = \frac{a_2^2}{3}$. *Then* $x = \frac{a_2}{3} - \sqrt[3]{\left(\frac{a_2}{3}\right)^3 - a_0}$ *or* $x = \frac{a_2}{3} + \sqrt[3]{a_0 - \left(\frac{a_2}{3}\right)^3}$.

[59] "*Cum volueris, in cubo et numero æqualibus censibus per unum valorem habere reliquum scias quod uterque valor est pars numeri censuum quæ quadrata et ducta in residuum producit numerum æquationis*", see [13, Chapter XXXVI, page 350].

[60] "*Prima regua Ludovici de Ferrariis quam invenit geometrica demonstratione*", see [13, Chapter XXXIX, Paragraph 1, page 352].

[61] In any case, we can easily recover it from the equality

$$\left(x + \frac{a_2}{3}\right)^3 = \left(\frac{a_2}{3}\right)^3 + a_0.$$

Again, no justification follows.[62] Cardano adds that "this solution falls on three kinds". If $\frac{a_2}{3} > \sqrt[3]{\left(\frac{a_2}{3}\right)^3 - a_0}$, then both solutions [are positive, non zero and they] go under the shape of the [cubic] *binomium* and *recisum* of the 1st [or 4th] type. If $\frac{a_2}{3} < \sqrt[3]{\left(\frac{a_2}{3}\right)^3 - a_0}$, then [only the second solution is positive, non zero and] it goes under the shape of the [cubic] *binomium* of the 2nd [or 5th] type.[63] We observe that Cardano made a similar, even if less detailed, remark in (AMA XXIV.i).

One should expect these rules to be "general", according to Cardano's classification in Chapter XX (see above, on page 215). According to my weaker interpretation on page 212, we are allowed to call these two rules "general" since they lead to "general solutions". Indeed, in (AMA XXXIX.1) and in (AMA XXXIX.2) the solutions go under the standard form of the cubic *binomia* and *recisa*.

Then, Cardano goes on with some other rules.

AMA XXXIX.3. *Consider* $x^3 + a_2x^2 + a_1x = a_0$.

Then, x^3 *and* a_1x *are "of the same nature" [that is, speaking of* binomia *and* recisa, *of the same type]. Moreover, the* recisum *of the 2nd [and 5th] type can be a solution.*

We remark that this equation has already been considered in (AMA XXXIX.1). Cardano justifies this proposition in a way similar to (AMA XXII.i). A few examples follow.

AMA XXXIX.4. *The* binomium *and* recisum *of the 1st and 4th type and the* recisum *of the 2nd and 5th type can be solutions of the equation* $x^3 + a_2x^2 + a_0 = a_1x$.

AMA XXXIX.5. *The* binomium *and* recisum *of the 1st and 4th type and the* binomium *of the 2nd and 5th type can be solutions of the equation* $x^3 + a_1x + a_0 = a_2x^2$.

---

[62] In any case, we can easily recover it from the equality

$$\left(x - \frac{a_2}{3}\right)^3 = a_0 - \left(\frac{a_2}{3}\right)^3.$$

[63] See above, on page 206.

AMA XXXIX.6. *The* binomium *of the 2nd and 5th type can be solutions of the equation* $x^3 = a_2x^2 + a_1x + a_0$.

For each one of the above three propositions, only a few examples follow after the statement.

By the classification in Chapter XX, one should expect these rules to be "particular", but in fact they are not – even according to our weaker interpretation. Indeed, Cardano says that "it is clear that all this chapters are very general except the last one, that is the cube equal to squares, things and a number. Indeed, this case, although general, is nevertheless very involute".[64] We remark that, consistently with our weaker interpretation on page 212, there is no reason why (AMA XXXIX.3)-(AMA XXXIX.6) should not be considered as general as (AMA XXXIX.1) or (AMA XXXIX.2). Then, we could suppose an omission or a misprint in Chapter XX. But this does not explain why (AMA XXXIX.6) should have a different status.

Cardano then affirms that "the chapters with four terms depend on [the chapters with] three [terms]".[65] More precisely,

AMA XXXIX.iii. *The equation* $x^3 + a_2x^2 + a_1x = a_0$ *depends on*

- $x^3 + a'_1x = a'_0$ *(no solution in terms of* binomia *and* recisa*),*[66]
- $x^3 + a''_2x^2 = a''_0$*; they both have the* recisum *of the 2nd and 5th type as solution.*[67]

AMA XXXIX.iv. *The equation* $x^3 + a_2x^2 + a_0 = a_1x$ *depends on*

- $x^2 + a'_0 = a'_1x$*; they both have the* binomium *and* recisum *of the 1st and 4th type as solution,*[68]

[64] "*Manifestum est igitur quod hæc capitula valde generalia sunt excepto ultimo, videlicet quando cubus æquatur censibus, rebus et numero, nam tale caputilum, etsi si generale, attamen est valde involutum*", see [13, Chapter XXXIX, Paragraph 6, page 355].

[65] "*Pendent autem capitula quaternaria ex ternariis*", see [13, Chapter XXXIX, Paragraph 6, page 355].

[66] See above, after (AMA XXII.ii).

[67] See above (AMA XXII.iv).

[68] At the beginning of Chapter XXII Cardano deals with the shape of the solutions of quadratic equations.

- $x^3 + a_0'' = a_1''x$; *they both have the* binomium *and* recisum *of the 1st and 4th type and the* recisum *of the 2nd and 5th type as solution.*[69]

AMA XXXIX.v. *The equation* $x^3 + a_1x + a_0 = a_2x^2$ *depends on*

- $x^2 = a_1'x + a_0'$; *they both have the* binomium *of the 2nd and 5th type as solution,*[70]
- $x^3 + a_0'' = a_2''x^2$; *they both have the* binomium *and* recisum *of the 1st and 4th type and the* binomium *of the 2nd and 5th type as solution.*[71]

AMA XXXIX.vi. *The equation* $x^3 = a_2x^2 + a_1x + a_0$ *depends on*

- $x^3 = a_1'x + a_0'$; *they both have the* binomium *of the 2nd and 5th type as solution,*[72]
- $x^3 = a_2''x^2 + a_0''$ *(no solution in terms of* binomia *and* recisa*).*[73]

These propositions seem to be a more elaborated version of the corresponding (AMA XXXIX.3)-(AMA XXXIX.6).

AMA XXXIX.7. *The equation* $x^3 + a_0 = a_2x^2 + a_1x$ *depends on*

- $x^3 + a_0' = a_1'x$; *they both have the* binomium *and* recisum *of the 1st and 4th type and the* recisum *of the 2nd and 5th type as solution,*[74]
- $x^3 + a_0'' = a_2''x^2$; *they both have the* binomium *and* recisum *of the 1st and 4th type and the* binomium *of the 2nd and 5th type as solution.*[75]

AMA XXXIX.8. *The equation* $x^3 + a_1x = a_2x^2 + a_0$ *depends on*

[69] See above (AMA XXII.i).
[70] See above, Footnote 68.
[71] See above (AMA XXII.iii).
[72] See above (AMA XXII.ii).
[73] See above, after (AMA XXII.iv).
[74] See above (AMA XXII.i).
[75] See above (AMA XXII.iii).

- $x^2 + a'_0 = a'_1 x$; *they both have the* binomium *and* recisum *of the 1st and 4th type as solution,*[76]

- $x^3 = a''_2 x^2 + a''_0$ *(no solution in terms of* binomia *and* recisa*).*[77]

We remark that this equation had already been considered in (AMA XXXIX.2).

AMA XXXIX.9. *The equation* $x^3 + a_2 x^2 = a_1 x + a_0$ *depends on*

- $x^2 = a'_1 x + a'_0$; *they both have the* binomium *of the 2nd and 5th type as solution,*[78]

- $x^3 = a''_1 x + a''_0$; *they both have the* binomium *of the 2nd and 5th type as solution.*[79]

We could then complete the Table 4.1 on page 211 as follows in Table 4.2 on page 230.

Referring to the above propositions, Cardano uses "to depend [*pendeo*]", "to be equivalent [*æquivaleo*]", and "to compose [*compono*]" as synonyms, with a preference for the last term. But he does not explain in which sense a complete equation "depends" on, or "is equivalent" to, or "is composed" by two other equations. I cannot figure out a sensible meaning for that.[80] Moreover and in any case, no explanation on the different status of (AMA XXXIX.6) is given.

[76] See above, Footnote 68.

[77] See above, after (AMA XXII.iv).

[78] See above, Footnote 68.

[79] See above (AMA XXII.ii).

[80] We cannot always pass from a complete equation to three-term equations by transformations, since also quadratic equations are involved. Or otherwise, if we consider the system formed by the three-term equations, it is not true that it leads to the corresponding complete equation. Or finally, we could remark that the three-term equations and the corresponding complete equation have a similar outfit so that, if we sum the three-term equations, we get the corresponding complete equation. Then, if a number is a solution of one of the three-term equations, it will also be a solution of the complete equation. But in this way we are assuming a relationship between the solutions of the three-term equations, which is not the case.

Table 4.2. – Shapes of irrational solutions to cubic equations in *Ars Magna Arithmeticæ*, Chapter XXXIX.

Equation	Shape of the irrational solution			
	binomia and/or *recisa*			*trinomia*
	(normal)	cubic	mixed cubic	
$x^3 + a_1x = a_0$		$\sqrt[3]{a} - \frac{k}{a}\sqrt[3]{a^2}$		
$x^3 = a_1x + a_0$	$\sqrt{a} + b$	$\sqrt[3]{a} + \frac{k}{a}\sqrt[3]{a^2}$		
$x^3 + a_0 = a_1x$	$a \pm \sqrt{b}$, $\sqrt{a} - b$			
$x^3 = a_2x^2 + a_0$				$a\sqrt[3]{a^2} + \sqrt[3]{a} + a$
$x^3 + a_2x^2 = a_0$	$\sqrt{a} - b$			$a\sqrt[3]{a^2} + \sqrt[3]{a} - a$
$x^3 + a_0 = a_2x^2$	$a \pm \sqrt{b}$, $\sqrt{a} + b$			
$x^3 + a_2x^2 + a_1x = a_0$	$\sqrt{a} - b$		$a - \sqrt[3]{b}$	
$x^3 + a_1x = a_2x^2 + a_0$	$a \pm \sqrt{b}$		$a + \sqrt[3]{b}$, $\sqrt[3]{a} \pm b$	
$x^3 + a_2x^2 = a_1x + a_0$	$\sqrt{a} + b$			
$x^3 = a_2x^2 + a_1x + a_0$	$\sqrt{a} + b$			
$x^3 + a_0 = a_2x^2 + a_1x$	$a \pm \sqrt{b}$, $\sqrt{a} \pm b$			
$x^3 + a_1x + a_0 = a_2x^2$	$a \pm \sqrt{b}$, $\sqrt{a} + b$			
$x^3 + a_2x^2 + a_0 = a_1x$	$a \pm \sqrt{b}$, $\sqrt{a} - b$			
Condition	$k \in \mathbb{Q}$			

This last part is quite confused. Since there are no definitively valid arguments for that, I prefer to trust Chapters XX and XXIV and consider that only $x^3 + a_2x^2 + a_1x = a_0$ and $x^3 + a_1x = a_2x^2 + a_0$ have a "general" rule.[81]

Summing up, we remark again that the hypothesis that the *Ars Magna* comes from a further development of the *Ars Magna Arithmeticæ* is not at all unlikely. Especially regarding cubic equations, the *Ars Magna Arithmeticæ* is at an earlier stage, since the treatment of these equations is less systematic. Concerning depressed equations and equations lacking in the first degree term, we basically find (less refined) all the results of the *Ars Magna*. The cubic formulae are not always explicitly written, but for the most of the non-complete equations they are already there *in nuce* (except maybe for $x^3 + a_0 = a_2x^2$ in Chapter XXXVI, of which we anyway know a *binomium* shape for one of its solutions, see (AMA XXXVI.iii)). The major difference with the *Ars Magna* is that here there are no proofs. Moreover, the ordering of the results is different so that the dependence of each equation lacking in the first degree term from a depressed one is clearer.

As in the *Ars Magna*, Tartaglia is mentioned only in relation to the equation $x^3 + a_1x = a_0$.

Concerning complete cubic equations, there is no longer correspondence between the *Ars Magna Arithmeticæ* and the *Ars Magna*. Indeed, even if Cardano knows the substitution $x = y \pm \frac{a_2}{3}$ by (AMA XXI.9), he does not use it, so that he never gets to an equivalent of the cubic formula as in the *Ars Magna*.

We have then the correspondences in Table 4.3.

Finally, we remark that the *Ars Magna Arithmeticæ* is again linked to the opposition between "general" and "particular", as it appears from the title of all the considered chapters. Since Cardano does

[81] We anyway remark that [57] includes Chapter XXXIX among the "particular" rules. Nevertheless, Pietro Cossali also affirms that, at Cardano's time, there was probably no agreement on considering these rules "general" or "particular", see [57, Volume II, Chapter II, Soluzioni speciali, page 183] and above, Footnote 3.1.4 on page 100.

Table 4.3. – Links between "general" propositions in the *Ars Magna Arithmeticæ* and *Ars Magna*.

Ars Magna	Equation	*Ars Magna Arithmeticæ*
(AM XI)	$x^3 + a_1x = a_0$	(AMA XXVIII)
(AM XII)	$x^3 = a_1x + a_0$	(AMA XXX)
(AM XIII), (AM XIII bis)	$x^3 + a_0 = a_1x$	(AMA XXXIV.i), (AMA XXXIV.iii)
(AM XIV)	$x^3 = a_2x^2 + a_0$	(AMA XXIX.ii)
(AM XV)	$x^3 + a_2x^2 = a_0$	(AMA XXXII.ii)
(AM XVI), (AM XVI bis)	$x^3 + a_0 = a_2x^2$	(AMA XXXVI.i), (AMA XXXVI.ii)

not state here whether "particular" is referred to the methods or to solutions, I assume the most natural choice, that is, that "general" is referred to the methods employed. Anyway, this interpretation clashes in two points. Firstly, it seems that a "general" method can bear "exceptions", as in the titles of Chapters XXX and XXXII. Secondly, while speaking of complete cubic equations, I could not account for Cardano's use of "general" but by interpreting "general" as referred to solutions.

4.2.6. "Particular" solving methods

In this section and when needed, we refer to the functions of the coefficients in *Ars Magna*, Chapter XXV using f', g', since my aim here is to explain the propositions concerning "particular" rules in the *Ars Magna Arithmeticæ* through the corresponding ones in the *Ars Magna.*

We find Cardano's "particular" solving methods for cubic equations in Chapters XXXI, XXXIII, XXXV, and XXXVII.

In **Chapter XXXI** "On the chapter and particular rules of the cube equal to the things and a number [*De capitulo et regulis particularibus cubi æqualis rebus et numero*]", Cardano states the following propositions.

AMA XXXI.i. *Consider* $x^3 = a_1 x + a_0$.
If $a_1 \sqrt[3]{f^3} = f^3 + a_0$, *then* $x + f$ *divides* $x^3 + f^3$.
If $a_1 \sqrt[3]{f^3} = f^3 - a_0$ *[with* $f^3 > a_0$*], then* $x - f$ *divides* $x^3 - f^3$.

No further explanation is given, except the example $x^3 = 16x + 21$ (with $\Delta_3 < 0$). Cardano explicitly remarks that this example cannot "be solved by the preceding chapter [*non potest solui per præcedens capitulum*]". Indeed, we remind[82] that Chapter XXX deals with the "general" case of $x^3 = a_1 x + a_0$ which nevertheless has "one exception". Then, as in (AM XII), it seems that a "general" rule or method can bear "exceptions".

We remark that this rule is (AM XXV.6) (with the same example) and (PA LI.26).

AMA XXXI.ii. *Consider* $x^3 = a_1 x + a_0$.

$$\text{If } \begin{cases} a_1 = f + g \\ a_0 = f\sqrt{g} \end{cases}, \text{ then } x = \sqrt{f + \tfrac{g}{4}} + \tfrac{\sqrt{g}}{2}.$$

We remark that this is (AM XXV.1) and also one of the examples which follow $x^3 = 20x + 32$ (with $\Delta_3 < 0$) is the same.

AMA XXXI.iii. *Consider* $x^3 = a_1 x + a_0$.

$$\text{If } \begin{cases} f = \sqrt{a_1 + g} \\ a_0 = fg \end{cases}, \text{ then } x = \sqrt{a_1 + g}.$$

We remark that this is (AM XXV.2) and also one of the examples which follow $x^3 = 32x + 24$ (with $\Delta_3 < 0$) is the same.

---

[82]See above, Section 4.2.5 on page 219.

In **Chapter XXXIII** "On the particular rules of the cube and squares equal to a number [*De regulis particularibus cubi et censuum æqualium numero*]", Cardano states the following propositions.

AMA XXXIII.i. *Consider* $x^3 + a_2x^2 = a_0$.
If $\begin{cases} a_2 = f + g \\ a_0 = fg^2 \end{cases}$, *then* $x = \sqrt{f\left(g + \frac{f}{4}\right)} - \frac{f}{2}$.

We remark that this is (AM XXV.14) and also one of the examples which follow $x^3 + 20x^2 = 72$ (with $\Delta_3 < 0$) is the same.

Then, Cardano states in a less detailed way the proposition (AMA XXXII.ii), and this time explicitly specifies that $*$ is (AMA XXX).

Finally, he remarks that the solutions of $x^3 + a_2x^2 = a_0$ are opposite in sign to the ones of $x^3 + a_0 = a_2x^2$.[83]

In **Chapter XXXV** "On the particular rules of the cube and number equal to the things [*De regulis particularibus cubi et numeri æqualium rebus*]", Cardano states the following propositions.

AMA XXXV.i. *Consider*[84] $x^3 + a_0 = a_1x$.
If $\begin{cases} \frac{a_1}{4} < g < \frac{a_1}{3} \\ g \text{ divides } a_0 \\ a_1 + \frac{a_0}{2\sqrt{g}} = 4g \end{cases}$, *then* $x = \sqrt{g} \pm \sqrt{a_1 - 3g}$.

Note that the condition $g < \frac{a_1}{3}$ grants that $a_1 - 3g > 0$, that is, that all the solutions are real. The condition $\frac{a_1}{4} < g$ grants that $\sqrt{g} - \sqrt{a_1 - 3g} > 0$. We remark[85] that this is (AM XXV.9) taking $f' = 4g - a_1$, and then $2\sqrt{g} = \sqrt{a_1 + f'}$.

AMA XXXV.ii. *Consider* $x^3 + a_0 = a_1x$.

---

[83] See (AM I.8 ii).

[84] At the end of the chapter (see [13, Chapter XXXV, pages 349-50]), Cardano adds a few words to explain why $\sqrt{a_1 - 3g}$ is not rational. Since they are really obscure, I do not comment them.

[85] Or, equivalently, this is (AM VIII.2 i), taking $G = -2\sqrt{g}$, $F = a_1 - 4g$, $\alpha_1 = -a_1$, and $\alpha_0 = a_0$.

$$If \begin{cases} a_1 = f + g \\ a_0 = f\sqrt{g} \end{cases}, \; then \; x = \sqrt{f + \tfrac{g}{4}} - \tfrac{\sqrt{g}}{2}.$$

Cardano remarks that this proposition corresponds to (AMA XXXI.ii). Indeed, if we write a_1'', a_0'' for the coefficients and f'', g'' for the functions of the coefficients in (AMA XXXI.ii), then it is enough to take $f = f''$ and $\sqrt{g} = -\sqrt{g''}$, in order to verify that $a_1 = a_1''$ and $a_0 = -a_0''$.

We remark that this is (AM XXV.9), taking $f = -f'$ and $\sqrt{g} = -\sqrt{a_1 + f'}$.

AMA XXXV.iii. *Consider* $x^3 + a_0 = a_1 x$.
If $a_0 - a_1 f = f^3$, *then* $x - f$ *divides* $x^3 + a_0 - a_1 f$.

Cardano remarks that this proposition corresponds to (AMA XXXI.i).

We remark that this is (AM XXV.9), taking $f = -f'$ and $\sqrt{g} = -\sqrt{a_1 + f'}$, and also (AM XXV.10) (with the same example) and (PA LI.26).

In **Chapter XXXVII** "On the particular rules of the cube and number equal to the squares [*De regulis particularibus cubi et numeri æqualium censibus*]", Cardano reminds that the *binomium* or *recisum* of the 1st or 4th type $a \pm \sqrt{b}$ and the *binomium* of the 2nd or 5th type $\sqrt{a} + b$ can be solutions of $x^3 + a_0 = a_2 x^2$ (see (AMA XXII.iii)). More precisely, he says that, when a solution is the *binomium* of the 1st or 4th type, then also its *recisum* is a solution and that, when a solution is an integer, then also the *binomium* of the 2nd or 5th type is a solution. Then, he wants to show that "the type of the *binomium* and *recisum* is not enough to meet each solution of this chapter".[86] But unfortunately his justification seems to be totally unclear.

AMA XXXVII.i. *Consider* $x^3 + a_0 = a_2 x^2$ *and let* x_1 *be a solution. Then*

$$x_2 = \sqrt{f x_1 + \left(\frac{f}{2}\right)^2} + \frac{f}{2}$$

[86] "*[O]portet ostendere quod genus binomij et recisi non est sufficiens satisfacere omni æquationi huius capituli*", see [13, Chapter XXXVII, page 350].

is another solution.

We remark that this is (AM XVI bis) and also (AMA XXXVII.iii), taking $f = a_2 - x_1$.

AMA XXXVII.ii. *Consider* $x^3 + a_0 = a_2 x^2$.

If $\begin{cases} \frac{a_2}{4} \leq g \leq \frac{a_2}{3} \\ \sqrt{\frac{a_0}{g}} \in \mathbb{Q} \\ a_2 + \frac{1}{2}\sqrt{\frac{a_0}{g}} = 4g \end{cases}$ *, then* $x = 2g \pm \sqrt{4g(a_2 - 3g)}$.

Cardano remarks that this proposition corresponds to (AMA XXXV.i).

We remark that this proposition corresponds to (AMA XXXIII.i). Indeed, if we write a''_1, a''_0 for the coefficients and f'', g'' for the functions of the coefficients in (AMA XXXIII.i), then it is enough to take $f'' = -4g$ and $g'' = 4g - a_2$, in order to verify that $a_2 = -a''_2$ and $a_0 = -a''_0$.

We moreover remark that this is (AM XXV.15) taking $g = f'$ and also one of the example is the same. Or, equivalently, this is (AM VIII.2 ii), taking with $G = -\frac{1}{2}\sqrt{\frac{a_0}{g}}$, $F = 4g$, $\alpha_2 = -a_2$, and $\alpha_0 = a_0$.

Finally, Cardano states in a less detailed way the proposition (AMA XXI.4).

Summing up, all the above "particular" propositions can also be found in the *Ars Magna*, mainly in Chapter XXV, as it can be seen in Table 4.4 on page 237. Out of the above numbered propositions, only one is not also in Chapter XXV. It is (AMA XXXVII.i), which is the same as (AM XVI bis) and as (AMA XXXVI.ii). We remark by the way that there is no equivalent of (AM XIII bis).

Cardano says that these rules are "known in a particular way".[87] Comparing to the use of this term in the *Ars Magna*, Cardano does not state here whether "particular" is referred to the methods or to solutions. Since Cardano provides no further specifications, I assume that, as in the *Ars Magna*, "particular" is here referred to the methods employed. Indeed, only the families of equations that can

[87] See [13, Chapter XX, page 326] or above, on page 214.

Table 4.4. – Links between "particular" propositions in *Ars Magna*, Chapter XXV and in the *Ars Magna Arithmeticæ*.

$x^3 = a_1x + a_0$		$x^3 + a_0 = a_1x$	$x^3 + a_2x^2 = a_0$		$x^3 + a_0 = a_2x^2$
{ (AM XXV.1) (AMA XXXI.ii)	⟷	{ (AM XXV.9) (AMA XXXV.i,ii)	{ (AM XXV.14) (AMA XXXIII.i)	⟷	{ (AM XXV.15) (AMA XXXVII.ii)
{ (AM XXV.2) (AMA XXXI.iii)	⟷	(AM XXV.17)	(AM XXV.18)		
(AM XXV.3)	⟷	(AM XXV.13)			
(AM XXV.4)	⟷	(AM XXV.11)			
(AM XXV.5)		(AM XXV.12)			
{ (AM XXV.6) (AMA XXXI.i) (PA LI.26)	⟷	{ (AM XXV.10) (AMA XXXV.iii) (PA LI.26)			

have $\Delta_3 < 0$ are considered, and moreover all the twenty-six provided examples have as a matter of fact $\Delta_3 < 0$. Then, I interpret the above propositions to be "particular" in the sense that in concrete terms they only work if one can find such f, g, that is, a solution of the considered equation. In this sense, the term "particular" applies also to (AMA XXXVII.i), which teaches how to find a second solution of $x^3 + a_0 = a_2x^2$, provided that one already found a first one.

4.2.7. Summing up

All along this section, we have remarked that there is a path linking the *Ars Magna*, the *Practica Arithmeticæ*, and the *Ars Magna Arithmeticæ* and that the direction is from the *Practica Arithmeticæ* to the *Ars Magna*, passing through the *Ars Magna Arithmeticæ*. Comparing the *Practica Arithmeticæ* to the *Ars Magna Arithmeticæ*, we find in the latter more and by far more developed topics on cubic equations. It is passing from the *Practica Arithmeticæ* to the *Ars Magna Arithmeticæ* that the role of equations radically changes from being side effects of proportions to being (or at least, starting to be) a subject of enquiry by itself.[88]

Comparing instead the *Ars Magna Arithmeticæ* to the *Ars Magna*, we observe that a remarkable part of the *Ars Magna*'s arguments for equations are already present in the *Ars Magna Arithmeticæ*, and this is especially true concerning Cardano's heuristic conceptions and transformations of equations. In both works, Cardano deals with them virtually in the same way, whereas other topics undergo different treatments. Namely, it is the case of the study of the shape of irrational solutions (of equations with rational coefficients), which is still characterised by the use of the Euclidean terminology for irrational numbers and which almost disappear in the *Ars Magna*. We moreover remark that at this stage, Cardano has only a partial treatment for cubic equations available, as Table 4.5 on page 239 shows. For most of the non-complete equations (with maybe the exception of $x^3 + a_0 = a_2x^2$), Cardano has a cubic formula available, or a

[88]See [65].

Table 4.5. – Correspondences in the treatment of cubic equations between the *Ars Magna* and the *Ars Magna Arithmeticæ*.

Ars Magna	Equation	*Ars Magna Arithmeticæ*
(AM XI)	$x^3 + a_1x = a_0$	(AMA XXVIII)
(AM XII)	$x^3 = a_1x + a_0$	(AMA XXX), (AMA XXXI) "particular"
(AM XIII)	$x^3 + a_0 = a_1x$	(AMA XXXIV), (AMA XXXV) "particular"
(AM XIV)	$x^3 = a_2x^2 + a_0$	(AMA XXIX)
(AM XV)	$x^3 + a_2x^2 = a_0$	(AMA XXXII), (AMA XXXIII) "particular"
(AM XVI)	$x^3 + a_0 = a_2x^2$	(AMA XXXVI), (AMA XXXVII) "particular"
(AM XVII)	$x^3 + a_2x^2 + a_1x = a_0$	(AMA XXXIX) "particular"
(AM XVIII)	$x^3 + a_1x = a_2x^2 + a_0$	(AMA XXXIX) "particular"

procedure that eventually can lead to it. No proof is given. For some non-complete equations (the ones that can have $\Delta_3 < 0$), Cardano also provides some "particular" rules. His treatment of complete cubic equations, instead, is sometimes quite disconcerting and as a matter of fact he does not have any formula. He only provides two "particular" rules. As seen, Cardano's treatment of cubic equations is tightly linked to the opposition between "general" and "particular", which anyway does not seem to be part of a completely stable terminology. The only significant rest of this opposition will be found in *Ars Magna*, Chapter XXV while speaking of the case $\Delta_3 < 0$. Moreover, the order in which the chapters are displayed in the *Ars Magna Arithmeticæ* highlights a dependency of the equations lacking in the first degree term from the depressed equations, which in the *Ars Magna* gets lost.

No treatment for quartic equations is available at all.

We would like now to draw for the *Ars Magna Arithmeticæ* a diagram to summarise the links that bind the cubic equations, similar to the one for the *Ars Magna* on page 164. I keep the same notations. Note that now the dotted lines correspond to some propositions in Chapter XXI. In particular, I have dashed the link between $x^3 = a_1x + a_0$ in (AMA XXX)) and $x^3 = a_2x^2 + a_0$ in (AMA XXIX.i)), since the substitution is not only stated, but also referred to those equations (among others) in (AMA XXI.9). We observe that the *Ars Magna*'s core of the bindings is already there. Anyway, we do not have the full system of the cubic formulae and – as said – the complete cubic equations are missing.

We remind that the only printed edition that we have is in the posthumous 1663 edition of the *Opera Omnia*. These remarks on the links between the *Ars Magna Arithmeticæ* and the *Ars Magna* could also shed light on why Cardano chose not to publish the *Ars Magna Arithmeticæ*. Indeed, in [65] it is argued that two of the contributing factors are the problem posed by the *casus irreducibilis* and Cardano's willing to compose the *Opus Arithmeticæ Perfectum*. Gavagna affirms that the editorial plan of the *Opus Arithmeticæ Perfectum* raised the need to reallocate the contents of the *Ars Magna Arithmeticæ* in two blocks. The first of them should have been on irrational quantities,

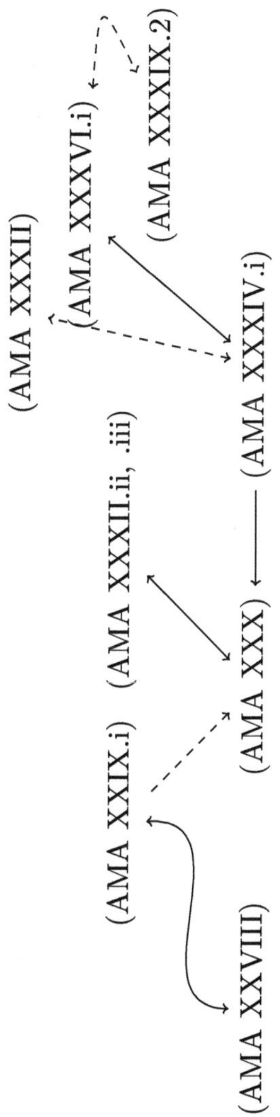

Figure 4.4. – Inter dependencies between the cubic equations in the *Ars Magna Arithmeticæ*.

while the second should have been included in the algebra volume and make up the core of the subsequent *Ars Magna.* This last part are the selected chapters of the *Ars Magna Arithmeticæ* that we have dealt with in detail in this section.

5. Strategies to possibly overcome the *casus irriducibilis*: the *De Regula Aliza*

We are now ready to consider the *De Regula Aliza*.

5.1. Getting acquainted with the *De Regula Aliza*

We have two editions available of this treatise. The first one is the *Hieronymi Cardani mediolanensis, civisque bononiensis, medici ac mathematici præclarissimi, de aliza regula, libellus, hoc est operi perfecti sui sive algebraicæ logisticæ, numeros recondita numerandi subtilitate, secundum geometricas quantitate inquirendis, necessaria coronis, nunc demum in lucem editæ*. It dates back to 1570 and was printed *in folio* in Basel by Officina Henricpetrina together with the *De Proportionibus*[1] and with the second edition of the *Ars Magna*. The second edition, which is shortly entitled *De regula aliza libellus*, is posthumous and placed in the fourth volume of the 1663 edition of Cardano's *Opera Omnia*, printed in Lyon by Ioannis Antonii Huguetan and Marci Antonii Ravaud. Some miscalculations and typos are corrected and sometimes the punctuation is improved, but as a whole it contains no major change compared to the first edition.

[1]The *De Proportionibus* is quite a big work, composed by 233 propositions over 271 *folia*. As the title suggests, the proportions are the common thread of the book. Apart from the general rules of calculation with proportions, we do not only find proportions applied to mathematics (geometry and arithmetic) or physics (statics and dynamics), but also (even though to a lesser extent) to navigation, medicine, geography, mechanics, optics, gnomonics, and astronomy.

Several copies of the 1570 edition can be found in Europe and America (Denmark, France, Germany, Italy, the Netherlands, Russia, Spain, Sweden, Switzerland, UK, but also Canada and USA). In Italy, there is a copy of the 1570 edition in some university libraries (Naples, Pisa, Rome), in national libraries (Florence, Naples, Rome, Turin), in public libraries (Mondovì, Lugo, Piacenza, Perugia, Ravenna), and in the diocesan library in Todi. In France, there is a copy of the 1570 edition at the National Library in Paris. As far as I know,[2] we do not have any manuscript of the *Aliza*.

Even though Cardano claims that none of his works had been rewritten less than three times,[3] the *Aliza* seems not to enjoy this benefit. Probably, Cardano barely proofread the text or proofread it very quickly. Indeed, apart from the widespread recurring typos, it happens that sometimes the figures are completely missing (as in Chapter II, see page 324) or that we find some oversights, like the unlikely multiplication of two areas (as in Chapter LX, see Footnote 102 on page 336). As said, Cardano's Latin sentences suffer from a high complexity in their structure, where hypotaxis rules. But even more important, the structure of the book – as said in Section 2.2 – is quite hard to detect, or there is rather no global structure at all. Therefore, it is more useful to consider *Aliza* as a miscellany and search for the chapters, or for their parts, that show a certain coherence. In this way, we manage to identify one general leading idea, which concerns the algebraic shapes of the cubic equations and from which we can try to go back up to the *casus irreducibilis*.

The *Aliza*'s miscellany is composed by sixty highly patchy chapters, each one with its own title. They fill up 111 *folia* in the 1570 edition. Their length is very variable, from half a *folio* up to six or seven *folia*. We sometimes find an inner organisation, like numbered paragraphs, corollaries, *scholia*, or *paradoxa*. Each chapter virtually contains at least a diagram or a table where a calculation is displayed; there are eighty-nine diagrams and forty-nine tables in all. In the margins

[2]See the website [44].

[3]"*[N]ullus liber minus quam ter scriptum est*", see [6, page 78].

and in the text, we find a good number of references, in most of the cases referring to the *Aliza* itself or to the *Ars Magna* and *De Proportionibus*, which were published – as said – in the same joined edition. There are also many references to the *Elements.*[4]

The *Aliza* is written using the abbreviations and accents common to the Latin of the Renaissance.[5] We should also consider the abbreviations that characterise Cardano's mathematics. I only recall that, acting as a man of his time, Cardano used to state propositions in words, so that only a loose kind of formalism – or rather, some stenographic expressions – are involved. I refer to the preliminary Chapter Notations and more for all the details. Without a fully developed formalism, it is not surprising that Cardano followed the habit of his time (which lasted until the 17th century and part of the 18th century) of using a particular example – like an equation with numerical coefficients – to make a general discourse.[6] Limit cases as

[4]For what edition of the *Elements* Cardano could have read, see page 409.

[5]Note that, whenever I transcript parts of the *Aliza*'s text, I dissimilate 'u' and 'U' from the original text in 'u', 'v' and 'U', 'V', while I assimilate 'i', 'j' and 'I', 'J' from the original text in 'i' and 'I'. Note moreover that, during the 16th century, the letter 's' is usually written as 'ſ' , except at the end of a word and when it is double ('ſs'). In any case, I transcribe it as 's'.

Concerning the accents and other special types that were common in the Renaissance typography, I performed the following substitutions in order to make the text more easily readable: 'ã' becomes 'am' or 'an', 'ë' and 'ę' become 'æ', 'ẽ' becomes 'em' or 'en', '&' becomes 'et', 'õ' becomes 'om' or 'on', 'p̃' becomes 'præ', 'ꝑ' becomes 'pre', 'ꝓ' becomes 'pro', 'q̃' becomes 'qua', 'qȝ' becomes 'que', 'ꝗ' becomes 'qui', 'q̊' becomes 'quo', 't̃' becomes 'tur', and 'ũ' becomes 'um' or 'un'. Exceptionally, 'aũt' becomes '*autem*', 'em̃' becomes '*enim*', and 'tñ' becomes '*tunc*'. Moreover, I have removed the grave accents that according to the common usage were employed to mark the adverbs and prepositions.

Concerning the abbreviations, all the words that were abbreviated in the original text using a point and that are employed as a technical jargon have been kept abbreviated. Instead, the abbreviated words that belong to the common language have been replaced by their full writing. For the same sake of readability, I have adapted to the modern usage the punctuation, the capital letters, and the italics.

[6]As Karine Chemla says concerning another case, a statement can be general without being the most abstract, see [54].

well are not usually considered.

As said, Cardano did not write a plain Latin. In general, the flowing of the sentences is burdensome. To this, we must add the fact that speaking about mathematics in full (or abbreviated) words turns out to be to a certain amount and in the vast majority of the cases equivocal. In the main, it is much easier to understand Cardano's arguments when at least a numerical example goes with them. Moreover, he usually created his own mathematical terms, when needed. For instance, Cardano used the following original terminology concerning the relations between the solids that compose the cube constructed on a divided segment. Let us consider a segment x divided into two parts such that $y+z=x$. Then Cardano called y^2z and yz^2 'mutual [*mutua*]' parallelepipeds. Moreover, a parallelepiped yz^2 is "opposite [*adversum*, *alternum*, *altrinsecum*, *remotum*]" to the cube y^3, and a parallelepiped y^2z is "connected [*coherens*, *proximum*]" to the cube y^3. Furthermore, Cardano talked about the 'solid root [*radix solida*]',[7] the 'reduplicated ratio [*proportio reduplicata*]',[8] 'wild quantities [*quantites sylvestres*]',[9] or about the 'tetragonical side'.[10]

[7]Cardano never defined what a "solid root" is. Anyway, from the examples in *Aliza*, Chapter L, we can infer that the "solid root" in the ratio of f to g is $f^3+2f^2g+fg^2$, see [9, Chapter L, Paragraphs 2-3, page 93].

[8]Cardano said that "I call reduplicate ratio when the ratio of the parts is the square [of the ratio] of the remainders" or "*[v]oco proportionem reduplicatam cum fuerit proportio partium ut residuorum duplicata*", see [9, Chapter XXXII, pages 62-63].

[9]Cardano said that "a wild quantity, namely [a quantity] that is not in any kind of roots, not even composed by those, nor left by subtraction, as the quantity the square root of which multiplied by the remainder from 12 produces 2" or "*in quãtitate sylvestri, scilicet quæ non sit in aliquo genere radicum, nec composita ex illis, nec per detractionem relicta, velut quantitas cuius R ducta in residuum ad 12 producat 2*", see [9, Chapter X, page 20].

[10]Again, Cardano never explicitly defined what the "tetragonical side" is. Anyway, from the examples in *Aliza*, Chapter IV, we can infer that the "tetragonical side" of a number is its fourth root, see [9, Chapter IV, page 8].

5.1.1. Overview of the *De Regula Aliza*'s contents

Trying to classify the *Aliza*'s chapters according to their contents is an extremely thorny issue. First of all – as said in Section 2.3 – these chapters have probably been conceived over a period of about fifteen years, and it does not seem that Cardano significantly proofread and uniformed them before the publication, in order to give some sort of global coherence to his treatise. There are chapters that are fully understandable, others that can be reconstructed to a greater or a lesser extent, and others that seem to be completely unintelligible. What immediately leaps out is that cubic equations are pervasive in the *Aliza*. We do not only have chapters that deal with $x^3 = a_1x + a_0$ (even if the number of pages devoted to this equation is strikingly huge compared to the others), but also the equations $x^3 + a_1x = a_0$ and $x^3 + a_0 = a_1x$ are taken into account as a proper subject of inquiry, and less frequently the equations $x^3 = a_2x^2 + a_0$ and $x^3 + a_0 = a_2x^2$. This does not mean, however, that we have any hope to organise the *Aliza* around these equations as it was the case for the *Ars Magna Arithmeticæ* or the *Ars Magna*, since the concerned chapters are scattered all along the book.

In this section, we build a first – though rough – picture of the *Aliza*, which is going to be developed more in the sections to come. I follow a two-folded approach: I firstly list the topics that are developed in each chapter, and then provide a classification of (some of) the chapters according to the general themes, or patterns, that can be identified. Recall anyway that this is only a sketch of the contents of a treatise that is much more complex.

Chapter I. On the supposed [things] and on the ways Foreword; how many parts should be in the solution(s) of $x^3 = a_1x + a_0$; cube of a binomial $(y+z)^3 = y^3 + 3y^2z + 3yz^2 + z^3$; the seven easier splittings (A I.1)-(A I.7) and the composed, anomalous, and useless splittings.

Chapter II. On the special rules on the cube equal to some things and a number of Chapter XXV of the *Ars Magna* How the propositions (AM XXV.1)-(AM XXV.8) for $x^3 = a_1x + a_0$ can be derived (or not) from the splittings.

Chapter III. On the way to discover the quantities that help to the chapters by the product of one part by the other and by the square of the difference of the parts Which shape can have the solution(s) and some auxiliary quantities for $x^3 + a_0 = a_2x^2$; how to find a_0 in $x^3 + a_0 = a_2x^2$ depending on the solutions and some auxiliary quantities.

Chapter IV. On the way to reduce all the quantities that are called 'first sides' by the tenth [Book] of Euclid in short Explanation of the Euclidean terminology for irrational numbers (*Elements*, Book X); how to perform calculations on irrational numbers and some properties.

Chapter V. On the consideration of the *binomia* and *recisa* that contain a rational figure, whence on the value of the chapters Which *binomia* and *recisa* cannot be solution of a non-complete cubic equation with rational coefficients; which shape can have the solution(s) and some auxiliary quantities for $x^3 + a_0 = a_1x$; how to find another solution of $x^3 + a_0 = a_1x$ given a first one (see also (AM XIII)).

Chapter VI. On the operations plus and minus according to the common usage Usual sign rules.

Chapter VII. On the examination of the values assumed by the second and third rules of the first chapter Algebra (that is, collection of calculation rules) of the splitting (A I.2) (Cossali's guess: how to pass from one splitting to another).

Chapter VIII. On the nature of the sides of the parallelepipeds (In)commensurability relationships of yz^2 depending on the (in)commensurability of y, z.

Chapter IX. In what way two parallelepipeds are constructed from any line [which is] not bigger than the fourth part of the cube of the given line Relationship between a surface and a height of a parallelepiped (unclear).

Chapter X. In what way the parts with the proposed line are appropriate to the parallelepiped Which shape can have (or cannot have) the solution of $x^3 = a_1x + a_0$ (using the splittings (A I.2)-(A I.4)).

Chapter XI. How many and what parts of the cube, and on the necessity of those, and what incommensurable [parts are] How many parts should be in the solution of $x^3 = a_1x + a_0$ (using terminology from *Elements*, Book X).

Chapter XII. On the way to geometrically prove the value of a cube and a number equal to some squares Proof for the existence of a solution of $x^3 + a_0 = a_2x^2$ through the intersection of a parabola and hyperbola.

Chapter XIII. On the discovery of the parts of the cubic *trinomium*, which produces the cube with two only cubic parts Which *trinomia* can (or cannot) be a solution of the family of equations $x^3 = a_1x + a_0$.

Chapter XIV. On the discovery of the kind of the value How to find the coefficients of an equation of the families $x^3 + a_0 = a_1x$ and $x^3 = a_1x + a_0$ given the shape of one of its solutions.

Chapter XV. On the discovery of the parts of the thing through the parts of the cube Calculations and properties of the terms of the expansion of the cube of a binomial; how to find y, z given some of the terms $y^3, 3y^2z, 3yz^2, z^3$ of the expansion (unclear).

Chapter XVI. That a *quadrinomium* from cubic roots is reduced to three parts, two of which are only cubic roots, or by far many Which *quadrinomia* can (or not) be a solution of the family of equations $x^3 = a_1x + a_0$.

Chapter XVII. In how many ways the number can be produced by the non-number How a rational number can be obtained by performing operations on irrationals.

Chapter XVIII. That the last division of the cube does not satisfy the given chapter Again on (A I.7) (unclear); on dividing x^3 respectively "according to the square root" (which means according to the following decomposition $x^3 = (y_1+y_2)^2+(z_1+z_2)^2$), "according to the cubic root" (maybe $x^3 = (y_1+y_2)^3+(z_1+z_2)^3$), and "according to the solid root" (namely $x^3 = (y+z)^3 = (y^3+2y^2z+yz^2)+(z^3+2z^2y+zy^2)$ in order to analyse the internal structure of the parts that compose x^3 and possibly to count out some of the decompositions (unclear, see also Chapter L).

Chapter XIX. That where the value [of the solution] satisfy, in no matter which way the cube is divided, it satisfies; otherwise not On the parts in the solution of the family of equations $x^3 = a_1x + a_0$ (unclear: very short, with few examples).

Chapter XX. Given a line, in what way it is divided into two parts so that the ratio of one to the product of the whole by the other is given in four ways [11] To find y, z, given (the value of $x = y + z$

[11]The title of this chapter is "*Data linea quomodo quadrifariam dividatur in duas partes ut sit proportio unius ad productum totius in alteram data*".

and) respectively the values of $\frac{2y+z}{z}$, $(y+z)\frac{z}{y^2}$, $(y+z)\frac{z}{y}$, and $\frac{y^3}{(y+z)z^2}$ or $\frac{y^3}{(y+z)^2 z}$ (not very clear).

Chapter XXI. Proof that shows the necessity of the solution Interpretation in a geometrical environment of the last proposition[12] of Chapter XX in order to argue that y, z cannot be found (not very clear).

Chapter XXII. On the contemplation of plus and minus, and that minus times minus makes minus, and on the causes of these according to the truth Alternative rule to multiply to negative numbers: minus times minus makes minus.

Chapter XXIII. On the examination of the chapter of the cube and a number equal to some things Order relationships between x, a_1, a_2 in each of the family of equations $x^3+a_0 = a_2x^2$, $x^3+a_0 = a_1x$, and $x^3 = a_1x + a_0$; proof in a geometrical interpretation of the case $x^3 + a_0 = a_1x$ (the aim of the proof is not very clear).

Chapter XXIV. Demonstration that shows that no chapter except the discovered [ones] can be generally known No equation of degree higher than four can be reduced to another of a lesser degree; on the shape of the solutions of the family of equations $x^3 = a_1x + a_0$.

Chapter XXV. On the examination of the third rule of Chapter XXV of the *Ars Magna* Comparison between the propositions (AM XXV.3) and (AM XII).

Chapter XXVI. On the ratio of the cube equal to some squares and a number to the cube with a number equal to some squares Given $x^3 = a_2x^2 + a_0$ and $y^3 + a_0 = a_2y^2$ one has that $\frac{y-a_2}{a_2-x} = \frac{x^2}{y^2}$; its proof in a geometrical interpretation and some examples.

[12] Which is: to find y, z given (the value of $x = y + z$ and) the value of $\frac{y^3}{(y+z)z^2}$.

Chapter XXVII. On the given value so that the number of the equality is found In non-complete cubic equations, given the value of x and one value among a_0 and either a_1 or a_2, how to find the remaining coefficient.

Chapter XXVIII. That in the purpose of Chapter XXVI one comes to the cube and some things equal to a number Special *versus* general rules; link between the family of equations $x^3 + a_0 = a_1 x$ and $y^3 + b_0 = b_2 y^2$; (title and aim not very clear).

Chapter XXIX. On the comparison of the chapters of the cube and some things equal to a number and of the cube and a number equal to as many things (Unclear, too many typos in the lettering of the diagram).

Chapter XXX. What [is] the equality of the cubes of the parts of a divided line How to factor the sum of power $y^3 + z^3$ and its proof in a geometrical environment.

Chapter XXXI. On the general value of the cube equal to some things and a number, [which is] called 'solid', and on its operations Another definition of 'solid roots' as a stenography in accordance to (AM XXV.3) (see also Chapters LIII, LVII); operations on it (unclear).

Chapter XXXII. On the comparison of two quantities according to the ratio of the parts First mention of the "reduplicated ratio";[13] unclear definition;[14] use of the "major" and "n-th reduplicated ratio" to compare two quantities (unclear).

---

[13]This will be the main topic for the next seven chapters, so that Chapters XXXII-XXXIX form a coherent block.

[14]Anyway, later on, especially from *Aliza*, Chapter XXXV, we understand that this is the usual duplicated ratio and it always appears in proportions of the kind $\frac{a}{b} = \frac{c^2}{d^2}$.

Chapter XXXIII. On the double order of four homologous parts of the same ratio to two others A proposition on proportions.

Chapter XXXIV. On the triple division of two quantities in a mutual reduplicated [ratio] Three propositions on proportions (two of them contain a reduplicated ratio); consequences on the structure of the *binomia* and *recisa* (namely, on the (in)commensurability of their parts).

Chapter XXXV. On the six mutual reduplicated ratios that are originated from the addition of one quantity to another and on the two useless [ratios] List of all the six possible proportions that contain a reduplicated ratio and whose terms are $n \in \mathbb{N}, x, x+n$, and again x.

Chapter XXXVI. On dividing two lines equal according to a given mutual reduplicated ratio (Diagram missing); on a property of the duplicated ratio (not very clear).

Chapter XXXVII. On the six comparisons of four quantities of a reduplicated ratio Six propositions of the kind: given the proportion $\frac{d}{b} = \frac{a^2}{c^2}$, some quantities among a, b, c, d and some of their sums, to find the remaining quantities.

Chapter XXXVIII. On the blurry comparison of the mutual quantities in a reduplicated ratio On a property of the reduplicated ratio (unclear).

Chapter XXXIX. On dividing two given lines according to the mutual reduplicated ratio in agreement with the given parts Link between $x^3 + a_0 = a_2x^2$ and $y^3 = a_2y^2 + a_0$ through the proportion with a reduplicated ratio $\frac{a_2-x}{y-a_2} = \frac{y^2}{x^2}$; (the remaining, very long part seems to be composed by many unclear propositions, especially concerning the diagram to which they refer).

Chapter XL. On three necessary [things] that is necessary to put before the discovery On the shape of the solutions of $x^3 + a_0 = a_1 x$ (very long, completely unclear); also some quadratic equations appear; on the family of equations $x^3 = a_1 x + a_0$ in relation to a certain "orthogonal triangle" (unclear).

Chapter XLI. On the very difficult problem that is very easily seen Statement of a problem that leads to a quartic equation.

Chapter XLII. On the double equality to compare in the chapter of the cube and a number equal to some things Five propositions on the family of equations $x^3 + a_0 = a_1 x$ (unclear).

Chapter XLIII. On the comparison of the number of the value to the parts of the number of the things (Completely unclear: geometrical proof disconnected from its interpretation in terms of equations).

Chapter XLIV. In what way a given line is divided according to the ratio that has a medium and two extrema in the solids From a problem on dividing a line to a cubic equation.

Chapter XLV. In what way the parts of a divided line are reciprocally compared to the solids and to the squares Rule to find the common solution to $x^3 + a_0 = a_1 x$ and $y^3 = b_1 y + b_0$ whenever it exists (the rule is a kind of double splitting that also involves (A I.3)).

Chapter XLVI. In what way the proposed rectangle and the cubes of its sides we have the whole cube Rule similar to (A I.1); method of the "separation of the mutual [parallelepipeds]"[15] when $x^3 = a_1 x +$

[15]This is a variation on (A I.3), where the system becomes

$$\begin{cases} a_0 = 2(y+z)yz \\ a_1 - \frac{a_0}{2} = y^3 + z^3 \end{cases}.$$

a_0 falls into the *casus irreducibilis*; corollary on how to derive a solution of a particular sixth degree equation.

Chapter XLVII. That a divided surface or solid has the sides bigger than the side of the whole Order relation between $y+z$ and respectively $\sqrt{y(y+z)}+\sqrt{z(y+z)}$, $\sqrt[3]{y(y+z)}+\sqrt[3]{z(y+z)}$ when $y, z > 0$; link with $x^3 = a_1x + a_0$ (unclear).

Chapter XLVIII. On the quantities of the squares and on the known mutual solids To (A I.1) and (A I.3) correspond respectively two different divisions of x in $y+z$ and $y'+z'$; hint to some other never mentioned splittings; recall (AM XIII); similarities between the rule in Chapter XLV and the splitting (A I.3) (not very clear).

Chapter XLIX. On certain equalities and on the ways out of the order How to deal with complete equations of the family $x^3 + a_1x = a_2x^2 + a_0$.

Chapter L. On the solid roots and on their handling Definition of 'solid root in the proportion of f to g' and ten propositions with some of its properties, applications, and examples of calculations.

Chapter LI. A certain special rule and likewise the way of the subtle handling How to calculate $(a\sqrt{b}+\sqrt{ab})^2$ (one can also derive a sixth degree equation from there); the rule mentioned in the title could be A I.1 (this chapter seems rather an exhibition of a tricky calculation).

Chapter LII. On the way of all the operations given some quantities in the middle way Examples of calculation with fractions that have an irrational number in the denominator.

Chapter LIII. On the accurate consideration of [the things] said above in Chapter VII Comparing the splittings (algebra of the splittings).

Chapter LIV. On the perpetual addition of some quantities Analysis in a geometrical interpretation of the equality $b(a+2b)-(a+b)^2 = b^2 - a(a+b)$.

Chapter LV. Very general question by which we came from three universal conditions to one special quantity, and it is admirable Four questions that involve continued ratios where the unknown appears and the properties of continued ratios (all of them not very clear), the first of which leads to a quadratic equation.

Chapter LVI. On two wonderful, but not pertinent questions Given $a, b, c \in \mathbb{Q}$, to find $a_1, a_2, b_1, b_2, c_1, c_2$ such that $a = a_1+a_2$, $b = b_1+b_2$, $c = c_1+c_2$, $\frac{a_1}{b_1} = \frac{b_1}{c_1}$, and $\frac{a_2}{b_2} = \frac{b_2}{c_2}$; on the equality $(a-b)(a+b) = a^2-b^2$ and its applications.

Chapter LVII. On the treatment of the general value of the cube equal to some things and a number Rewriting the equation $x^3 = a_1x + a_0$ as $x = \sqrt{a_1 + \frac{a_0}{x}}$ and new stenography to speak about the shape of a solution.

Chapter LVIII. On the common quantity from two incommensurable [quantities], in how many ways it is said Possible shape for the solutions of the family of equations $x^3 = a_1x + a_0$.

Chapter LIX. On the order and on the examples in the second and fifth *binomia* Methods to solve particular equations of the family $x^3 = a_1x + a_0$.

Chapter LX. General demonstration of the chapter of the cube equal to some things and a number Geometrical interpretation of the family of equations $x^3 = a_1x + a_0$; method similar to the splittings.

Generally speaking, we now manage to identify some recurring patterns, which are resumed in Table 5.1. Note that these patterns can of course overlap – or better, they are meant to do so. Among these, the pattern on the splittings (for $x^3 = a_1x + a_0$) is by far the most significant, as it will appear in the next sections. Taking into account the equation $x^3 = a_1x + a_0$ and performing the substitution $x = y + z$, Cardano gets the equality $y^3 + 3y^2z + 3yz^2 + z^3 = a_1(y + z) + a_0$. His strategy is then to arrange the left side $y^3 + 3y^2z + 3y\,z^2 + z^3$ in the sum of two parts, both of which are going to be matched either to $a_1(y + z)$ or to a_0 in the right side, and then (if possible) directly to one of the given coefficients a_1, a_0. I call 'splitting'[16] all these combinations, since the left side of the equality is split in two parts. This thread is peculiar to the *Aliza* and represents the original feature of this treatise. Indeed, even though in quantitative terms the number of chapters devoted to it is not so impressive, it is the true theoretical core around which some of the other threads can be (directly or indirectly) organised.

The pattern on the possible shapes of the solutions refers to the shapes in which an irrational solution of a cubic equation can be written, assuming that the equation has rational coefficients. In the main, Cardano was interested in how numbers can be expressed rather than in the numbers themselves. This is what I call a 'shape', that is, an expression written in a certain stenography, in this case in Cardano's one. We have already come across this topic in the *Ars Magna Arithmeticæ*, and indeed in the *Aliza* Cardano uses some techniques that remind of the former treatise. Nevertheless, the focus in the *Aliza* on analysing the possible shapes of a solution (when it is a matter of $x^3 = a_1x + a_0$) can also be explained by the fact that the splittings themselves originate from a reflection on how many

[16]From a word-by-word translation of Pietro Cossali's '*spezzamenti*', see [57] and [60].

Table 5.1. – Recurring patterns in the *Aliza*.

Pattern	Chapters
Splittings (for $x^3 = a_1x + a_0$)	I, II, VII, X, XVIII, XLV, XLVI, XLVIII, LIII, LX
Possible shapes of the solutions	I, III, X, XI, XIII, XVI, XIX, XXIV, XL, LVII, LVIII
The equation $x^3 = a_1x + a_0$	I, II, X, XI, XIII, XIV, XVI, XXIV, XIX, XXIII, XXIV, XXV, XL, XLV, XLVI, XLVII, LVII, LVIII, LIX, LX
The equation $x^3 + a_0 = a_2x^2$	III, XII, XXIII, XXVI, XXVIII, XXXIX
The equation $x^3 + a_0 = a_1x$	V, XIV, XXIII, XXVIII, XL, XLII, XLV
General properties of cubic equations	V, XXIV, XXVII, XLIV
Sign rules	VI, XXII
(Ir)rationality and (in)commensurability	IV, V, VIII, XI, XVII, XXXIV, LII
Calculation rules	I, IV, VII, IX, XV, XVII, XVIII, XX, XXX, XXXI, XXXII, XXXIII, XXXIV, XXXV, XXXVI, XXXVII, XXXVIII, XXXIX, XLVII, L, LI, LII, LIV, LV, LVII

parts compose a solution (since there the substitution $x = y + z$ is of the utmost importance). Therefore, it appears how, when Cardano addressed the problem entailed by the *casus irreducibilis*, he turned to these algebraic shapes rather than to the numerical values of the solutions, or to the structural relationships between them (as Lagrange, Ruffini, Abel, and Galois will subsequently do). In the end, he was trying to provide a general theory of the shapes under which certain numbers can be written. Indeed, a recurring strategy in Cardano's mathematical treatises is to establish whether these shapes can convey a solution of an equation or not.

The equation $x^3 = a_1x + a_0$ is the most represented one in the *Aliza*, from the very beginning until the last chapter of the treatise. Which is not astonishing: since in the *Ars Magna* Cardano had already shown that most of the solving methods for the other cubic equations come down to solve this equation and since this family of equations is the first one to fall into the *casus irreducibilis*, it is quite natural to firstly address the problem in this context. Then, this pattern is obviously linked to all the chapters that are about the splittings and also about the possible shapes of the irrational solutions for the equation at issue.

In terms of number of chapters, the following two patterns on the equations $x^3 + a_0 = a_2x^2$ and $x^3 + a_0 = a_1x$ are sufficiently well-represented, even though they seem to be quite disconnected from the preceding patterns.

Then comes the pattern on general properties of cubic equations. Only a few chapters of the *Aliza* handle cubic equations from the viewpoint of the properties that all the families share. This does not mean that Cardano's justifications hold independently of the considered family of equations, but rather that he proves case by case a statement that is common to a certain number of cases.

In the *Aliza*'s framework, the pattern on sign rules is a minor topic. Nevertheless, the alternative rule to multiply two negative numbers had drawn at best the attention of some scholars (like Commandino and Harriot) during the centuries. With regard to the declared aim of the *Aliza*, that is, to deal with and eventually overcome the problem entailed by the *casus irreducibilis*, this rule (if ever it could exist) would have solved the question at the bases. Indeed, if the product of two negative numbers had been another negative number, the square root of a negative number would have been a negative number.

The study of particular kinds of irrational numbers, according to the terminology derived from Book X of the *Elements*, is one of the basis of Cardano's analysis of cubic equations with rational coefficients. The pattern on (ir)rationality and (in)commensurability is therefore linked with the study of the possible shapes of an irrational solution. Moreover, it is connected to the splittings, since the substitution $x = y + z$ leads to systems in which some particular kinds of irrational numbers appear.

Finally, there is the pattern on the calculation rules. This is meant to cover a wide variety of subjects, such as the expansion of the cube of a binomial, the proportions, the reduplicated ratios, and the solid roots. Here and there, a good number of pages of the *Aliza* is devoted to a variety of calculation rules, recalling in this way the feature of the older *abaco* treatises.

As said, some of these recurring patterns lead to identify some common threads of the *Aliza*. These constitute the core of my analysis of the book. More precisely, I proceed with the following plan. In Section 5.2, I present the main thread that can be identified in the *Aliza*: the splittings for the equation $x^3 = a_1x + a_0$. Cardano deals with them in the very first chapter, which is a real cornerstone for us. My hypothesis is that, through the splittings, Cardano was searching for some different cubic formulae that do not fall into the *casus irreducibilis*. Indeed, it seems that he hoped that there exists a correspondence between the *casus irreducibilis* and the structure of the concerned equation, and thus he would like to discover a handling that leads to cubic formulae that only involve real numbers. But he did not managed to – and, as we know, he could not.[17] Still, Cardano wanted to publish the book and this could have also partially given rise to the obscurity of the *Aliza*.

---

[17]See Footnote 26 to Chapter 2 from page 15.

I thus organise a considerable number of chapters around these splittings and under different viewpoints. Basically, I try to unravel where this idea comes from and how one could employ the splittings in an unconventional, alleged way. In Section 5.3.1, I retrieve the origins of the substitution $x = y + z$ (which is fundamental in the splittings and in deriving the solving method for $x^3 = a_1x + a_0$) as a side-effect of the study of the shape of some irrational numbers that are solutions of cubic equations with rational coefficients. It turns out, indeed, that there are only two possible irrational shapes, and both of them are binomials. In Section 5.3.2, I make a second hypothesis on the possible origin of the splittings. This time I highlight the correlation with the *Ars Magna*, and, in particular, with some rules from Chapter XXV that are intended to rephrase an equation in such a way that it is easier to guess one of its solutions. We recall that the link between the *Aliza* and *Ars Magna*, Chapter XXV had already been suggested in *Ars Magna*, Chapter XII. In Section 5.3.3, I present another hypothesis on the origin of the splittings, based on my interpretation of the last chapter of the *Aliza*. It appears that Cardano's researches could have been inspired by a certain kind of loosely interpreted geometry. In Section 5.3.4, I deal with a last topic concerning the splittings, this time from a different viewpoint. Indeed, the splittings can also be considered as a subject of inquiry by themselves. Cardano establishes some calculation rules on the splittings. These, in turn, could possibly also be useful in solving equations that fall into the *casus irreducibilis*, if one is willing to relax his requirements on the cubic formula.

Then, in Section 5.4, I complete the picture by further providing a few technical results (concerning irrational numbers and the echoes of (AM VIII.2)) that are exploited in the previously analysed chapters.

In Section 5.5, I deal with a single chapter of the *Aliza*, where a different kind of geometry is exceptionally employed to deal with a problem that in the end involves a cubic equation. A comparison with Eutocius' and al-Khayyām's solutions to the very same problem is useful in order to identify Cardano's position in the negative.

Finally, in Section 5.6, I present another of the *Aliza*'s threads, which at that period is only a rough outline. As we will see, it could consist in preserving the cubic formula (together with the square roots of negative numbers that appear in the *casus irreducibilis*) and trying to develop an arithmetic of these nasty numbers. In this case, one needs first of all to deal with the sign of those numbers. Cardano made a (failed) attempt in one controversial chapter of the *Aliza*.

As said, such analysis cannot obviously account for the *Aliza* as a whole. It only concerns twenty-four chapters out of sixty (which are Chapters I, II, III, IV, V, VI, VII, X, XI, XII, XIII, XIV, XVI, XVII, XXII, XXIV, XXV, XXX, XL, LIII, LVII, LVIII, LIX, and LX), that is a little more than one third of the whole treatise. It happens that most of these chapters are placed at the two extremities of the book. Given the fact that the *Aliza* is a messy miscellany rather than a uniform work, I doubt that it will be possible to finally settle each and every single chapter in the same coherent logical structure. The chapters that I do not consider in the following can be divided into two families. Those of one of them are negligible. Those of the other obscure. The chapters of the first family are those that explain some technical points, either directly concerning equations or not. More precisely, they are Chapters VIII, IX, XV, XXVII, and XLI. The second family gathers the thirty-one remaining chapters. In turn, we can further split this family into the chapters that are completely obscure (like Chapters XVIII, XIX, XXI, XXVI, XXIX, XXXI, XXXII, XXXIV, XXXVI, XXXVII, XXXVIII, XXXIX, XLII, LIV, LVI, and the first part of Chapter XL) and the chapters in which we understand the particular calculations, but not their general aim (like Chapters XX, XXIII, XXVIII, XXXIII, XXXV, XLIII, XLIV, XLV, XLVI, XLVII, XLVIII, XLIX, L, LI, LII, and LV). Unfortunately, among the chapters that are completely obscure, we count for instance Chapter XXXI "On the general value of the cube equal to some things and a number, which is called solid, and on its operations [*De æstimatione generali cubi æqualis rebus et numero solida vocata, et operationibus eius*]". As the title itself shows, it sounds very interesting, but I could not work it out.

5.1.2. Again on dating the *Aliza*'s miscellany

Finally, I would like to shortly come back to the issue of the *Aliza*'s date. As it appears from Section 2.3, it is almost impossible to precisely determine when the book had been composed. At best, and considering it as a whole, I can provide the overall estimate that it was around 1550. But we should not forget that this book is a miscellany. Here one may want to try to establish an inner relative chronology of its chapters, starting from the hints that can be found in the text. Again, this is extremely to be achieved, but a few points on this topic deserve to be clarified.

The *Aliza* contains many references to the *Ars Magna* and *De Proportionibus*, which seems very reasonable since the three books were published together. Unfortunately, these references are completely ineffective in establishing a relative chronology of the *Aliza*'s chapters. Indeed, even though the *Aliza* had probably not been carefully proofread, their labels could have been in any case added at a second stage in the passages where Cardano employs the corresponding results. In a few cases, I believe nonetheless that it is possible to give an estimate. First, the opening paragraphs of Chapter I convey the strong feeling that they could have been a late (maybe the last) addition. Indeed, they really seem to have the role of a sort of short introduction to (a considerable part of) the book, as we will see in Section 5.2.

Second, there is a handful of chapters, namely Chapters V, X, XIII, XVI, and especially Chapter XI, where Cardano studied the shapes for the irrational solutions of $x^3 = a_1x + a_0$, with a_1, a_0 rational. As we will see in Section 5.3.1, the fact that he counted out some shapes that actually are possible shapes makes us suppose that he did not still have the cubic formula available. Third, as we will see on page 312, it seems that in Chapter XXIV Cardano did not have available Ferrari's method for quartic equations. Hence this chapter has probably been composed before 1540 (see here on page 401). The same also holds for Chapter XLI, which I will not mention in the following.

5.2. The method of the splittings in *Aliza*, Chapter I

As we have said, the equation $x^3 = a_1x + a_0$ is a central subject matter in the *Aliza*. This equation appears from the very beginning of the book, where its **Chapter I** "On the supposed [things] and on the ways [*De suppositis ac modis*]" is entirely devoted to it. This chapter, and especially its start, stands out, since it is the only place where we can find a few general, introductory words in a programmatic spirit – a sort of foreword. Moreover, this chapter gives rise to a very common pattern, which will be recurring in the rest of the book, so that I can organise a sizeable number of other chapters around what we call the 'splittings' in Chapter I. Note that all these chapters, which share the same core, might be the remainders of the original "*aliza* problem", of which Cardano spoke in the *Ars Magna*, Chapter XII (see above, on page 19).

It is therefore fully worthwhile to have an closer look at the very first paragraph of Chapter I:

> [s]ince I [Cardano] have already demonstrated in the *Ars magna* that all the chapters are transformed, provided that the two principal [chapters] had been discovered [to be] general and not by transformation, it is plain that, having discovered another general chapter in addition to the chapter of the cube and some things equal to a number and [the chapter] of the cube equal to some squares and a number, which is deduced by transformation from the preceding [one], even if [the other general chapter] was general, all the chapters either of three or of four terms would not only be known in general, but also demonstrated, provided that this very same [chapter] is discovered by a demonstration.[18]

[18]"*Cum iam in Arte magna demonstraverimus omnia capitula converti, modo duo principalia nec iam ex conversione inventa generalia fuerint, manifestum est, invento alio capitulo generali, præter capitula cubi et rerum æqualium numero et*

As we have said, this is the only place in the whole *Aliza* where a sort of declaration of intent is made. Here, Cardano explains an overall strategy to deal with the *casus irreducibilis.* As a matter of fact, however, it is not shining as bright as one may hope for. What did Cardano want to convey here? First he tells us a story that we know very well, also kindly giving the reference. As we have seen in Section 3.5, the *Ars Magna* shows that the solving methods of all the cubic equations are linked one to the other, and that all of them stand, in particular, in a special connection either with the solving method of the equation $x^3 + a_1x = a_0$ or with the one of the equation $x^3 = a_1x + a_0$. More precisely, all the proofs (except those for $x^3 + a_1x = a_0$ and $x^3 = a_1x + a_0$) employ some kind of substitution to drive each equation back to the solving methods either of $x^3 + a_1x = a_0$ or of $x^3 = a_1x + a_0$. We have also remarked in Sections 3.2.1 and 3.2.2 that both the proofs for $x^3 + a_1x = a_0$ and for $x^3 = a_1x + a_0$ follow a very similar pattern, though never making reference one to the other. Then, it is not surprising that here Cardano maintains that all the "chapters" or the families of equations[19] were transformed (by substitution) into two principal ones (which are $x^3 + a_1x = a_0$ and $x^3 = a_1x + a_0$), which in turn were generally solved without substitutions. But we have also seen that, since the solving methods of all the equations are linked one to the other in such a way, if $x^3 = a_1x + a_0$ falls into the *casus irreducibilis*, then also all the equations the solving methods of which appeal to $x^3 = a_1x + a_0$ fall into it. In this way, the problem entailed by the *casus irreducibilis* spreads.

Cardano's approach is the following. He would like to find a certain family of equations such that it is at the same time "general" (but different from the family of equations of $x^3 + a_1x = a_0$ and $x^3 = a_2x^2 + a_0$) and solved by a method proved by a demonstration. Then, all the

cubi æqualis quadratis et numero, quod ex priore per conversionem deducitur, et si generale sit, omnia capita seu ex tribus seu ex quatuor nominibus generaliter non solum cognita esse, sed et demonstrata, modo hoc ipsum demonstratione inventum sit", see [9, Chapter I, page 1].

[19] See above, Footnote 3.

families of equations will also be solved. To get a full understanding of this plan, it is now essential to interpret the adjective "general". It is one of Cardano's favourite expressions – as we have already had the opportunity to verify (see for instance Sections 3.1.4, 4.2.4, and 4.2.6). We saw that Cardano related this term either to the shape of a solution of a cubic equation, or to the rule that gives a solution, or finally to the solving method of an equation. In the opening passage of the *Aliza* "general" is referred to a family of equations, and I am going to make the comparison with the meaning of "general" referred to methods (on page 84). Indeed, a family of equations is "general" (or can be generally solved) if its solving method is general, that is, if we can solve every given equation that belongs to that family by it.[20] In our particular case, this means that a family of equations is "general" if the concerned cubic formula (or the procedure from which the formula is derived) has no conditions. Under this viewpoint, the families of $x^3 + a_1x = a_0$ and of $x^3 = a_2x^2 + a_0$ are "general", since any arbitrary numerical equation belonging to those families can be solved through their cubic formulae (since $\Delta_3 \geq 0$ and one can always get a real solution without running into imaginary numbers). By the way, note that there exist also some substitutions (see here, Propositions (AM VII.2-3) on page 69, (AM VII.6-7) on page 71, and (AMA XXI.1) on page 192) that link $x^3 = a_2x^2 + a_0$ to $x^3 + a_1x = a_0$ and *vice versa*, so that Cardano could say that the solving method for one equation is deduced from the other, even though both equations are "general". But not all the solving methods of the other cubic equations can be reduced to one of those two without loss of generality. Indeed, there are some solving methods, more precisely those that resort to $x^3 = a_1x + a_0$, that require to previously know how to solve this equation. But the solving method that gives the cubic formula of

[20] Anyway, we have seen as well that Cardano's terminology is not always completely stable. Indeed, in the *Ars Magna*, the Chapter XII on $x^3 = a_1x + a_0$ is called "general", but it is actually subject to a condition – this is the nitty-gritty of the issue! – and, in the *Ars Magna Arithmeticæ*, the same happens for Chapters XXX and XXXII respectively on $x^3 = a_1x + a_0$ and $x^3 + a_2x^2 = a_0$ (see here, on pages 110 and 214).

the former equation stands independently from all the others, which means that it cannot be derived neither from that of $x^3 + a_1x = a_0$ nor from that of $x^3 = a_2x^2 + a_0$. These one-way connections can be grasped looking at the diagram on page 164. I claim that the "another chapter" that Cardano was referring to in the quoted passage is the one of $x^3 = a_1x + a_0$. Cardano's plan is then to find a family of equations such that it is "general", that is, to find a solving method for $x^3 = a_1x + a_0$ that does not have any condition. This means that he has to face the fact that sometimes square roots of negative numbers appear in the cubic formulae. To give consistency to my claim we need to address the rest of Chapter I.

It is indeed very reasonable to maintain that Cardano wanted to focus on $x^3 = a_1x + a_0$. On the one hand, the problem of the *casus irreducibilis* arises first in this equation so that, if one manages to impose no conditions on its formula, the problem will automatically be fixed also for the other equations. On the other hand, this equation is one of the most studied in the *Aliza*, and especially in Chapter I. In the remaining part of the opening passage, Cardano says that the solving methods for the cubic equations with three terms are easier to be discovered than the ones for the complete cubic equations. Moreover, among the solving methods for cubic equations with three terms, the method for $x^3 = a_1x + a_0$ is the easiest one, and that – as Cardano argues – is due to the following three reasons. First, this solving method is for the most part covered by the rule in (AM XI). Second, it is discovered by the very same reasoning. Third, certain other solving methods are obtained starting from it by transformation.[21] But then, why should the solving method for $x^3 = a_1x + a_0$ be

[21] "But among these [chapters] that consist of three terms the easiest one is the chapter of the cube equal to some things and a number, the greatest part of which is already obtained by the first rule [(AM XI)], and which stands according to the reasoning of the chapter already discovered [again (AM XI)], and also which we will show [to be] from that to the others not contrary to transformation" or "*[h]orum autem quæ tribus nominibus constant facillimum est capitulum cubi æqualis rebus et numero, cuius pars iam maxima ex prima regula habetur, et quod secundum rationem capituli iam inventi se habet; et etiam quod ex illo in alia non contra conversionem ostenderimus*", see [9,

easier than the one for $x^3 + a_1x = a_0$? Indeed, the above three reasons do not especially characterise $x^3 = a_1x + a_0$, since they also apply to $x^3 + a_1x = a_0$. Cardano was rather saying that the solving method for $x^3 = a_1x + a_0$ is the easiest one among those for the cubic equations with three terms that can have $\Delta_3 < 0$. Accordingly, only the equations that can have $\Delta_3 < 0$ can be reduced by transformation to the solving method for $x^3 = a_1x + a_0$.

In the second paragraph, Cardano shows what substitution he was going to use, namely $x = y + z$, and why he needed a substitution of this kind. I intend to deal with this topic in the next section, since there is a handful of chapters that can be grouped around it.

5.2.1. The splitting (A I.1)

Here, Cardano introduces what Pietro Cossali called '*spezzamenti*', and what we are going to call 'splittings' according to Cossali's semantic clue. Cardano briefly recalls the formula for the cube of a binomial, saying that he had already dealt with it (though he gives no precise reference; possibly he implicitly refers to (AM VI.6), see above on page 101). Even if he does not explicitly say so, it is clear that Cardano applies the substitution $x = y + z$ (where y, z are unknown to be further determined) to $x^3 = a_1x + a_0$, and uses the formula for the cube of a binomial to develop the calculations. In this way, he gets

$$y^3 + 3y^2z + 3y\,z^2 + z^3 = a_1(y + z) + a_0.$$

Afterwards, his strategy is to arrange the left side $y^3+3y^2z+3y\,z^2+z^3$ in the sum of two parts, both of which are going to be matched at first to $a_1(y + z)$ or to a_0 in the right side, and then (if possible) directly to one of the given coefficients a_1, a_0. Cardano states that this could be done in "seven easiest ways [*septem modis facilioribus*]" – what we call the 'splittings', since the left side of the equality is split in two parts.

Chapter I, page 1]. I interpret 'the first rule' as referring to the rule in Chapter XI of the *Ars Magna*.

A I.1. *Consider* $x^3 = a_1x + a_0$. *Write* $x = y + z$ *with* y, z *two [real, positive] numbers.*
If y, z *can be chosen such that the splitting according to which "the number is assigned to the cubes [...] and the things themselves are equal to the six parallelepipeds that remain [*numerus tribuatur cubis [...] et res ipsæ æquantur parallelipedis sex*]", that is*

$$\begin{cases} a_0 = y^3 + z^3 \\ a_1x = 3y^2z + 3yz^2 \end{cases}, \quad \textit{or again} \quad \begin{cases} a_0 = y^3 + z^3 \\ a_1 = 3yz \end{cases},$$

holds, then the condition

$$\begin{cases} a_0 \geq \frac{1}{4}x^3 \\ [a_1x \leq \frac{3}{4}x^3] \end{cases}$$

follows.

Let us call $AB = y$ and $BC = z$. The first line in the condition is stated as follows.

> But since the cubes of AB and BC can never be smaller than one-fourth of the whole cube AD, and indeed this does not happen except when AC will be divided into equal [parts] by B, therefore being the number smaller than one-fourth of the whole cube AD, it will not be able to be equal to the cubes of AB, BC.[22]

Cardano states that, if $x = y + z$, then $y^3 + z^3 \geq \frac{1}{4}x^3$, and justifies it by "the 9th [proposition] of the second [book] of Euclid and by some Dialectic rules [*[p]er 9 secundi Elementorum et regula Dialec.*]". Note that both the 1570 and 1663 editions have "*regula*". Since the preposition "*per*" takes the accusative case, we would have expected to find either '*regulam*' or '*regulas*', which is not the case. In these

[22] "*Sed quoniam cubi ab et bc nunque possunt esse minores quarta parte totius cubi ad, et hoc etiam non contingit nisi cum fuerit ac divisa per æqualia in b. Cum igitur numerus fuerit minor quadrante cubi totius ad, non poterit æquari cubis ab, bc*", see [9, Chapter I, page 2].

uncertain cases I prefer to leave the interpretation as loose as possible and I have then chosen to translate the term in plural.[23] This mention of the "Dialectic rules" seems to be a blurry reference to some classical logical rules. Indeed, Cardano also wrote a *Dialectica*, printed in 1566 in Basel (in the second volume of [8]), which indeed contains some logical rules (and by the way also some mathematical examples to expound them). Anyway, since *Elements* II.9 implies[24] that $y^2 + z^2 \geq \frac{1}{2}x^2$, we are searching for one or more rules that enable to pass from the inequality with the square to the inequality with the cube, but I could retrieve none in the *Dialectica*. It is nevertheless not hard to derive[25] the searched inequality from *Elements* II.9. Note that, in

[23]Note that Pietro Cossali interpreted instead the term '*regula*' to be a singular word. He affirms then that the "Dialectic rule" reminds of adding in an equality the same quantity on both sides of the equal sign, leaving thus the equality unchanged, but no real evidence supports his view. Cossali argued as follows: "*[q]uanto alla Regola Dialettica, si tratta della nota proprietà delle equazioni secondo la quale è possibile aggiungere o sottrarre ad ambo i membri dell'equazione una stessa espressione, proprietà che consente di spostare un termine da un membro all'altro dell'equazione*", see [60, Chapter I, Paragraph 3, footnote 20, page 30]).

[24]According to Campano, "*[s]i linea in duo æqualia duoque inæqualia dividitur: quæ fiunt ex ductu utriusque inæe sectionum in seipsam pariter accepta: duplum sunt utrisque pariter acceptis quæ quidem ex dimidia æque quæ utrique sectioni interiacet quadratis describuntur*" and, according to Zamberti, "*[s]i recta linea secetur in æqualia et non æqualia: quæ ab inæqualibus totius segmentis fiunt quadrata dupla sunt eius quod a dimidia et eius quod a medio sectionum fit quadratorum*", see [32, pages 27r-27v]. In our case, this means that $y^2 + z^2 = 2\left(\left(\frac{y+z}{2}\right)^2 + \left(\frac{y+z}{2} - z\right)^2\right)$.

[25]Let us assume that $y^2 + z^2 \geq \frac{1}{2}(y+z)^2$. We write

$$\begin{aligned}\frac{1}{4}(y+z)^3 = \frac{1}{2}(y+z)^2\,\frac{1}{2}(y+z) &\leq (y^2+z^2)\,\frac{1}{2}(y+z)\\ &= \frac{1}{2}(y^3 + y^2z + yz^2 + z^3)\\ &\leq y^3 + z^3.\end{aligned}$$

Cossali also gave two proofs of this, the second of which is an anachronistic attempt to reconstruct Cardano's reasoning (see [60, Chapter I, Paragraph 3, pages 28-31] and [57, Chapter VIII, Paragraph 5, page 439]).

order to verify this inequality, we must assume both y and z (real) positive.[26]

Then, once obtained $y^3 + z^3 \geq \frac{1}{4}(y+z)^3$, or $a_0 \geq \frac{1}{4}x^3$ because of the splitting (A I.1), it is immediate to get by the formula for the cube of a binomial that $3y^2z + 3yz^2 \leq \frac{3}{4}(y+z)^3$, or $a_1 x \leq \frac{3}{4}x^3$.

Cardano says that, since this splitting has a condition, the family of equations is not "general" – and this agrees with the above interpretation of "general". Moreover (but not surprisingly), it turns out that this condition implies a condition on the sign of the discriminant. Indeed, the system

$$\begin{cases} a_0 \geq \frac{1}{4}x^3 \\ a_1 x \leq \frac{3}{4}x^3 \end{cases}$$

entails that

$$\begin{cases} a_0 \geq \frac{1}{4}x^3 \\ a_1 \leq \frac{3}{4}x^2 \end{cases}$$

and in turn that

$$\begin{cases} \frac{1}{4}a_0^2 \geq \frac{1}{4}\left(\frac{1}{4}x^3\right)^2 \\ a_1^3 \leq \left(\frac{3}{4}x^2\right)^3 \end{cases},$$

that is $\frac{1}{27}a_1^3 \leq \frac{1}{4^3}x^6 \leq \frac{1}{4}a_0^2$. Then, $\frac{1}{4}a_0^2 - \frac{1}{27}a_1^3 \geq 0$, or $\Delta_3 \geq 0$.

We finally observe that, if one knows the splitting (A I.1) for $x^3 = a_1 x + a_0$, they should also be able to derive its cubic formula,

---

In modern terms, the searched inequality is a generalised mean inequality. We take an integer $n \geq 0$, and some real $p > 0$ and $x_1, \ldots, x_n \geq 0$. Consider the following generalisation of the arithmetic, geometric, and harmonic means

$$M_p(x_1, \ldots, x_n) = \left(\frac{1}{n}\sum_{i=1}^{n} x_i^p\right)^{\frac{1}{p}},$$

called 'generalised mean' or 'Hölder mean'. Then, for $p > q$, we have the generalised mean inequality $M_p \leq M_q$. Taking $n = 2$, $p = 1$, $q = 3$, we have that $\frac{1}{2}x \leq \sqrt[3]{\frac{1}{2}(y^3 + z^3)}$, hence the inequality that Cardano wanted to prove.

[26] Let us suppose by the absurd that $y \geq 0$ and $z \leq 0$, for example $y = 1$ and $z = -2$. Then, $-7 < -\frac{1}{4}$, which contradicts $y^3 + z^3 \geq \frac{1}{4}x^3$.

since there are no major conceptual gaps to be filled: a calculation is enough. We write the splitting in the following way

$$\begin{cases} a_0 = y^3 + z^3 \\ \frac{a_1}{3} = yz \end{cases}.$$

We take $y^3 = Y$ and $z^3 = Z$. Then, considering the real roots, we have

$$\begin{cases} a_0 = Y + Z \\ \frac{a_1}{3} = \sqrt[3]{Y}\sqrt[3]{Z} \end{cases},$$

that is

$$\begin{cases} a_0 = Y + Z \\ \frac{a_1^3}{27} = YZ \end{cases}.$$

Solving this system is equivalent to solve the quadratic equation $Y^2 + \frac{a_1^3}{27} = a_0 Y$. Its solutions are $Y_1, Y_2 = \frac{a_0}{2} \pm \sqrt{\left(\frac{a_0}{2}\right)^2 - \left(\frac{a_1}{3}\right)^3}$ and, because of the symmetry of the system, we have that $Y_1 = Z_2$ and $Y_2 = Z_1$. We recover then the cubic formula

$$x = \sqrt[3]{Y_1} + \sqrt[3]{Y_2} = \sqrt[3]{\frac{a_0}{2} + \sqrt{\left(\frac{a_0}{2}\right)^2 - \left(\frac{a_1}{3}\right)^3}} + \sqrt[3]{\frac{a_0}{2} - \sqrt{\left(\frac{a_0}{2}\right)^2 - \left(\frac{a_1}{3}\right)^3}}$$

(where, as said, I have only considered the real roots of $\sqrt[3]{Y_1}$, $\sqrt[3]{Y_2}$). We recall that this is the same method used by Cardano in *Ars Magna*, Chapter XII (see here Section 3.2.2, page 113) to derive the cubic formula. It seems extremely unlikely to me that Cardano was not aware of this link with the cubic formula. Cossali maintained, very reasonably, that this splitting directly comes from Tartaglia's poem (see Appendix A from page 405) and that, once discovered it, Cardano looked for other similar splittings.[27]

[27] "*Giusta il metodo di Tartaglia questa equazione [$y^3+3y^2z+3yz^2+z^3-p(y+z)-q=0$] si spezza nelle due $y^3+z^3=q$, $3y^2z+3yz^2=p(y+z)$. È egli di natura sua generale questo spezzamento? È egli l'unico? E qui stendendo Cadano lo*

5.2.2. The splitting (A I.2)-(A I.7)

Given this first splitting, the one that is related to it in the simplest way is the following.

A I.2. *Consider* $x^3 = a_1x + a_0$. *Write* $x = y + z$ *with* y, z *two [real, positive] numbers.*
If y, z *can be chosen such that the splitting according to which "all the parallelepipeds are given to the number and the cubes to the things [*parallelipeda omnia dentur numero et cubi rebus*]", that is*

$$\begin{cases} a_0 = 3y^2z + 3yz^2 \\ a_1x = y^3 + z^3 \end{cases}, \quad \text{or again} \quad \begin{cases} a_0 = 3y^2z + 3y\,z^2 \\ a_1 = y^2 - yz + z^2 \end{cases},$$

holds, then the condition

$$\begin{cases} [a_0 \leq \frac{3}{4}x^3] \\ a_1x \geq \frac{1}{4}x^3 \end{cases}$$

follows.

This splitting is completely parallel to (A I.1), since the arrangement of the terms coming from the substitution $x = y + z$ in the equation $x^3 = a_1x + a_0$ is the same. The only difference is that the way in which the coefficients are assigned to the two arrangements of terms is swapped: $3y^2z + 3y\,z^2$ is now given to a_0 and $y^3 + z^3$ to a_1x. Also the condition is similarly stated,[28] except that it cannot be translated

sguardo su le varie combinazioni de' termini, moltissimi gli si presentarono alla mente i supposti possibili a farsi", see [60, Chapter I, Paragraph 3, page 27-28].

[28] "And because the parallelepipeds cannot be bigger than three-fourths of the whole cube, because the cubes taken together cannot be smaller than one-fourth, and therefore this chapter cannot be general, since, when the number will be bigger than three-fourths of the whole cube, it cannot be assigned to the parallelepipeds" or "*[e]t quia parallelipeda non possunt esse maiora dodrante totius cubi, quia cubi non possunt pariter accepti esse minores quadrante, ideo nec hoc capitulum potest esse generale, quoniam cum numerus fuerit maior dodrante totius cubi, non poterit tribui parallelipedis*", see [9, Chapter I, page 2].

in a condition on the discriminant (since the discriminant is defined in connection with the cubic formula that derives from (A I.1)). Note that Cardano used *De Proportionibus*, Proposition 146,[29] to state[30] the proportion $y^3+z^3 : 3y^2z+3yz^2 = y^2+z^2-yz : 3yz$, which indeed allows[31] to rewrite $a_1x = y^3+z^3$ as $a_1 = y^2-yz+z^2$.

Cardano then observes that neither (A I.1) nor (A I.2) are "general" because of their conditions, but that, if they are considered together, they are so, since their conditions, taken together, cover all the possible values of a_0, a_1 (see below, figures 5.1 and 5.2).

A I.3. *Consider $x^3 = a_1x + a_0$. Write $x = y + z$ with y, z two [real, positive] numbers.*
*If y, z can be chosen such that the splitting according to which "four parallelepipeds are given to the number, the remaining two with the cubes to the things. But in these bodies this is particular, that the ratio of those bodies is as [the ratio] of the squares of the parts simultaneously joined to twice the product [*quatuor parallelipeda numero dantur, reliqua duo cum cubis rebus. Est autem hoc inter corpora illa præcipuum, quod proportio ipsorum corporum est ut quadratorum

---

[29] *De Proportionibus*, Proposition 146: "The body that is made by a divided line times the surface equal to the squares of both parts having subtracted the surface of one part times the other, is equal to the aggregate of the cubes of both parts" or "*[c]orpus quod fit ex linea divisa in superficiem æqualem quadratis ambaru partium detracta superficie unius partis in alteram, est æquale agggregato cuborum ambarum partium*", see [11, page 140r]. This means that $y^3+z^3 = (y+z)(y^2+z^2-yz)$.

[30] "Or one may see by [the things] demonstrated in the book on Proportions that the proportion of the aggregate of the cubes to the aggregate of the six parallelepipeds is as [the proportion] of the aggregate of the squares AB and BC of the parts, subtracted the product of AB times BC from three times the product or the surface of AB times BC, or three times the surface AE" or "*[l]icet autem videre ex demonstratis in Libro de proportionibus quod proportio aggregati cuborum ad aggregatum sex parallelipedum est, veluti aggregati quadratorum ab et bc partium detracto producto ab in bc ad triplum producti seu superficiei ab in bc, seu triplum superficiei ae*", [9, Chapter I, page 2].

[31] Indeed, the proportion says that $3yz(y^3+z^3) = 3yz(y+z)(y^2+z^2-yz)$.

partium simul iunctorum ad ambo producta]", *that is*[32]

$$\begin{cases} a_0 = 2y^2z + 2yz^2 \\ a_1x = y^3 + y^2z + yz^2 + z^3 \end{cases}, \quad \textit{or again} \quad \begin{cases} a_0 = 2y^2z + 2yz^2 \\ a_1 = y^2 + z^2 \end{cases},$$

holds, then the condition

$$\left[\begin{cases} a_0 \leq \frac{1}{2}x^3 \\ a_1x \geq \frac{1}{2}x^3 \end{cases}\right]$$

follows.

Note that Cardano only states the proportion $y^3 + z^3 + y^2z + yz^2 : 2y^2z + 2yz^2 = y^2 + z^2 : 2yz$, from which it follows[33] that $a_1x = y^3 + y^2z + yz^2 + z^3$ can be rewritten as $a_1 = y^2 + z^2$. This time he does not state any condition (though he does not even say that the splitting is "general", as he will do whenever it is so). Nevertheless, this splitting has the above condition, and Cossali identified it in the following way.[34] Recall that Cossali gave two proofs of the condition for (A I.1) (see above, Footnote 25 on page 270). Unlike the second one, the first proof does not attempt to reconstruct Cardano's reasoning: still, it is even more general. There, in an intermediate step, Cossali showed that the maximum value of $y^2z + yz^2$ is $\frac{1}{4}x^3$ (and is attained

[32]The first sentence in the description identifies three possible splittings, namely

$$\begin{cases} a_0 = 2y^2z + 2yz^2 \\ a_1x = y^3 + y^2z + yz^2 + z^3 \end{cases} \text{or} \begin{cases} a_0 = y^2z + 3yz^2 \\ a_1x = y^3 + 2y^2z + z^3 \end{cases} \text{or} \begin{cases} a_0 = 3y^2z + yz^2 \\ a_1x = y^3 + 2yz^2 + z^3 \end{cases}$$

The second sentence definitely points at the first splitting since only the following proportion holds

$$(y^3 + y^2z + yz^2 + z^3) : (2y^2z + 2yz^2) = y^2 + z^2 : 2yz.$$

[33]Indeed, the proportion says that $2yz(y^3 + z^3 + y^2z + yz^2) = 2yz(y + z)(y^2 + z^2)$.

[34]See [60, Chapter I, Paragraph 4, pages 32-33] and [57, Chapter IX, Paragraph 2, pages 445-446].

when $y = z$).[35] This lets us easily gather that $2y^2z + 2yz^2 \leq \frac{2}{4}x^3$, or $a_0 \leq \frac{1}{2}x^3$ because of the splitting (A I.3), and then by the formula for the cube of a binomial that $y^3 + z^3 + y^2z + yz^2 \geq \frac{2}{4}x^3$, or $a_1x \geq \frac{1}{2}x^3$.

A I.4. *Consider $x^3 = a_1x + a_0$. Write $x = y + z$ with y, z two [real, positive] numbers.*
*If y, z can be chosen such that the splitting according to which "the number is given only to the two mutual parallelepipeds [*numerus datur duobus tantum parallelipedis mutuis*]", that is*

$$\begin{cases} a_0 = y^2z + yz^2 \\ a_1x = y^3 + 2y^2z + 2yz^2 + z^3 \end{cases}, \quad \left[\text{or again} \begin{cases} a_0 = y^2z + yz^2 \\ a_1 = y^2 + yz + z^2 \end{cases} \right],$$

holds, then the condition

$$\begin{cases} a_0 \leq \frac{1}{4}x^3 \\ [a_1x \geq \frac{3}{4}x^3] \end{cases}$$

follows.

The condition is immediately justified by the above considerations.

Let us resume through the diagrams in Figures 5.1 and 5.2 the conditions given by these first four splittings. Note that the condition in (A I.1) together with each one of the conditions in (A I.2)-(A I.4) covers all the possible values of a_0 and a_1. More precisely, the conditions in (A I.1) and (A I.4) delimit a complementary range of

[35] Since $x = y + z$, Cossali writes $y = \frac{1}{2}x + t$ and $z = \frac{1}{2}x - t$, with $t > 0$, and then he calculates $y^2z + yz^2 = yz(y + z)$. He gets to $y^2z + yz^2 = \frac{1}{4}x^3 - xt^2$, which means that $\frac{1}{4}x^3$ is the maximum value (attained at $t = 0$), since all the numbers involved are positive. See [60, Chapter I, Paragraph 3, pages 28-29] and [57, Chapter VIII, Paragraph 5, page 339].

Otherwise, one can simply verify that $y^2z + yz^2$ (taking $y = x - z$) attains a global maximum by deriving (twice) with respect to z and then calculating it.

We remark that Cardano refers to the maximum value of $y^2z + yz^2$ elsewhere. Indeed, in *Aliza*, Chapter XXV we find to this end a reference to *De Proportionibus*, Proposition 209. For more details, see below, Footnote 98 on page 329.

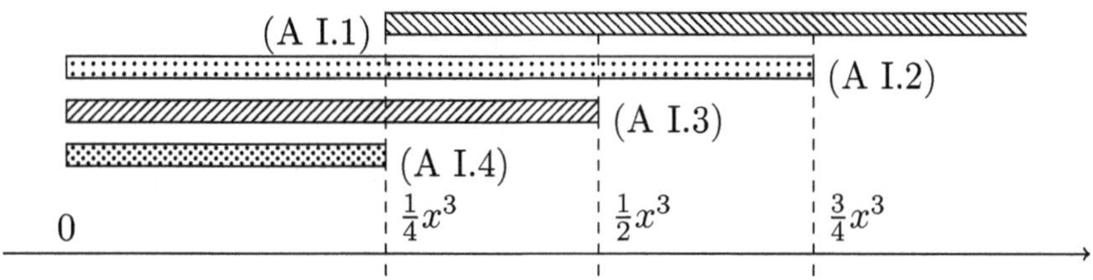

Figure 5.1. – Conditions on a_0 in the splittings (A I.1)-(A I.4).

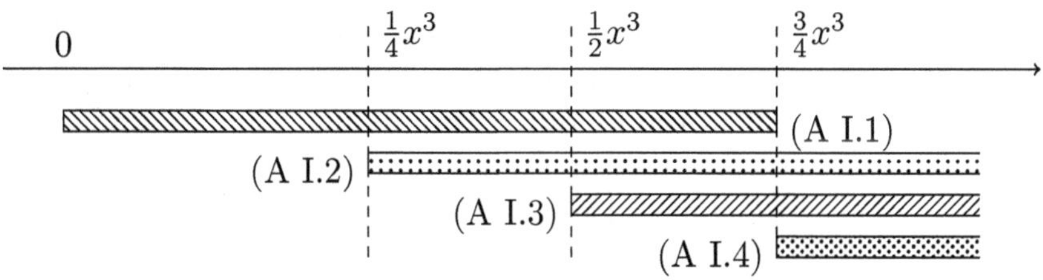

Figure 5.2. – Conditions on $a_1 x$ in the splittings (A I.1)-(A I.4).

validity, while the others overlap. It seems that Cardano was eventually looking for a complementary solving method, that is, for a solving method for the same equation that holds when one cannot apply the one in (A I.1) or, equivalently, in *Ars Magna*, Chapter XII. The preferred way to discover this new solving method is passing by another splitting. Hopefully this could have also led to a complementary formula, as it already was the case for (A I.1).

We moreover recall that there exist the connections between (A I.2) and (A XXV.5), between (A I.3) and (A XXV.3), and between (A I.4) and (A XXV.4), which we will analyse while dealing with Chapter II (see below, on page 323).

Let us now come back to the other splittings.

A I.5. *Consider* $x^3 = a_1 x + a_0$. *Write* $x = y + z$ *with* y, z *two [real, positive] numbers.*
If y, z *can be chosen such that the splitting according to which "we give the number to one cube, the remaining seven bodies to the things*

*[*damus numerum uni cubo, reliqua septem corpora rebus*]", that is*[36]

$$\begin{cases} a_0 = y^3 \\ a_1x = 3y^2z + 3yz^2 + z^3 \end{cases} \quad or \quad \begin{cases} a_0 = z^3 \\ a_1x = y^3 + 3y^2z + 3yz^2 \end{cases}$$

*holds, then no [non-trivial] condition follows. This splitting is "the worst of all [*deterius omnibus*]".*

Cardano's statement is in this case less clear. Obviously it is not true that – strictly speaking – no condition on the coefficients is entailed, since there is at least the trivial condition given by the splitting itself. It can be rewritten[37] in the following way

$$\begin{cases} a_0 \leq x^3 \\ a_1x \leq x^3 \end{cases},$$

which is quite a useless condition, since it always holds when x is a solution of the equation. Indeed, we have that $x^3 = a_1x + a_0$ and both the coefficients a_1, a_0 and the solution x are positive, so that it is clear that a_0 and a_1x are smaller than, or equal to, x^3. Therefore, we account for Cardano's "no condition" as "no [non-trivial] condition".

Cardano then affirms that this splitting is the "the worst of all" since "it falls back on a chapter of four terms, thence from that to the first [*redit ad capitulum quatuor nominum inde ex eo ad primum*]".

[36]We remark that, as long as one does not put any further condition on y, z, they are *a priori* interchangeable in Cardano's phrasing. For example, the expression 'one cube' addresses both y^3 and z^3, or 'one cube and one opposite parallelepiped' addresses both $y^3 + yz^2$ and $z^3 + y^2z$. A few splittings, namely the ones where the assignation for a_0 (and for a_1x) are symmetrical in y, z, are defined in a univocal way. This is the case for instance in (A I.1)-(A I.4). For the sake of clarity, I choose in the other cases to explicitly write all the combinations of y, z.

[37]Indeed, the first line of the condition is obtained by $y^3 \leq (y + z)^3 = x^3$, since all y, z are positive. The second line is obtained by studying $3y^2z + 3yz^2 + z^3$, once substituted $y = x - z$, on the compact $[0, x]$ with respect to the variable z. Its maximum is x^3 when $y = 0$ and $z = x$.

Indeed, if we try to solve the system in y, z as we did to derive the cubic formula in (A I.1), we bump into the complete equation

$$z^3 + \sqrt[3]{a_0}z^2 + (\sqrt[3]{a_0^2} - a_1)z - a_1\sqrt[3]{a_0} = 0$$

(which nevertheless can have $\Delta_3 < 0$). Deleting the term of degree two thanks to the substitution $z = w - \sqrt[3]{a_0}$ (which Cardano knew, see (AM IV.2) on page 81 and (AMA XXI.9) on page 195), we recover again $w^3 = a_1 w + a_0$, which was the equation from which we started.

A I.6. *Consider $x^3 = a_1 x + a_0$. Write $x = y + z$ with y, z two [real, positive] numbers.*
*If y, z can be chosen such that the splitting according to which "we give the number to one cube and three parallelepipeds that are made by the side of that cube times the square of the side of the other cube and the remaining four bodies, that is the cube with three parallelepipeds, to the things [*demus numerum uni cubo et tribus parallelipedis quæ fiunt ex latere cubi illius in quadrata lateris alterius cubi et reliqua quatuor corpora, scilicet cubum cum tribus parallelipedis, rebus*]", that is*

$$\begin{cases} a_0 = y^3 + 3yz^2 \\ a_1 x = 3y^2 z + z^3 \end{cases} \quad \textit{or} \quad \begin{cases} a_0 = 3y^2 z + z^3 \\ a_1 x = y^3 + 3yz^2 \end{cases}$$

*holds, then no [non-trivial] condition follows. This splitting is "mis-shaped [*difforme*]".*

As before, the splitting entails[38] the same trivial condition

$$\begin{cases} a_0 \le x^3 \\ a_1 x \le x^3 \end{cases}.$$

[38]Indeed, the first line of the condition is obtained by studying $y^3 + 3yz^2$, once substituted $y = x - z$, on the compact $[0, x]$ with respect to the variable z. Its only critical point is the inflection point in $z = \frac{1}{2}x$. Then, it is trivially bounded by x^3. The same holds for the second line, since the splitting is symmetric in y, z.

Cardano remarks that $(3y^2z + z^3) - (y^3 + 3yz^2) = (z - y)^3$ and provides an example that eventually leads to both equations $x^3 = \frac{158}{7}x + 185$ and $x^3 = \frac{185}{7}x + 158$, since the arrangement of the terms coming from the substitution $x = y + z$ in the equation $x^3 = a_1x + a_0$ is symmetric. It is exactly because of this symmetry that Cardano calls the splitting "misshaped [*difforme*]". Indeed, he observes[39] that "similar [*simila*]" aggregates, like $y^3 + 3yz^2$ and $3y^2z + z^3$, are made equal to "dissimilar [*dissimila*]" ones, like a_0 and a_1x. This in particular means that a_0 and a_1x are to be considered "dissimilar" under a certain viewpoint. I believe that this viewpoint should be geometrical and, more precisely, should concern a dimensional argument. Indeed, one would only say that a_0 and a_1x are "dissimilar" if x were interpreted as a segment and not as a number (both a_0 and a_1 being numbers).[40] On the other hand, Cardano calls the aggregates $y^3 + 3yz^2$ and $3y^2z + z^3$ "similar", and indeed they both represent some three-dimensional solids, namely a cube with the three "opposite" (see above on page 246) parallelepipeds. We observe nevertheless that such a dimensional argument also holds for the other splittings. The difference with this case is that here the two aggregates $y^3 + 3yz^2$ and $3y^2z + z^3$ are symmetric. This remark by Cardano could be derived from an early geometrical (dimensional) inspiration.

---

[39] "Since what is made equal is the similar, namely the cubes with the three opposite parallelepipeds, to which the dissimilar is compared, indeed one aggregate is made equal to the number, the other [is made equal] to the things" or "*quoniam quod æquatur est simile scilicet cubi cum parallelipedis tribus adversis, cui æquatur dissimile, nam unum aggregatum æquatuor numero, aliud rebus*", see [9, Chapter I, page 3].

[40] One could also argue that a_0 and a_1x are dissimilar, still a_0, a_1, x being all numbers. Indeed, this really happens in the *Aliza*, and more precisely in the second paragraph of the same Chapter I (see below, Section 5.3.1). There, Cardano says that a_0 and a_1x are of a "different kind", since one is rational and the other is irrational.

Anyway, I do not believe that it is the case here. Indeed, that second paragraph is also a later addition that makes reference to an issue on cubic equations with rational coefficients, which is expounded in a bunch of subsequent chapters. Therefore I am inclined to make a separation between the introductory opening of this chapter and the following part on the splittings.

A I.7. *Consider $x^3 = a_1 x + a_0$. Write $x = y + z$ with y, z two [real, positive] numbers.*
*If y, z can be chosen such that the splitting according to which "we give the number to the aggregate from the cube and two connected parallelepipeds with the opposite [one] and the things to the remaining four bodies (*numerum damus aggregato ex cubo et duobus cohærentibus parallelipedis cum uno adverso et res reliquis quatuor corporibus*)", that is*

$$\begin{cases} a_0 = y^3 + 2y^2 z + yz^2 \\ a_1 x = y^2 z + 2yz^2 + z^3 \end{cases}, \quad \left[or \begin{cases} a_0 = y^2 z + 2yz^2 + z^3 \\ a_1 x = y^3 + 2y^2 z + yz^2 \end{cases} \right],$$

or again

$$\begin{cases} a_0 = y^3 + 2y^2 z + yz^2 \\ a_1 = yz + z^2 \end{cases}, \quad \left[or \begin{cases} a_0 = y^3 + 2y^2 z + yz^2 \\ a_1 = y^2 + yz \end{cases} \right]$$

*holds, then no [non-trivial] condition follows. This splitting is "misshaped [*difforme*]".*

As before, the splitting entails[41] the condition

$$\begin{cases} a_0 \leq x^3 \\ a_1 x \leq x^3 \end{cases}.$$

Cardano remarks that $y^3 + 2y^2 z + yz^2 : y^2 z + 2yz^2 + z^3 = y : z$ holds and gives an example, which eventually leads to the equations $x^3 = 35x + 98$ and $x^3 = 14x + 245$. As before, due to the symmetry of the splitting, Cardano finds it "misshaped [*difforme*]".

We remark that, according to Cardano's reasoning, since none of the splittings in(A I.5)-(A I.7) has a condition, if one could conceive a solving method starting from one of them, this would be general and could eventually replace the one in (A I.1).

---

[41]Indeed, the first line of the condition is obtained by studying $y^3 + 2y^2 z + yz^2$, once substituted $y = x - z$, on the compact $[0, x]$ with respect to the variable z. Then, it is trivially bounded by x^3. The same holds for the second line, since the splitting is symmetric in y, z.

Nevertheless, this is possible for no splitting except (A I.1). We have already observed that, if we try to work out (A I.5) with the same method as in (A I.1), we fall back again on a cubic equation (namely, the same from which we started). And it gets not better with the other splittings. Let us consider the degree of the systems in the splittings, which is always nine. Sometimes, it happens that it can be lowered. This is the case for (A I.1)-(A I.4) and (A I.7), where the degree decreases to six, so that also the equations associated to those systems have at most degree six. It is not always an easy job to derive these equations, even when the degree lowers. In the context of the study of equations and of the deep structure of their coefficients during the 18th and 19th centuries, some elimination methods had been conceived. They enable one to derive the equation associated to a two-equation system (later extended to more than two) in one or more unknowns (usually two, but later extended to more than two) *via* the elimination of one term or of one of the unknowns.[42] The most influential characters were Newton, Euler, Cramer, Bézout, and Lagrange, but also Cossali got involved in the topic. Interestingly, Cossali applied one of his previous studies[43] on elimination methods to the systems in the splittings at issue in order to derive the associated equations. He showed[44] that the systems in (A I.1)-(A I.4) are usually associated to an equation of degree six. In particular, whereas the splitting (A I.4) leads to an equation lacking in the terms of degree five and three, and the splittings (A I.2) and (A I.3) lead to equations lacking in the term of degree five, the splitting (A I.1) leads to an equation in which only the terms of degree six, three, and zero appear.[45] This particular feature of the associated equation is what

[42]For a brief account, see [92, pages 131-145].

[43]See [59].

[44]See [60, Chapter I, Paragraph 7, pages 49-57] and [57, Chapter IX, Paragraph 5, pages 454-465].

[45]As it often happens, Cossali's calculations contain little mistakes. He claims that the equation $-27y^6 + 27a_0y - a_1^3 = 0$ is associated to the system

$$\begin{cases} a_0 = y^3 + z^3 \\ a_1x = 3y^2z + 3yz^2 \end{cases}$$

opens the door to derive a cubic formula, since the equation can be solved as a quadratic one. The systems in the splittings (A I.5) and (A I.6) are instead associated to an equation of degree nine. Again, thanks to the particular structure of the equation associated to (A I.5) (only the terms of degree nine, six, three, and zero appear and the coefficients are appropriated), one is enabled to reduce the polynomial to the cube of a binomial, which is $(a_0 - y^3)^3$. Unfortunately, this leads back to the starting point. Finally, the system in the splitting (A I.7) is associated to a complete cubic equation, which as such can have $\Delta_3 < 0$. Then, if Cardano's hope was to derive another cubic formula for $x^3 = a_1 x + a_0$, the splittings (A I.2)-(A I.7) are useless (and indeed any other possible splitting will be useless as well).

The splittings in the *Aliza* recall *Ars Magna*, Chapter XXV (see here, Section 3.1.4 on page 84). In particular, we have a straight correspondence between (A I.2) and (AM XXV.5), between (A I.3) and (AM XXV.3), and between (A I.4) and (AM XXV.4) (see respectively the systems (AM XXV.5 bis) on page 88, (AM XXV.3 bis) on page 86, and (AM XXV.4 bis) on page 87). Considering the shift between the 1545 and 1570 (and then 1663) editions in Cardano's remarks at the end of *Ars Magna*, Chapter XII (see here, on page 114), we expected this link. Indeed, we will see that Cardano explicitly mentions this link in the *Aliza* (see below, Section 5.3.2 on page 323). Nevertheless, the two methods are presented with different purposes in the two books. They both concern the problem entailed by the *casus irreducibilis*, but in the *Ars Magna* – as we have already noticed on page 97 – the main gain was to help intuition to find a solution of the considered equation thanks to a rephrasing of the equation *via* a more friendly system. This enables one to lower the degree of the equation by polynomial division. In the *Aliza* instead, Cardano wanted to fly

in the splitting (A I.1). Actually, the equation is $-27y^6 + 27a_0y^3 - a_1^3 = 0$ (which can be solved as a quadratic equation). Indeed, if we solve the quadratic equation in z in the second line of the system, we get $z_1 = -y$ and $z_2 = \frac{a_1}{3y}$. The first value leads to a contradiction, since from the substitution in the first line we have $-a_0 = 0$. The second value leads to the above equation.

higher.[46] The aim is no longer to reach a case-by-case method, still subject to uncertainty, but an algorithmic one that hopefully works each time. This agrees with the fact that Cardano called "particular" the propositions in *Ars Magna*, Chapter XXV, while in the *Aliza* he is looking for a "general chapter". There (see above, page 98), although the propositions made reference to an *a priori* "general" method, they could only be applied in some blessed cases, thus in practice one cannot solve every given problem by them. Here, Cardano is aspiring to develop a truly applicable "general" method, that is to find an algorithm or a formula. It is hard to say whether the *Aliza* is a reworking of the *Ars Magna* or *vice versa*. For sure, Cardano mentions the *Ars Magna* in the *Aliza* and not the other way around – this is the only evidence that we have. It is likely that Cardano firstly discovered the "particular" rules in *Ars Magna*, Chapter XXV and then transformed some of them into the splittings. Indeed, only a few rules of the *Ars Magna* are compatible with the substitution $x = y + z$ in the *Aliza* and these could have been the core from which Cardano deduced the other splittings by analogy. Moreover, since no trace of similar propositions can be found either in the *Ars Magna Arithmeticæ* or in the *Practica Arithmeticæ*, it makes sense to suppose that the particular rules in *Ars Magna*, Chapter XXV were developed after the cubic formulae and the contact with Tartaglia.

5.2.3. The remaining splittings

According to Cardano's criteria (that is, always considering positive coefficients and writing no equality with zero), there are sixty-two possible splittings.[47] To count them all, we remark that splitting $y^3 + 3y^2z + 3yz^3 + z^3$ in two parts and assigning one to a_0 completely determines the choice for the other part (assigned to a_1x). We consider

[46] But note the little inconsistency in the *Aliza*, Chapter I when Cardano provides the following examples $x^3 = 29x + 140$, $x^3 = \frac{185}{7}x + 158$, $x^3 = \frac{158}{7}x + 185$, $x^3 = 35x + 98$, $x^3 = 14x + 245$ (which lead to equations with $\Delta_3 > 0$).

[47] See also Cossali's account [60, Chapter I, Paragraph 7, pages 47-49] and [57, Chapter IX, Paragraph 5, pages 452-454].

then all the possible combinations of the eight terms in $y^3 + 3y^2z + 3yz^3 + z^3$. We have two choices for each of y^3, z^3 (since we can have one or none of them) and four choices for each of y^2z, yz^2 (since we can have three, two, one, or none of them and we cannot distinguish them three by three). Moreover we have to count out the two limit cases, where there are all the terms or no term. This makes sixty-two possibilities.

The most effective way to classify the combinations is on the basis of how many terms are associated to a_0 as in Table 5.2. Note that the dotted line is the y, z-symmetry axis. For the sake of brevity, in Table 5.2 I omit the terms assigned to a_1x. We remark anyway that one can also directly read them in the table. For instance, when one term of $y^3 + 3y^2z + 3yz^3 + z^3$ is associated to a_0, seven terms remain for a_1x, and they are exactly the ones listed in the seven terms column for a_0. In the same way, the two terms column is associated to the six terms column, the three terms column is associated to the five terms column, and the four terms column is associated with itself.

Aside from the "easiest (*faciliora*)" splittings (A I.1)-(A I.7), Cardano says that the remaining ones are either "anomalous [*anomali*]" or "useless [*inutiles*]":

> [t]he remaining compositions are anomalous, as if we had given the number to one parallelepiped, to three [parallelepipeds], to five [parallelepipeds], to two [parallelepipeds] non-mutual, to four [parallelepipeds] two of which are non-mutual, or again to one cube and one parallelepiped, [to one cube and] to two [parallelepipeds] or [to one cube and] to three [parallelepipeds] not of the same kind. The others are useless, as if we had given the number to the aggregate from both cubes and two or four parallelepipeds in whatever way.[48]

[48]"*Reliquæ autem compositiones, aut sunt anomale, velut si daremus numerum uni parallelipedo, vel tribus vel quinque vel duobus, non mutuis aut quatuor, ex quibus duo mutua non essent, aut uni cubo et uni parallelipedo, vel duobus vel tribus non eiusdem generis. Aliæ sunt inutiles, velut si daremus numerum*

Table 5.2. – The sixty-two possible splittings in *Aliza*, Chapter I classified according to the number of terms associated to a_0.

Number of terms associated to a_0 in the splittings			
1-term	2-term	3-term	4-term
	y^3+z^3 (A I.1)		$y^3+y^2z+yz^2+z^3$
	y^2z+yz^2 (A I.4)		$2y^2+2yz^2$ (A I.3)
		$y^3+y^2z+yz^2$	$y^3+2y^2z+yz^2$ (A I.7)
		$y^3+y^2z+z^3$	$y^3+2y^2z+z^3$
		y^3+2y^2z	y^3+3y^2z
	y^3+yz^2	y^3+2yz^2	$y^3+y^2z+2yz^2$
y^3 (A I.5)	y^3+y^2z	y^2z+2yz^2	y^3+3yz^2 (A I.6)
y^2z	$2y^2z$	$3y^2z$	y^2z+3yz^2
yz^2	$2yz^2$	$3yz^2$	$3y^2z+yz^2$
z^3 (A I.5)	y^2z+z^3	$2y^2z+yz^2$	$3y^2z+z^3$ (A I.6)
	yz^2+z^3	$2y^2z+z^3$	$2y^2z+yz^2+z^3$
		$2yz^2+z^3$	$3yz^2+z^3$
		$y^3+yz^2+z^3$	$y^3+2yz^2+z^3$
		$y^2z+yz^2+z^3$	$y^2z+2yz^2+z^3$ (A I.7)

Number of terms associated to a_0 in the splittings		
5-term	6-term	7-term
	$y^3+2y^2z+2yz^2+z^3$	
	$3y^2z+3yz^2$ (A I.2)	
$y^2+3y^2z+z^3$		
$y^3+y^2z+2yz^2+z^3$		
$y^3+y^2z+3yz^2$		
$y^3+3y^2z+yz^2$	$y^3+3y^2z+yz^2+z^3$	
$3y^2z+2yz^2$	$y^3+2y^2z+3yz^2$	$y^3+3y^2z+2yz^2+z^3$
$y^3+2y^2z+2yz^2$	$y^3+3y^2z+2yz^2$	$y^3+3y^2z+3yz^2$
$2y^2z+2yz^2+z^3$	$2y^2z+3yz^2+z^3$	$3y^2z+3yz^2+z^3$
$2y^2z+3yz^2$	$3y^2z+2yz^2+z^3$	$y^3+2y^2z+2yz^2+z^3$
$y^2z+3yz^2+z^3$	$y^3+y^2z+3yz^2+z^3$	
$3y^2z+yz^2+z^3$		
$y^3+2y^2z+yz^2+z^3$		
$y^3+3yz^2+z^3$		

In the above quotation, Cardano lists some of them in a compact way. Among the "anomalous" splittings Cardano mentions, in this order, the splittings determined by the following assignations:

$$
\begin{aligned}
a_0 = y^2 z, &\quad a_0 = yz^2,\\
a_0 = 3y^2 z, &\quad a_0 = 3yz^2,\\
a_0 = y^2 z + 2yz^2, &\quad a_0 = 2y^2 z + yz^2,\\
a_0 = 3y^2 z + 2yz^2, &\quad a_0 = 2y^2 z + 3yz^2,\\
a_0 = 2y^2 z, &\quad a_0 = 2yz^2,\\
a_0 = 2y^2 z, &\quad a_0 = 2yz^2,\\
a_0 = y^2 z + 3yz^2, &\quad a_0 = 3y^2 z + yz^2,\\
a_0 = y^3 + y^2 z, &\quad a_0 = yz^2 + z^3,\\
a_0 = y^3 + yz^2, &\quad a_0 = y^2 z + z^3,\\
a_0 = y^3 + y^2 z + yz^2, &\quad a_0 = y^2 z + yz^2 + z^3,\\
a_0 = y^3 + 2y^2 z, &\quad a_0 = 2yz^2 + z^3,\\
a_0 = y^3 + 2y^2 z, &\quad a_0 = 2yz^2 + z^3,\\
a_0 = y^3 + 2yz^2, &\quad a_0 = 2y^2 z + z^3,\\
a_0 = y^3 + 2y^2 z + yz^2 \text{ (A I.7)}, &\quad a_0 = y^2 z + 2yz^2 + z^3 \text{ (A I.7)},\\
a_0 = y^3 + y^2 z + 2yz^2, &\quad a_0 = 2y^2 z + yz^2 + z^3.
\end{aligned}
$$

Note that I interpret "four [parallelepipeds] two of which are non-mutual" as pointing only at the splittings determined by $a_0 = y^2 z +$

aggregato ex ambobus cubis, et duobus parallelipedis aut quatuor quomodocumque, nam si numerus cum parvus sit, non sufficit aggregato cuborum, quomodo sufficit eidem si addantur parallelipeda", see [9, Chapter I, page 4]. Note that the correlative of the first '*aut*' has been omitted and was – in my understanding – originally referred to '*inutiles*' in the next sentence. Indeed, it was common at that time to mix up '*vel*' and '*aut*' especially in listing, which happened in the two lists referring to the anomalous and useless splittings.

$3yz^2$ and $a_0 = 3y^2z + yz^2$, but not at the one in (A I.3) determined by $a_0 = 2y^2z + 2yz^2$. Indeed, taking into account four parallelepipeds, the expression 'two of which are non-mutual' is superfluous since Cardano had only available six parallelepipeds three by three of the same kind. In my interpretation, Cardano added this expression to distinguish these splittings from the one in (A I.3). Moreover, in "[as if we had given the number to one cube and] to two [parallelepipeds] or [to one cube and] to three [parallelepipeds] not of the same kind" I loosely interpret "not of the same kind" to be referred only to "three [parallelepipeds]" but not to "two [parallelepipeds]". We finally remark that, since the splittings in (A I.7) also appear above, the labels 'misshaped' and 'anomalous' are not exclusive. According to Cardano's description, also the splittings determined by the assignations $a_0 = y^3 + y^2z + 2yz^2$ and $a_0 = 2y^2z + yz^2 + z^3$ are misshaped. We have no hint on why Cardano called these splittings 'anomalous'. I would tend towards a literal interpretation of '*anomalus*' as 'without name', meaning that it was not worthwhile to give a name to those splittings. Of course, this interpretation leaves the question open why Cardano decided not to give a name to these splittings.

Among the "useless" splittings Cardano mentions, in this order, the splittings determined by the following assignations:

$$a_0 = y^3 + y^2z + yz^2 + z^3,$$

$$a_0 = y^3 + 2y^2z + z^3, \quad a_0 = y^3 + 2yz^2 + z^3,$$
$$a_0 = y^3 + 3y^2z + yz^2 + z^3, \quad a_0 = y^3 + y^2z + 3yz^2 + z^3,$$
$$a_0 = y^3 + 2y^2z + 2yz^2 + z^3.$$

With regard to these splittings, Cardano adds a crucial observation. Indeed, in the last sentence of *Aliza*, Chapter I, he argues that

> if it [the number a_0] is not enough for the aggregate of the cubes since the number is small, in what way is it enough

for the same if the parallelepipeds are added?[49]

This is a key sentence to make sense out of the splittings. In the *Ars Magna* (and in (A I.1)), Cardano assumed $a_0 = y^3 + z^3$, and as a consequence he needed to put the condition on $y^3 + z^3$ – or, more precisely, on a_0 – in which he asked this number to be big. In the *Aliza*, Cardano was trying to fix this difficulty. As we have argued, it is very reasonable that he was looking for some splittings where the condition was not verified, that is, where the number a_0 is small, hoping that they could lead to another solving method (either to complement or to replace the one in (A I.1)). This is the reason why Cardano calls the above four splittings useless: there, the number a_0 is bigger than in (A I.1) since it is equal to the two cubes y^3 and z^3 plus some other parallelepipeds (and we recall that $a_0 = y^3 + z^3$ was already too big). In the list of the useless splittings, Cardano does not include the ones determined by the following assignments:

$$a_0 = y^3 + y^2z + z^3, \quad a_0 = y^3 + yz^2 + z^3,$$

$$a_0 = y^3 + y^2z + 2yz^2 + z^3, \quad a_0 = y^3 + 2y^2z + yz^2 + z^3,$$

$$a_0 = y^3 + 3y^2z + z^3, \quad a_0 = y^3 + 3yz^2 + z^3,$$

$$a_0 = y^3 + 3y^2z + 2yz^2 + z^3, \quad a_0 = y^3 + 2y^2z + 3yz^2 + z^3.$$

Actually there is no doubt that they should have been included since also there the number is made equal to the two cubes y^3 and z^3 plus some other parallelepipeds.

According to the above interpretation, in *Aliza*, Chapter I, forty-eight splittings are mentioned in all. In the end, the following fourteen splittings remained unmentioned:

$$a_0 = y^3 + 3y^2z, \quad a_0 = 3yz^2 + z^3,$$

$$a_0 = y^3 + y^2z + 3yz^2, \quad a_0 = 3y^2z + yz^2 + z^3,$$

[49] "*[N]am si numerus cum parvus sit, non sufficit aggregato cuborum, quomodo sufficit eidem si addantur parallelipeda*", see [9, Chapter I, page 4].

$$a_0 = y^3 + 3y^2z + yz^2, \quad a_0 = y^2z + 3yz^2 + z^3,$$
$$a_0 = y^3 + 2y^2z + 2yz^2, \quad a_0 = 2y^2z + 2yz^2 + z^3,$$
$$a_0 = y^3 + 2y^2z + 3yz^2, \quad a_0 = 3y^2z + 2yz^2 + z^3,$$
$$a_0 = y^3 + 3y^2z + 2yz^2, \quad a_0 = 2y^2z + 3yz^2 + z^3,$$
$$a_0 = y^3 + 3y^2z + 3yz^2, \quad a_0 = 3y^2z + 3yz^2 + z^3.$$

5.2.4. Summing up

In *Aliza*, Chapter I, Cardano is looking for another cubic formula for the equation $x^3 = a_1x + a_0$, in which the square roots of negative numbers do not appear when all the solutions are real and distinct. We have remarked that the splitting (A I.1) leads to the cubic formula that we know, since solving the system that characterises (A I.1) is equivalent to solve a cubic equation. We have also remarked that Cardano was almost for sure aware of this connection.

It is indeed very reasonable that Cardano was looking for another cubic formula or for a complement to it *via* one of the other splittings, since this device (that is, splitting $(y + z)^3$) already worked in (A I.1). There, the method employed has a straight connection with the solving method in *Ars Magna*, Chapter XII, and therefore with Tartaglia's poem (see Appendix A from page A.2). The other splittings can be seen as a sort of generalisation of it. The conditions in (A I.2)-(A I.4) provide, for each splitting, a range of validity complementary to the one for (A I.1), so that a complementary solving method could have possibly been derived. The splittings (A I.5)-(A I.7) have no condition, so that a solving method that replaces the one in (A I.1) could have possibly been derived. Unfortunately, the difficulty with the splittings (A I.2)-(A I.7) – and actually with any other splitting – is exactly that one cannot conceive any solving method for the considered equation starting from them. Indeed, they all lead to equations (namely, to the equations in one unknown associated to the systems in the splittings) that are never of a lesser degree than three.[50] Therefore they cannot

[50]See Footnote 26 on page 15.

be used in order to derive a solution for the family of equations $x^3 = a_1x + a_0$, as it was instead the case for (A I.1).

Still, some questions remain open. We recall that Cardano says that the splittings (A I.1)-(A I.7) are the "easiest" ones. I have not managed to find a pattern to justify this claim. Moreover, the criteria according to which Cardano listed the remaining splittings are not completely clear.

5.3. Analysing the splittings

In the following sections I deal with all the chapters of the *Aliza* that can be grouped around the thread given by the splittings. In particular, I use these connections in order to try to retrieve some possible origins for the splittings (including a geometrical one).

5.3.1. On the origin of the substitution $x = y + z$: a hypothesis on the general shapes for irrational solutions

In the preceding section, I have suggested that the splittings can be seen as a sort of generalisation of the solving method in *Ars Magna*, Chapter XII (or of Tartaglia's poem). The tricky part in deriving the splittings is to get the substitution $x = y + z$ and – as we will see – Cardano thought hard and for a long time over it.

In Section 4.2.4, we have expounded Cardano's researches in the *Ars Magna Arithmeticæ* about the algebraic shape that a "general" solution of a cubic equation with rational coefficients can have. A late cross reference to this approach appears moreover in *Ars Magna*, Chapter IV, where nevertheless the coefficients are no longer assumed to be rational (see above, Section 3.1.4). We remark that, in [60, Chapter II, Paragraph 1, pages 59-67] and in [57, Chapter VII, Paragraph 6, pages 399-405], Cossali strove for recapitulating all of Cardano's results concerning irrational solutions (in the *Ars Magna Arithmeticæ* and in the *Ars Magna*, as well as in the *Aliza*). Unfortunately, his account is far from being faithful to the spirit of Cardano's texts.

The topic of the algebraic shape of an irrational solution of an equation with rational coefficients occurs also in the *Aliza.* At the beginning of **Chapter I** (see above on page 268), even before mentioning the splittings, Cardano discusses the substitution $x = y + z$ for $x^3 = a_1x + a_0$. He shows that x consists at least of two incommensurable[51] parts, by an argument that appeals to "generality".

> Being therefore the cube equal to two quantities of a different kind (otherwise this would not be general, if it was extended to the numbers alone and to their parts), it is necessary both that it is unravelled in just two parts, one of which is a number and [is] equal to an assigned [one], the other [part] contains as many parts different in nature as there are in the things equal to these. Therefore it is necessary that the cube consists of at least two different parts, and therefore its side or the things, and indeed several quantities of different kinds cannot be made by a nature of a single kind through a multiplication however often repeated, as it is demonstrated by Euclid in the tenth Book (Proposition 20). Truly, if two quantities alien to the number were contained in the thing, it would be necessary that they were incommensurable one to the other, otherwise they would have been equivalent to one [quantity]. But in this way it is necessary that the cube, which contains three parts, a number and two irrationals, is made in order that it can be equal to some things and the number.[52]

[51] For the definition of '(in)commensurable', see below, Footnote 61.

[52] "*Cum ergo cubus æqualis sit duabus quantitatibus diversi generis (aliter non esset hoc generale, si ad solos numeros et eorum partes extenderetur), necesse est ut et ipse in duas tamtum partes resolvatur, quarum una numerus sit, et assignato æqualis: alia totidem partes contineat natura varias, quot in rebus continentur de eis æquales. Quo circa necesse est cubum saltem ex duabus partibus constare natura diversis, igitur et latus eius seu res, neque enim ab unius generis natura plures per multiplicationem quotiescunque repetitam plures quantitates diversorum generum fieri possunt, ut ab Euclide in decimo libro*

Firstly, let us consider Cardano's remark in the parentheses. He states that, in the equation $x^3 = a_1x + a_0$, the terms a_1x and a_0 must be of a "different kind [*diversi generis*]", since he is aiming to a "general" method, that is, a method that works, for instance, also when x^3 is irrational. Right before, at the beginning of the second paragraph of Chapter I,[53] Cardano said that a_1 and a_0 are (rational) numbers. Then, his remark in the parentheses implies that x can be irrational, so that a_1x, and accordingly x^3, can be irrational too. Indeed, let us suppose that a_1x and a_0 were of the same "kind". Then, a_1x will be rational, since a_0 is rational. But this cannot be, since $x^3 = a_1x + a_0$ and x^3 could be irrational. This way of reasoning recalls the study of the shape of irrational solutions that we have seen in the *Ars Magna Arithmeticæ* (see Section 4.2.4). Having cleared this point, Cardano comes to the conclusion that x^3 must be composed (at least) of two parts – let us call them Y and Z, with the following constraints:

- Y must contain "as many parts different in nature [*totidem partes contineat natura varias*]" as there are in a_1x,
- Z must be rational and equal to the number a_0,
- ($Y + Z$ is not a rational number).

Thereby, Cardano says that also x must be composed of (at least) two parts – what we have called y and z. Indeed, if x is composed only of one part, its cube x^3 is also composed only of one part, but this

demonstratum est. Verum si in re contineantur duæ quantitates a numero alienæ, necesse est ut inter se sint incommensæ, aliter æquivalerent uni. At ex eiusmodi necesse est cubum fieri, qui tres partes contineat, numerum et duas rhete, ut rebus ac numero possit coæquari", see [9, page 1].

[53] "[I]n the chapter of the cube equal to some things and a number, two [numbers] are proposed in particular, the number of the equality and also the number of the things. But more generally let now the cube of a certain line or quantity be equal to the proposed numbers, the simple [one] and [the one] of the things" or "*[i]n capitulo igitur cubi æqualis rebus et numero duo proponuntur speciatim, numerus æquationis, numerus etiam rerum. Generaliter autem, quod cubus nunc alicuius lineæ seu quantitatis, propositis numeris simplici et rerum æqualis est*", see [9, Chapter I, page 1].

cannot be according to the above argument. Moreover, if x is the sum of two, or more than two, irrational (and no rational appears in the sum), then (at least two of) these irrational must be incommensurable between them. Indeed, by the absurd, let us suppose for the sake of simplicity that $x = y + z$ with y, z irrational and commensurable. This means that $\frac{y}{z}$ is rational, that is $y = kz$ with k rational. Then $x = z(k+1)$ with $k+1$ rational, which cannot be, since we assumed that x cannot be composed only of one part. This explains Cardano's reasoning to the least detail.

Let us now consider two examples. We make reference to the two "general" shapes (that is, those shapes that are expounded thanks to the *binomia* and *recisa* and that can encompass other more "particular" shapes for the solutions) mentioned in the *Ars Magna Arithmeticæ* as solutions of $x^3 = a_1x + a_0$ with a_1, a_0 rational. They are the *binomium* of the 2nd or 5th type $\sqrt{a} + b$ and the cubic *binomium* $\sqrt[3]{a} + \sqrt[3]{b}$ such that $\sqrt[3]{ab}$ is rational. We want to show that these two examples fit the above requirements on Y, Z. It is clear that both of them match the third one. They also match the first and the second ones. Indeed, when $x = \sqrt{a} + b$ is a *binomium* of the 2nd or 5th type, its cube is

$$x^3 = 3ab + b^3 + 3b^2\sqrt{a} + \sqrt{a^3} = (3b^2 + a)\sqrt{a} + 3ab + b^3.$$

and we can take for instance $Z = b^3$, which is rational, and $Y = (3b^2 + a)\sqrt{a} + 3ab$. Then, Y contains two parts (a rational and an irrational one) as well as a_1x contains a rational and an irrational part (since $x = \sqrt{a} + b$). When instead $x = \sqrt[3]{a} + \sqrt[3]{b}$ is a cubic *binomium* such that $c = \sqrt[3]{ab}$ is rational, its cube is

$$x^3 = 3\sqrt[3]{a^2b} + 3\sqrt[3]{ab^2} + a + b = 3c(\sqrt[3]{a} + \sqrt[3]{b}) + a + b$$

and we can take $Z = a + b$, which is rational, and $Y = 3c(\sqrt[3]{a} + \sqrt[3]{b})$. Then, Y contains two (irrational) parts, which are $3c\sqrt[3]{a}$ and $3c\sqrt[3]{b}$ (and they are incommensurable, hence they cannot be summed), as well as x.

We remark that in the above quotation Cardano pays particular attention[54] to the case in which neither y nor z are (rational) numbers, as in $x = \sqrt[3]{a} + \sqrt[3]{b}$. He specifies indeed that then y and z must be such that the equation $(y+z)^3 = a_1(y+z) + a_0$ can hold. Indeed, the cube of a sum of two or more terms usually has more terms than the sum itself (except when some condition like '$\sqrt[3]{ab}$ is rational' is imposed) and Cardano wants to count this possibility out.

As said, Cardano's account on the substitution $x = y + z$ immediately foreruns the last part of Chapter I on the splittings. If his aim is to justify the substitution, this account is not complete. Indeed, Cardano does not explicitly show that x has at most two parts. This is the missing passage in order to show that x exactly has two parts, which is the result that Cardano wants to use. Anyway, we will see in the rest of this section that there is a sizeable number of *Aliza*'s chapters (as usual, haphazardly ordered) that recall and complete the argument. This paragraph is a later addition. Indeed, on the one hand, it is the obvious continuation of the first paragraph, which seems to be a later addition too. On the other hand, considering the framework of Chapter I, it is likely that Cardano needs to justify the substitution that he is going to use to derive the splittings. Then, he

[54]"Truly, if two quantities alien to the number were contained in the thing, it would be necessary that they were incommensurable one to the other, otherwise they would have been equivalent to one [quantity]. But in this way it is necessary that the cube, which contains three parts, a number and two irrationals, is made in order that it can be equal to some things and the number. Having divided the thing into two parts, it is necessary that the cube produces three of this sort and that, if [the thing is divided] into three, it produces four and so on, and (as it is said) in order that in those the number [is] equal to the proposed number, [and] that actually the remaining parts are from the nature of the parts of the sides and [are] equal to the aggregates of these" or "*[v]erum si in re contineantur duæ quantitates a numero alienæ, necesse est ut inter se sint incommensæ, aliter æquivalerent uni. At ex eiusmodi necesse est cubum fieri, qui tres partes contineat, numerum et duas rhete, ut rebus ac numero possit coæquari. Cum ergo diviserimus rem in duas partes, oportet cubum tres eiusmodi progignere, et si in tres ut progignat quatuor atque ita deinceps, et (ut dictum est) ut in illis sit numerus numero proposito æqualis; reliquem partes vero ut sint ex natura partium lateris, et illarum aggregatis coæquales*", see [9, Chapter I, pages 1-2].

puts there just a few lines to give a semblance of uniformity, relying on the fact that more detailed arguments will follow elsewhere.

The study of the substitution $x = y+z$ possibly originated from former studies on the shapes of the irrational solutions of $x^3 = a_1x + a_0$, with a_1, a_0 rational, and, generally speaking, on irrational numbers. Concerning this last topic, I provide some technical details in Section 5.4.1. With regard to the first one, we find some chapters in the *Aliza* that recall or improve Cardano's results in the *Ars Magna Arithmeticæ* (see Section 4.2.4).

More precisely, at the beginning of **Chapter V** Cardano states that

A V.i. *The* binomia *and* recisa *of the 3rd or 6th type* $\sqrt{a} \pm \sqrt{b}$ *cannot be solution [of a non-complete cubic equation with rational coefficients].*

Cardano does not specify of which equation $\sqrt{a} \pm \sqrt{b}$ cannot be a solution. Nevertheless, recall that we have already met a slightly more general version (for any degree) of this proposition in (AMA XIX). Since in the *Aliza* in general and in the rest of this chapter Cardano takes into account only cubic equations, I have considered them as the implicit object of this proposition too (with the correction in the Footnote 29 on page 201). Moreover, Cardano's justification here corresponds to the missing part of the justification in the *Ars Magna Arithmeticæ* (see above on page 199), since he only justifies the proposition for $x^3 = a_2x^2 + a_0$.

> I say that the value by a *binomium* or *recisum* in which there is no number is not suitable in this case, because, [the value being] subtracted from the number, three non-composite quantities, a number and two roots, are left, and from those roots multiplied by themselves it is nothing but a number and a root of a number, then in the product they cannot erase each other.[55]

[55] "*Dico quod æstimatio in binomio vel reciso, in quo non est numerus, non est idonea in hoc casu, quia detracta a numero relinquit tres quantitates incom-*

Let us take by the absurd $x = \sqrt{a} + \sqrt{b}$ to be a solution of $x^3 = a_2x^2 + a_0$, with a_2, a_0 rational. Then, we get the contradiction $x^2(x-a_2) \neq a_0$, since

$$(a+2\sqrt{ab}+b)(\sqrt{a}+\sqrt{b}-a_2)=(a+3b)\sqrt{a}+(3a+b)\sqrt{b}-2a_2\sqrt{ab}-a_2(a+b)$$

cannot be rational. We remark that the justification sticks to the very same pattern (substituting by the absurd the (un)expected value and deriving a contradiction through calculations) as in the *Ars Magna Arithmeticæ*, though this time Cardano rewrites the equation as $x^2(x - a_2) = a_0$.

In **Chapter X** "In what way the parts with the proposed line are appropriate to the parallelepiped [*Quomodo conveniant partes cum linea proposita in parallelipedo*]" Cardano directly employs the splittings (A I.2)-(A I.4) to draw some more systematic conclusions on the shape of the solutions of $x^3 = a_1x + a_0$, with a_1, a_0 rational. Also Proposition (A V.i) is rephrased.

A X.1. *Consider the equation $x^3 = a_1x + a_0$ and take y, z such that $y + z = x$ is a solution.*
If moreover x is rational, then either y, z are rational or y, z are a binomium *and its* recisum.

Cardano does not further specify the type of the *binomium* and *recisum.*

A X.1 — *Proof.* Let a_1, a_0, and $x = y + z$ be given rational.

positas, numerum et duas radices, et ex radicibus illis in se ductis non sit nisi numerus, et una radix numeri, ergo in producto non poterunt se delere", see [9, Chapter V, page 13].

By (A I.2), Cardano knows that $a_0 = 3y^2z+3yz^2$. Then, $yz = \frac{a_0}{3(y+z)}$ is rational. Therefore, y, z are either both rational or they are a *binomium* and its *recisum*.[56]

The same is done using (A I.3) or (A I.4) and replacing the factor 3 in the denominator with 2 or 1. □

We remark that Cossali[57] gave a statement similar to the above one, though his proof completely departs from Cardano's one.

The following propositions are proved in an analogous way. Note that, since they all contain a negative statement, Cardano employs the same method by the absurd to contradict the fact that the coefficients are rational. I only give the proof of Proposition (A X.2) as an example.

A X.2. *Consider the equation $x^3 = a_1x + a_0$. Then, no irrational $\sqrt{a}$ can be a solution of the equation.*

A X.2 — *Proof.* Let a_1, a_0 be given rational and $x = y + z$ be given irrational. Suppose moreover that $x = \sqrt{a}$ and by the absurd let it be a solution of the equation.

As in the previous proof, Cardano gets to the system

$$\begin{cases} y + z = x \\ yz = \frac{a_0}{3(y+z)} \end{cases}.$$

In particular, $y, z = \frac{x}{2} \pm \sqrt{\left(\frac{x}{2}\right)^2 - \frac{a_0}{3(y+z)}}$ [being y, z the solutions of the associated quadratic equation]. Anyway, this cannot be, since a_1 is rational. Indeed, using the splitting (A I.2) and (A XXX), Cardano

[56] Since y, z are the solutions of the associated quadratic equation. Nevertheless, Cardano (as usual) justifies this last step only through an example, namely $x^3 = 30x + 36$ (where $\Delta_3 < 0$).

[57] "*La forma irrazionale $t \pm \sqrt{u}$ non soddisfa all'equazione $x^3 - px \mp q = 0$, che a condizione, che essa equazione abbia una radice razionale*", see [60, Chapter II, Paragraph 1, page 61] and [57, Chapter VII, Paragraph 6, pages 400-401].

infers from $a_1 x = y^3 + z^3$ that $a_1 = y^2 - yz + z^2$. [But, using the above values of y, z, one (obviously) finds that $y^2 - yz + z^2 = x^2 - \frac{a_0}{x}$, which is irrational since $\frac{a_0}{x}$ is so.]

The same can be done using (A I.3) or (A I.4) and replacing the factor 3 in the denominator with 2 or 1. □

We remark that Cardano could have replaced $x = \sqrt{a}$ in the equation to immediately derive a contradiction, but he does not so. He rather chooses to stick to the pattern of the previous proof by rephrasing $a_1 = x^2 - \frac{a_0}{x}$.

A X.3. *Consider the equation $x^3 = a_1 x + a_0$. Then, no irrational $\sqrt[3]{a}$ can be a solution of the equation.*

A X.4. *Consider the equation $x^3 = a_1 x + a_0$. Then, no irrational $\sqrt{a} \pm \sqrt{b}$ can be a solution of the equation.*

We remark that this proposition is a particular case of (A V.i). The proof given here, analogous to the one of (A X.2), fixes the uncertainties of the proof of (A X.3).

A X.5. *Consider the equation $x^3 = a_1 x + a_0$. Then, no irrational $\sqrt[4]{a}$ or $\sqrt[6]{a}$ can be a solution of the equation.*

A X.6. *Consider the equation $x^3 = a_1 x + a_0$. Then, no irrational $\sqrt[n]{a \pm \sqrt{b}}$ with $n \geq 2$ can be a solution of the equation.*

The conclusion of this chapter is puzzling:

> [t]herefore, it is necessary that the universal value of this sort is either under the *binomium* in which there is a number or not, or under the *trinomium* in which there is the number or not, or in more terms in which there is [the number] or not, or in a wild quantity, namely [a quantity] that is not in any kind of roots, not even composed by those, nor left by subtraction.[58]

[58] "*Necesse est igitur ut huiusmodi æstimatio universalis sit aut sub binomio, in quo sit numerus, aut in quo non sit, aut trinomio in quo sit numerus, aut in*

At this stage, Cardano recaps the situation as follows. "Universal" (or "general") values for the equation $x^3 = a_1 x + a_0$, with a_1, a_0 rational, can go under the shapes of any *binomium* (and *recisum*), any *trinomium* (and *recisum*), or any "wild [*sylvester*]" quantity – whatever a "wild" quantity is, since Cardano's subsequent explanation is completely unclear.[59] In this way, also the *binomia* (and the *recisa*) of the 3rd or 6th type $\sqrt{a} \pm \sqrt{b}$ are encompassed, though they had already been counted out on two occasions.

Concerning clearness, **Chapter XI** "How many and what parts of the cube, and on the necessity of those, and what incommensurable [parts are] [*Partes cubi quot et quæ, et de necessitate illarum, et quæ incommensæ*]" is by far one of the worst chapters. Its extreme obscurity is a good example of Cardano's bad style in the *Aliza*. Anyway, this chapter instantly draws our attention, since it starts by rephrasing the paragraph from Chapter I that I have quoted at the beginning of this section:

> [t]herefore we repeat and we say that, regarding the side of the cube the quantity of which is searched, if the cube must be equal to two [parts], some things and a number, it is necessary that the cube [is] divided into such a way

quo non sit, aut in pluribus nominibus in quo sit, aut in quo non sit, aut in quãtitate sylvestri, scilicet quæ non sit in aliquo genere radicum, nec composita ex illis, nec per detractionem relicta", see [9, Chapter X, Paragraph 6, page 20].

59 Cardano adds a further explanation obscure as well. "as the quantity the square root of which multiplied by the *residuum* to 12 produces 2, whence the chapter cannot be discovered" or "*velut quantitas cuius R ducta in residuum ad 12 producat 2, ubi capitulum inventum non esset*", see [9, Chapter X, Paragraph 6, page 20].

It seems then that a "wild quantity" s is such that $\sqrt{s}(12 - s) = 2$. The solutions of the equation $s^{\frac{3}{2}} + 2 = 12 s^{\frac{1}{2}}$ with $s \geq 0$ correspond to the solutions of the equation $p^3 + 2 = 12p$ with $p \geq 0$ *via* the substitution $p = s^{\frac{1}{2}}$. Since this last equation has $\Delta_3 < 0$, the formula contains imaginary numbers. Then, a "wild quantity" could maybe be related to imaginary numbers. This is the only occurrence of the term in the whole *Aliza*.

> at least into two, then its side. Indeed, from one it comes nothing but one. Then, at least in two.[60]

First of all, Cardano reminds that the unknown must be divided at least in two parts. We consider then $x = y + z$. In the following, using a very Euclidean terminology for irrational numbers, Cardano takes again into account the remaining possibilities (since the case 'y commensurable in length to z' has just been counted out). The steps follow such an awfully haphazard and tiresome layout that I had to considerably reorganise them.

According to the terminology of *Elements*, Book X, two numbers y, z can have different kinds of relationships. They can be commensurable or incommensurable,[61] or commensurable or incommensurable to the square power.[62] Indeed, we can easily generalise the definition of '(in)commensurable to the square power' to '(in)commensurable to the

[60] "*Repetamus igitur et dicamus quod latus cubi, cuius quantitas quæritur, si debet æquari cubus duobus rebus et numero, oportet ut cubus sic divisus in duo saltem, ergo latus eius, nam ex uno non provenit nisi unum. Ergo in duo saltem*", see [9, Chapter XI, page 20].

[61] According to Campano, "*[q]uantitates quibus fuerit una quantitas communis eas numerans: dicentur communicantes. Quibus vero non fuerit una communis quantitas eas numerans: dicentur incommensurabiles*" and, according to Zamberti, "*[c]ommensurabiles magnitudines dicuntur quas eadem mensura dimetietur. Incommensurabiles autem quæ sub nullius communis mensuræ dimensionem cadunt*", see [32, Book X, Definitions 1-2, page 133r]. Again, according to Campano, "*[o]mnium duarum quantitatum communicantium est proportio tanquam numeri ad numerum*" and, according to Zamberti, "*[c]ommensuurabiles magnitudines adinvicem rationem habent quam numerus ad numerum*", see [32, Book X, Proposition 5, page 135v]. The *vice versa* also holds, see [32, Book X, Proposition 6, pages 135v-136r]. Finally, according to Zamberti, "*[i]ncommensurabiles magnitudines adinvicem rationem non habent quam numerus ad numerum*", see [32, Book X, Proposition 7, page 136v]. Then, y, z are commensurable (respectively incommensurable) if and only if $\frac{y}{z}$ is rational (respectively irrational).

[62] According to Campano, "*[l]ineæ in potentia communicantes dicuntur quarum superficies quadratas una communis superficies numerat. Lineæ incommensurabiles in potentia dicuntur quarum superficies quadratas non numerat una communis superficies*" and, according to Zamberti, "*[r]ectæ lineæ potentia commensurabiles sunt quando quæ ab ipsis quadrata eadem area dimetitur. Incommensurabiles autem quando ea quæ ex ipsis quadrata nulla area communi*

3rd power' (and actually to '(in)commensurable to the nth power' for any natural n), encompassing then the definition of 'commensurable' in 'commensurable to the 1st power'.[63] This is not only a more comprehensive way of speaking, but as a matter of fact suits one of Cardano's needs. Indeed, when for the first time he says that two numbers are "commensurable to the power [*potestate commensæ*]", he does not specify to which power, but indeed his examples concern the pairs a, $\sqrt[3]{b}$, and $\sqrt[3]{a}$, $\sqrt[3]{b}$ (where a, b are rational such that their cubic roots are irrational), which are indeed commensurable only to the 3rd power. Nevertheless, referring to those kinds of numbers, Cardano never explicitly uses '(in)commensurable to the 3rd power', but rather a generic '(in)commensurable to the power', or a deceptive '(in)commensurable to the 2nd power'. For instance, he states that $\sqrt[6]{32}$ and $\sqrt{2}$ are commensurable to the 2nd power, but he actually shows that they are commensurable to the 3rd power. Then, we need the above extended terminology in order to clarify Cardano's one.

Cardano employs this pseudo-Euclidean terminology for two purposes. On the one hand, it gives a criterion on which one can build a classification for the irrational shapes of the solutions of a cubic equation with real coefficients. On the other hand, it conveys a distinction that Cardano will often use in his justifications. Indeed, it eventually highlights whether one can sum or not the terms in the cube of the binomial $x^3 = (y+z)^3 = y^3 + 3y^2z + 3yz^2 + z^3$. As we have already hinted to on page 294, if two numbers v, w are commensurable, then $\frac{v}{w}$ is rational. This means, for instance, that $v = kw$ with k rational, and therefore $v + w = w(k+1)$, where $k+1$ is rational. In this way,

mensura dimetitur", see [32, Book X, Definitions 3-4, page 133r]. Then, y, z are commensurable (respectively incommensurable) in square if and only if $\left(\frac{y}{z}\right)^2$ is rational (respectively irrational).

[63] Note that, fixed one n, the relation 'being incommensurable to the nth power' is the negation of the relation 'being commensurable to the nth power'. This last relation is transitive.

Note moreover that 'being commensurable to the nth power' implies 'being commensurable to the mth power' for each natural m that is a multiple of n, or equivalently 'being incommensurable to the nth power' implies 'being incommensurable to the mth power' for any natural m that divides n.

one manages to write a certain sum of two terms as one monomial. The point at issue clearly arises if we consider that in most of the cases Cardano uses numerical examples, where, to expound a generic argument, one automatically writes a sum such as the one above as a monomial. Then, when we take the cube of a certain numerical binomial, the (in)commensurability of the terms of the binomial says something about the number of terms of its cube. In the end, we recall that these surveys on the (in)commensurability relationships are linked to the method of the splittings. Indeed, knowing how many terms there are in the cube of the binomial $(y+z)^3$ is one of Cardano's key arguments to set the substitution $x = y + z$.

As a preliminary step, Cardano remarks that, when y, z are incommensurable to the 1st power, then $\frac{yz^2}{y^3}, \frac{y^2z}{z^3}, \frac{y^2z}{yz^2}$ (which are all equal to $\frac{y}{z}$) are irrational (and thus we cannot lower the number of the terms in the cube of the binomial). When instead y, z are commensurable to the 3rd power (and not to the 1st), then $\frac{y^3}{z^3}$ is rational (thus $y^3 + z^3$ can be written as one term) and $\frac{y^3}{y^2z}, \frac{y^2z}{yz^2}, \frac{y}{z}$ are irrational (thus we cannot lower the number of the terms). As said, Cardano uses the fact that certain terms in the cube of the binomial $x = y + z$ can be summed (or not) to count how many terms (and of which kind) remain. This can entail a contradiction when the cube x^3 is equated to $a_1x + a_0$, excluding thus some of the possible shapes for a solution. This general plan is quite clear, but unfortunately not its details. More or less following this overall ordering, but with a lot of repetitions, Cardano affirms that:

A XI.i when $y = \sqrt[3]{a}$, $z = \sqrt[3]{b}$, and y, z are incommensurable to the 1st power [and to the 2nd power[64]], then y^3, z^3 are rational,

[64] This last condition is automatically verified if y, z are incommensurable to the 1st power. Indeed, consider the polynomial $P(u) = u^3 - \frac{a}{b} \in \mathbb{Q}[u]$ and call $t = \sqrt[3]{\frac{a}{b}}$. Then, t satisfies $P(u)$, which is its minimal polynomial, since $P(u)$ (with no rational roots) is irreducible. Let us now consider another polynomial $Q(u) \in \mathbb{Q}[u]$ such that $Q(t) = 0$. Then, $P(u)$ divides $Q(u)$ and the degree of $P(u)$ is greater than, or equal to, 3. Suppose by the absurd that t^2 is

but not y^2z, yz^2.[65] Then, the sum $y+z$ cannot be a solution of $x^3 = a_1x + a_0$;[66]

A XI.ii when y, z are not cubic roots and are commensurable to the 3rd power [and incommensurable to the 1st power], then the ratio $\frac{y^2z}{yz^2}$ is irrational.[67] In particular, if $y = \sqrt[6]{32}$ and $z = \sqrt{2}$, then y^2z, yz^2 are irrational, and also y^3, z^3;[68]

A XI.iii when $y, z = \sqrt[3]{a \pm \sqrt{b}}$ [that is, y, z are incommensurable to the 1st, 2nd, and 3rd powers], then their sum cannot be a solution of

---

rational (that is, y, z are commensurable to the 2nd power). Then, the choice $Q(u) = u^2 - t^2$ is satisfied by t, but has degree 2, hence the contradiction.

[65] "Indeed, if AC and CB were not commensurable, then neither g and f, nor d and e [would be commensurable], but d and g are numbers, because [they are] cubes of cubic roots, therefore e and f cannot be numbers" or "*nam si non sint commensæ ac et cb, igitur nec g et f, nec d et e, sed d et g sunt numeri, quia cubi Rcu: igitur e et f non possunt esse numeri*", see [9, Chapter XI, page 21]. We have called $AC = y$, $CB = z$, $d = y^3$, $e = y^2z$, $f = yz^2$, and $g = z^3$.

[66] "Then AB cannot be composed by two incommensurable cubic roots, because the parallelepipeds were not numbers. [...] Then, in any case, if the number was not near, none of the two quantities can satisfy the parallelepipeds according to the number in order that the remaining [part] of the cube satisfy the things" or "*[n]on ergo potest esse ab composita ex duabus R cubicis incommensis, quia parallelipeda non essent numeri [...]. Nullæ ergo duæ quantitates aliquo modo si non adsit numerus possunt satisfacere parallelipedis pro numero, ut reliquum cubi satisfaciat rebus*", see [9, Chapter XI, page 21].

[67] "In truth if they were commensurable to the second [third] power [...], but, if they were not cubic roots, e will be to f as AC to CB, but AC is not commensurable to CB, therefore not even e to f" or "*[s]i vero commensæ potentia secunda [...], sin autem no essent Rcu: erit e ad f ut ac ad cb, sed ac non est commensum cb, ergo nec e cum f*", see [9, Chapter XI, page 22].

[68] "The said example is $\sqrt[6]{32}+\sqrt{2}$. According to the single parallelepipeds to avoid the effort, the cube is $\sqrt{72}+\sqrt[6]{2048}+\sqrt[6]{8192}$. Here it is agreed that no number is made, therefore it cannot be appropriate. I only say that no parallelepipeds can be a number under these hypotheses" or "*[e]xemplum dictum est Rcu:quad. 32 p: R 2 cubus secũdum simplicia parallelipeda ad laborem fugiendum est Rquad. 72 p: Rcu:quad. 2048 p: Rcu:quad. 8192. Hic constat nullum fieri numerum, ideò convenire non potest. Dico modo quòd nullum parallelipedum potest in his suppositis esse numerus*", see [9, Chapter XI, page 22].

the considered equation.[69]
Cardano's arguments are sometimes unclear. Unfortunately, he only provides the numerical example in (A XI.ii). Cardano's statements are quite deceptive, up to the point of simply being wrong due to a lack of accuracy. A blatant case is (A XI.i), where he does not mention the solution $\sqrt[3]{a}+\sqrt[3]{b}$, with $\sqrt[3]{ab}$ rational, which is a clear counter-example to the statement. Moreover, if we take a, b irrational and consider (A XI.iii) as a generalisation of (A XI.i), then the canonical shape for a solution given by the cubic formula is also not taken into account. The Euclidean terminology makes Cardano mislead. If this terminology enables Cardano to study the number of the terms in the cube of a binomial, it also prevents him from having a good understanding of the fact that the sum of two cubic roots can actually be a solution. In this last case, he merely did not choose the most befitting language.

Facing such an enigmatic chapter, I suggest the following interpretation. Consider $x = y + z$ and its cube $y^3 + 3y^2z + 3yz^2 + z^3$. We want to infer some information on the (in)commensurability to the nth power (for $n = 1, 2, 3$) of the terms in the cube depending on the relations between y and z. Then, we are able to count how many parts there are in the cube and to specify y, z such that x is a solution of $x^3 = a_1x + a_0$, with a_1, a_0 rational. More precisely, we want to study what happens when the ratios $\frac{y^3}{y^2z}, \frac{yz^2}{z^3}$ (both equal to $\frac{y}{z}$), $\frac{y^3}{yz^2}, \frac{y^2z}{z^3}$ (both equal to $\frac{y^2}{z^2}$), and $\frac{y^3}{z^3}$ are either rational, or not. There are five cases.

If $\frac{y}{z}$ is rational, then $\frac{y^2}{z^2}$ and $\frac{y^3}{z^3}$ are rational too. In this case, x has one (rational or irrational) part, and can be a solution of the equation. This is one of the cases that Cardano charges to be non-"general". If $\frac{y}{z}$ is irrational and $\frac{y^2}{z^2}$, $\frac{y^3}{z^3}$ are rational, then y^3, y^2z are incommensurable to the 1st power as well as yz^2, z^3 and y^2z, yz^2. Moreover, y^3, yz^2 are

[69] "Nor could these two parts be universal cubic roots. Since, [...] if [the parts are] incommensurable, there will be made four incommensurable parts in the cube, therefore one will be superfluous" or "*Neque poterunt hæ duæ partes esse RV:cu.. Quoniam [...] si incommensæ fient quatuor partes in cubo incommensæ, ergo una erit superflua*", see [9, Chapter XI, page 22].

commensurable to the 1st power as well as y^2z, z^3 and y^3, z^3. But this cannot be. Indeed, y^3, y^2z, yz^2, z^3 will be all commensurable to the 1st power by transitivity. If $\frac{y}{z}$ is irrational, $\frac{y^2}{z^2}$ is rational, and $\frac{y^3}{z^3}$ is irrational, then y^3 and y^2z, yz^2 and z^3, y^2z and yz^2, y^3 and z^3 are incommensurable to the 1st power, while y^3 and yz^2, and y^2z and z^3 are commensurable to the 1st power. In this case, x^3 has two parts. This is for instance what happens when $x = \sqrt{a} + b$ is a *binomium* of the 2nd or 5th type. If $\frac{y}{z}$ and $\frac{y^2}{z^2}$ are irrational and $\frac{y^3}{z^3}$ is rational, then y^3 and y^2z, yz^2 and z^3, y^2z and yz^2, y^3 and yz^2, y^2z and z^3 are incommensurable to the 1st power, while y^3 and z^3 are commensurable to the 1st power. In this case, x^3 has three parts. This is for instance what happens when $x = \sqrt[3]{a} + \sqrt[3]{b}$ with $\sqrt[3]{ab}$ rational. Finally, if $\frac{y}{z}$, $\frac{y^2}{z^2}$, and $\frac{y^3}{z^3}$ are irrational, then all the terms in the cube are two by two incommensurable to the 1st power. In this case, x^3 has four parts. But it cannot be equated to $a_1x + a_0$, since it only has three parts. Note that the relations of 'being (in)commensurable to the nth power' for $n > 3$ are not influential in this classification, since only rational numbers up to the third power appear in the cube.

Cardano's above statements fit this scheme: (A XI.i) (and partially (A XI.ii)) correspond to the fourth case, while (A XI.iii) corresponds to the fifth case. Moreover, in this chapter and in others, Cardano has already dealt with the first case.

At around the middle of this chapter, Cardano starts to recap:

A XI.iv The solution of the considered equation may not contain cubic roots at all (whether universal, or with numbers or not).[70] Take a, b, c rational such that their cubic roots are irrational. Then:

[70]"In the end it is deduced from this that that part of the chapter of the cube equal to some things and a number cannot consist of quantity composed by two simple or universal cubic roots or by the number and a cubic root" or "*[e]x quo tandem concluditur, partem illam capituli cubi æqualium rebus et numero non posse consistere in quantitate composita ex duabus R cubicis simplicibus aut universalibus, aut numero et R cubica*", see [9, Chapter XI, page 23].

- when $y = a$ and $z = \sqrt[3]{b}$, then their sum cannot be a solution of $x^3 = a_1x + a_0$ (by substitution and comparison);[71]
- when $y = \sqrt[3]{a}$ and $z = \pm\sqrt[3]{b}$, then their sum cannot be a solution of $x^3 = a_1x + a_0$;[72]

[71]"Indeed in the number and cubic root it will be necessary to give the cubes to the number, because [the cubes] will be numbers, then they will not satisfy a small number. Besides the parallelepipeds will be incommensurable and [will be] two cubic roots and in the thing there is nothing but one part that is a cubic root, therefore the number of the things will not be [given]'" or "*in numero et R cubica oportebit dare cubos numero, quia erunt numeri, ergo in numero parvo non satisfacient, præterea parallelipeda incommensa erunt et duæ* Rcu: *et in re non est nisi una pars quæ sit* Rcu: *igitur non erit numerus rerum*", see [9, Chapter XI, page 23].

[72]"And not even if both parts are cubic roots, since if you gave the parallelepipeds to the number, firstly they will not necessarily fit the cubes if [the cubic roots] are not commensurable, then the cubic roots will not be a number. Besides the cubes will be numbers, then they could not contain the thing according to the number, since the thing consists of two cubic roots multiplied by the number, they would produce the number. And we cannot even give both cubes, when they are positive, to the number, where the number was not smaller than the fourth part of the whole cube, as we taught. But where it is bigger or equal, we give [both cubes to the number] and that part of the chapter of the cubes equal to some things and a number that is already known is made. Therefore the remaining part in that equality never takes place. And we cannot even give the difference of the cubes to the number, as in a positive and a negative cubic root like $\sqrt[3]{6} - \sqrt[3]{2}$, because the negative parallelepipeds will be bigger and the positive [parallelepipeds] [will be] smaller. Then, since in the thing the positive cubic root is necessarily bigger than the negative cubic root, the things could in no way be contained in the parallelepipeds according to the number, but, joining well the positive [part] of the things with the negative [one] and the negative [part] with the positive [one] with the cube, the chapter of the cube and some things equal to a number is perfectly made" or "*[n]eque si ambæ partes sint* Rcu: *, quoniam si dederis parallelipeda numero primum non convenient cubi necessario, si non sint commensa, sint* Rcu: *ergo non numerus. Præterea cubi erunt numeri, ergo non poterunt res continere per numerum, cum res constet ex duabus* Rcu: *ductæ in numerum, producerent numerum. Neque possumus dare utrunque cubum p: numero ubi numerus sit minor quarta parte totius cubi, ut docuimus, ubi autem est maior vel æqualis damus, et fit illa pars capituli cubi æqualis rebus et numero, quæ iam nota est, igitur reliqua pars in hac æquatione nullum habet locum. Neque possumus dare differentiam cuborum numero, ut in* Rcu: *p: et m:, velut* Rcu: *6 m:* Rcu: *2,*

- the *trinomium* $\sqrt[3]{a}+\sqrt[3]{b}+\sqrt[3]{c}$ [where $\sqrt[3]{a}$, $\sqrt[3]{b}$, $\sqrt[3]{c}$ are two by two incommensurable to the 1st and 2nd powers and $\sqrt[3]{abc}$ is irrational], such as $\sqrt[3]{6}+\sqrt[3]{5}+\sqrt[3]{2}$, cannot be a solution of $x^3 = a_1x + a_0$ (by substitution and comparison);[73]

A XI.v when $y = \sqrt{a}$ and $z = b$, then their sum cannot be a general solution of $x^3 = a_1x + a_0$ [referring to Chapter I];[74]

A XI.vi when $y = a$ and $z = \sqrt{b \pm \sqrt{c}}$, then their sum cannot be a solution of $x^3 = a_1x + a_0$ (by substitution and comparison);[75]

---

quia parallelipeda m: erunt maiora et p: minora, ergo cum in re R*cu: p: sit maior* R*cu: m: necessario, nullo modo res poterunt contineri per numerum in parallelipedis, sed bene iungendo p: cum m: et m: cum p: rerum cum cubo fiet ad unguem capitulum cubi et rerum æqualium numero*", see [9, Chapter XI, page 23].

Cardano's argument is two-folded. First, take $x = \sqrt[3]{a}+\sqrt[3]{b}$ to be a solution of $x^3 = a_1x + a_0$ with a_1, a_0 rational. Consider $a + 3\sqrt[3]{a^2b} + 3\sqrt[3]{ab^2} + b = a_1\sqrt[3]{a} + a_1\sqrt[3]{b} + a_0$. It is evident that Cardano cannot take $3\sqrt[3]{a^2b}+3\sqrt[3]{ab^2} = a_0$ and $a + b = a_1\sqrt[3]{a} + a_1\sqrt[3]{b}$. But also the assignment $a + b = a_0$ and $3\sqrt[3]{a^2b} + 3\sqrt[3]{ab^2} = a_1\sqrt[3]{a} + a_1\sqrt[3]{b}$ does not always work, namely – Cardano argues making implicitly reference to (A I.1) – when $a_0 < \frac{1}{4}x^3$ (see here, on page 269). Cardano omits to mention the condition $\sqrt[3]{ab}$ rational, under which the assignment partially works.

Second, take $x = \sqrt[3]{a} - \sqrt[3]{b} > 0$ (with $\sqrt[3]{a} > \sqrt[3]{b}$) and consider $a - 3\sqrt[3]{a^2b} + 3\sqrt[3]{ab^2} - b = a_1\sqrt[3]{a} - a_1\sqrt[3]{b} + a_0$. Then, the assignment $a - b = a_0$ and $-3\sqrt[3]{a^2b} + 3\sqrt[3]{ab^2} = a_1\sqrt[3]{a} - a_1\sqrt[3]{b}$ only works for $x^3 + a_1x = a_0$, since $-3\sqrt[3]{a^2b}+3\sqrt[3]{ab^2} < 0$.

73 "Therefore let AB to be divided into three parts that all are incommensurable cubic roots and not in the same proportion and it is agreed that eight kinds of bodies are made [...] the cube will not be equal to some things under any number" or "*[i]am ergo ventum est necessario ad triarios, sit ergo ab divisa in tres partes, quæ omnes sint* R*cu: incommensæ, nec in eadem proportione, et constat quod fient octo genera corporum [...] cubus non poterit æquari rebus sub aliquo numero*", see [9, Chapter XI, pages 23-24].

74 "But the value cannot consist of a number and a square root in order to be general, indeed this is shown above" or "*[s]ed neque æstimatio potest constare ex numero et R quadrata, ut sit generalis, hoc enim est demonstratum supra*", see [9, Chapter XI, page 23].

75 "[The value] cannot consist of a number and a universal square root, because two incommensurable parts will be in the cube beside the number (because the universal root is not a first power commensurable to the number, see Chapter

A XI.vii when $y = a$ and $z = \sqrt[3]{b \pm \sqrt{c}}$, then their sum cannot be a solution of $x^3 = a_1x + a_0$ (by substitution and comparison).[76]

As said, this chapter is quite messy. Due to the Euclidean terminology employed, it stands a bit aside from the other chapters considered in this section. It is composed by three parts, respectively containing the preliminary remarks, the statements (A XI.i)-(A XI.iii), and the statements (A XI.iv)-(A I.7). It happens that sometimes one part contains a statement that is not consistent with a statement in another part. We can explain this contradiction saying that the three parts have been juxtaposed precisely because of the Euclidean terminology employed. The fact that Cardano employs a terminology with ancient roots and which is manifestly not up-to-date with his aims, together with the fact that he fails to recognise a shape for the irrational solutions of $x^3 = a_1x + a_0$ (that he would have not missed if he had had the cubic formula available), makes me suppose that Chapter XI attests one of the earliest inquiries by Cardano on the topic.

Chapter XIII "On the discovery of the parts of the cubic *trinomium*, which produces the cube with two only cubic parts [*De inventione partium trinomii cubici, quod cubum producit cum duabus partibus tantum cubicis*]" is devoted to the study of the *trinomia* as solutions of $x^3 = a_1x + a_0$, with a_1, a_0 rational. Cardano maintains the following statements.

III at the end) and only one [part] in the thing, then the number of the things does not correspond" or "*neque potest constare ex numero et RV: quadrata quia in cubo erunt duæ partes præter numerum incommensæ (quia RV: non est potentia prima commensa numero) et in re una tantum ergo non constabit numerus rerum*", see [9, Chapter XI, page 23].

[76]"And not even by a number and a universal cubic root. It will be necessary to give the cubes to the number and the parallelepipeds will be two incommensurable, then as before, being only one universal cubic root, the thing will not be able to be contained in the cube under any number" or "*[n]eque ex numero et RV: cu. quoniam oportebit dare cubos numero et parallelipeda erunt duo incommensa, ergo ut prius cum sit tantum una RV: cu. non poterunt res numero aliquo contineri in cubo*", see [9, Chapter XI, page 23].

A XIII.i. *Consider the equation* $x^3 = a_1x + a_0$. *Then, no irrational* $\sqrt[3]{a} + \sqrt[3]{b} + \sqrt[3]{c}$ *with* $\sqrt[3]{abc}$ *rational (or, equivalently,*[77] *such that* $\sqrt[3]{a} : \sqrt[3]{b} = \sqrt[3]{b} : \sqrt[3]{c}$*) can be a solution of the equation.*

A XIII.ii. *Consider the equation* $x^3 = a_1x + a_0$. *Then, no irrational* $\sqrt[3]{a} + \sqrt[3]{b} + c$ *[with* $\sqrt[3]{abc}$ *rational or, equivalently,] such that* $\sqrt[3]{a} : \sqrt[3]{b} = \sqrt[3]{b} : c$ *can be a solution of the equation.*

Cardano plainly justifies these statements by substitution and comparison, employing only a bit of the Euclidean terminology from Chapter XI. Here the (in)commensurability relationships are no longer the centrepiece: the classification of the shapes for the solution of the considered equation is no longer based on them. They are used only insofar as they enable to infer how many parts there are in the cube x^3, given a certain shape for the solution x.

For instance, in the case (A XIII.i), similarly to what was done in the third point of (A XI.iv), Cardano remarks[78] that, if by the absurd $x = \sqrt[3]{a} + \sqrt[3]{b} + \sqrt[3]{c}$, with $\sqrt[3]{abc}$ rational, had been assumed, then its cube would have been

$$x^3 = (a+b+c+6\sqrt[3]{abc})+3(\sqrt[3]{ab^2}+\sqrt[3]{a^2c}+\sqrt[3]{bc^2})+3(\sqrt[3]{a^2b}+\sqrt[3]{b^2c}+\sqrt[3]{ac^2}),$$

where the parentheses help to identify the groups of terms commensurable to the 1st power one to the other. Indeed, since $\sqrt[3]{abc}$ rational

---

[77] Indeed, if $\sqrt[3]{abc}$ is rational, then we can write $\sqrt[3]{c} = \frac{k}{\sqrt[3]{ab}}$, with k rational, and the proportion holds. *Vice versa*, if the proportion $\sqrt[3]{a} : \sqrt[3]{b} = \sqrt[3]{b} : \sqrt[3]{c}$ holds, then $\sqrt[3]{b^2} = \sqrt[3]{ac}$, which implies that $\sqrt[3]{abc}$ is rational.

[78] "And I only say that, if the cubic *trinomium* the parts of which multiplied produced the number was assumed, only two parts that are cubic roots would be produced, but there are three incommensurable parts in the thing – as it is said, therefore the parts of the cube cannot contain the parts of the things according to the number" or "*dico modo quod si assumatur trinomium cubicum, ex cuius ductu partium producatur numerus, quod producentur duæ partes tantum, quæ R cubicæ, sed in re sunt tres partes incommensæ, ut dictum est, igitur partes cubi non possunt continere partes rerum secundum numerum*", see [9, Chapter XIII, page 27].

implies that $\sqrt[3]{c} = \sqrt[3]{\frac{b^2}{a}}$, we can rewrite the cube as follows

$$x^3 = (a + b + c + 6\sqrt[3]{abc}) + 3(a + 2b)\sqrt[3]{\frac{b}{a}} + 3(2a + b)\sqrt[3]{\frac{b^2}{a^2}}.$$

Equivalently, let us call the irrational $\frac{\sqrt[3]{a}}{\sqrt[3]{b}} = \frac{\sqrt[3]{b}}{\sqrt[3]{c}} = k$. Then, k^3 is rational, $\sqrt[3]{b} = k\sqrt[3]{c}$, and $\sqrt[3]{a} = k^2\sqrt[3]{c}$. The cube can thus be rephrased as

$$x^3 = (k^6c + k^3c + c + 6k^3c) + 3c(k^3 + 1)k + 3c(k^3 + 2)k^2.$$

Cardano concludes that x^3 cannot be equated to $a_1x + a_0$ because the number and the kinds of cubic roots on the two sides of the equal are not the same.

In **Chapter XVI** "That a *quadrinomium* from cubic roots is reduced to three parts, two of which are only cubic roots, or by far many [*Quod quadrinomii ex radicibus cub. cubus ad tres partes quarum duæ sint tantum R cubæ reducitur, aut longe plures*]" Cardano finally takes into account the *quadrinomia* as solutions of $x^3 = a_1x + a_0$, with a_1, a_0 rational. This chapter is in the same style (justifications by substitution and comparison, with a little bit of the Euclidean terminology) as Chapter XIII.

A XVI.i. *Consider the equation* $x^3 = a_1x + a_0$. *Then, no irrational* $\sqrt[3]{a} + \sqrt[3]{b} + \sqrt[3]{c} + \sqrt[3]{d}$ *such that* $\sqrt[3]{a} : \sqrt[3]{b} = \sqrt[3]{b} : \sqrt[3]{c} = \sqrt[3]{c} : \sqrt[3]{d}$ *can be a solution of the equation.*

Note that the condition stated in the continued proportion implies that $\sqrt[3]{abc}$ is rational. More precisely, let us call the irrational $\frac{\sqrt[3]{a}}{\sqrt[3]{b}} = \frac{\sqrt[3]{b}}{\sqrt[3]{c}} = \frac{\sqrt[3]{c}}{\sqrt[3]{d}} = k$. Then, $\sqrt[3]{c} = k\sqrt[3]{d}$, $\sqrt[3]{b} = k^2\sqrt[3]{d}$, and $\sqrt[3]{a} = k^3\sqrt[3]{d}$, and therefore $\sqrt[3]{abc} = k^6d$ is rational, since k^3 is rational. We moreover remark that the *quadrinomium* is actually a *trinomium*, since $\sqrt[3]{a} + \sqrt[3]{b} + \sqrt[3]{c} + \sqrt[3]{d} = (k^3 + 1)\sqrt[3]{d} + k^2\sqrt[3]{d} + k\sqrt[3]{d}$.

Similar considerations also hold for the following two propositions.

A XVI.ii. *Consider the equation* $x^3 = a_1x + a_0$. *Then, no irrational* $\sqrt[3]{a} + \sqrt[3]{b} + c + \sqrt[3]{d}$ *such that* $\sqrt[3]{a} : \sqrt[3]{b} = \sqrt[3]{b} : c = c : \sqrt[3]{d}$ *can be a solution of the equation.*

A XVI.iii. *Consider the equation* $x^3 = a_1x + a_0$. *Then, no irrational* $\sqrt[3]{a}+\sqrt[3]{b}+\sqrt[3]{c}+\sqrt[3]{d}$, *with*[79] $\sqrt[3]{a^2bc}$ *or* $\sqrt[3]{bcd^2}$ *rational, can be a solution of the equation.*

A XVI.iv. *Consider the equation* $x^3 = a_1x + a_0$. *Then, no irrational* $\sqrt[3]{a} + \sqrt[3]{b} + \sqrt[3]{c} + \sqrt[3]{d}$, *with no relation between* $\sqrt[3]{a}, \sqrt[3]{b}, \sqrt[3]{c}, \sqrt[3]{d}$, *can be a solution of the equation.*

We remark that in this case there are many more cubic roots incommensurable to the 1st power contained in the cube than in the preceding propositions (namely, 16 cubic roots in the development of the cube are incommensurable one to the other). Then – Cardano ends – "no *quadrinomium* by cubic roots is suitable [*nullum quadrinomium ex R cubicis non est idoneum*]".

Chapters V, X, XI, XIII, and XVI are located in the first third of the *Aliza*. In the main, they are spaced out by chapters on different topics.[80] Then, a bit outdistanced, we find **Chapter XXIV** "Demonstration that shows that no chapter except the discovered [ones] can be known generally [*Demonstratio ostendens quod caput nullum præter*

[79]Cardano only says to take the four cubic roots such that " two [cubic roots] times the square of another produce a number[*duæ in quadratum alterius producant numerum*]". Indeed, considering that $\sqrt[3]{c} = k\sqrt[3]{d}$, $\sqrt[3]{b} = k^2\sqrt[3]{d}$, and $\sqrt[3]{c} = k^3\sqrt[3]{d}$, there are two possibilities. Indeed, either $\sqrt[3]{a^2bc} = k^9d\sqrt[3]{d}$ or $\sqrt[3]{bcd^2} = k^3d\sqrt[3]{d}$, both assuming that $\sqrt[3]{d}$ is rational, can only be rational (the other possibilities contain the powers of k that are not multiple of 3). Then, the *quadrinomium* is actually a *trinomium* one term of which is the number.

[80]See for instance Chapter VI on the sign rule (see below, on page 379) or Chapter XII on the geometrical proof for the existence of a (positive, real) solution of $x^3 = a_1x + a_0$ (see below, on page 361). Note that, despite the title, Chapter XIV teaches how to find the coefficients of an equation of the families $x^3 + a_0 = a_1x$ and $x^3 = a_1x + a_0$, once the shape of one of its solutions has already been given. Also Chapter XV, in which Cardano harks back on the splittings, does not concern the study of the possible shape for the solutions of $x^3 = a_1x + a_0$.

inventa generale sciri potest]". Unfortunately, it is another of the *Aliza*'s chapters in which Cardano's bad Latin overcomes clearness. The chapter is divided into two parts. In the first one, which corresponds to the first short paragraph, Cardano maintains that no equation with more than four different terms (regarding their powers, that is, any equation of degree higher than four) that cannot be reduced to another with fewer terms (that is, either to an equation of lower degree that Cardano is able to solve, or to an equation that can be solved by substitution, like the biquadratic ones) can be solved in general. When, instead, there are four terms up to the cube (that is, a complete cubic equation), one should – according to Cardano – get an equation with three terms,[81] and, from there, the equation $x^3 = a_1x + a_0$.[82]

[81]This (and in particular reducing a complete cubic equation to a depressed one) was a common strategy, achieved through some well known substitutions (see, for instance, the substitutions in Section 3.4 on page 3.4). And indeed, in his poem Tartaglia mentions only depressed cubic equations.

[82]The paragraph is quite tricky, then here the quotation follows so that the reader can judge by himself: "[i]t remains to show what has been proposed at the beginning, because of which we wrote this, namely that there is no other general chapter that can be known beyond those [ones] that are bequeathed. Since, more than four different kinds [of terms] cannot be reduced to fewer, either by a division, or [by taking the] root, or by a change, or [by] an own rule, or [by] suppressing, or for the sake of the origin, or by a Geometrical demonstration, because there are great inequalities in the single [terms] that can hardly be understood in four quantities and, if the perfection cannot be found in these, the less [it can be found] in those. Then on this, if there are four quantities up to the cube, it is now wise to reduce to three quantities and [to reduce] all the chapters of three quantities to the cube equal to some things and a number. Therefore, if I will show that this cannot be general, the purpose is also clear in the unknown part" or "*[r]eliquum est ut ostendamus quod ab initio propositum est, cuius causa hæc scripsimus, scilicet non esse capitulum aliud generale, quod sciri possit, ultra ea quæ tradita sunt, quoniam ultra quatuor diversa genera nisi possit reduci ad pauciora, vel per divisionem, vel radicem, aut per mutationem, aut regulam propriam, vel deprimendo, aut ob originem, aut per demonstrationem geometricam, cum in singulis sint magnæ inæqualitates, quæ vix possunt intellegi in quatuor quantitatibus, nec in eis potuerit invenire perfectio quanto minus in illis. De his ergo, si sint quatuor usque ad cubum, iam doctus est reducere ad tres quantitates, et capita trium quantitatum omnia ad cubum æqualem rebus et numero. Si igitur ostendero*

We remark that Cardano is suggesting again the same strategy as in the first paragraph of Chapter I. Note moreover that at this stage the quartic equations seem not to be included among the equations that Cardano is able to solve.

Cardano goes on observing that, according to (AM XXV.1) and to the third corollary to (AM XIII) (see above, respectively on pages 84 and 121), the two (positive, real) solutions of $x^3 = a_1 x + a_0$ (when $\Delta_3 < 0$ and the coefficients are rational) are a number and a *binomium* of the (2nd and) 5th type(s). Therefore, "it is necessary that the general value can be shared by the number and by the fifth *binomium* [*oportet æstimationem generalem posse communicari numero et quinto binomio*]". Cardano then focuses on the radical part of the *binomium* at issue. He firstly remarks that the cube of any *binomium* is again a *binomium* where the same radical appears. For instance, we have that $(\sqrt{a}+b)^3 = (a+b^2)+2b\sqrt{a}$. Then, taking this *binomium* as a solution and equating $(\sqrt{a}+b)^3 = a_1(\sqrt{a}+b)+a_0$, he finds some conditions on the inner structure (or shape) of the root $\sqrt{a}$. He considers the cases in which a is a perfect square and $a = \sqrt{a'}+a''$, or $a = \sqrt{a'}+\sqrt{a''}+a'''$, or $a = \sqrt[4]{a'}+\sqrt{a''}+a'''$, and by substitution and confrontation comes to the conclusion that only $a = \sqrt{a'}+a''$ is possible.

In particular, Cardano considers the equation $x^3 = 29x + 42$ (with $\Delta_3 < 0$), since he is able to write its coefficients as

$$\begin{cases} 29 = (18+\sqrt{72}) + (11-\sqrt{72}) \\ 42 = (18+\sqrt{72})\sqrt{11-\sqrt{72}} \end{cases}.$$

hoc non posse esse generale, etiam in parte ignota liquet propositum", see [9, Chapter XXIV, page 49]. Note that my understanding is based on interpreting 'four different kinds [*quatuor diversa genera*]' as 'four different kinds of terms' in an equation, and precisely as 'four different kinds of terms according to their power'. This interpretation is supported by the second sentence (in the Latin version), where the "four different kinds" becomes "four up to the cube [*quatuor usque ad cubum*]", which can subsequently be reduced to "three quantities [*tres quantitates*]", and in turn to $x^3 = a_1 x + a_0$.

The close of the above quotation is disconcerting, especially because this chapter actually ends by stating the generality of a solution.

Then, he applies (AM XXV.1), taking $f = 18+\sqrt{72}$ and $g = 11-\sqrt{72}$ (and $\sqrt{g} = \frac{3}{2} - \frac{\sqrt{2}}{2}$), to draw

$$x = \sqrt{\frac{83}{4} + \frac{18}{4}\sqrt{2}} + \frac{3}{2} - \frac{\sqrt{2}}{2}.$$

In this case, x has (almost) the shape of the considered *binomium* with another *binomium* under the square root. Then $x = 6$, since he knows that $\sqrt{\frac{83}{4} + \frac{18}{4}\sqrt{2}} = \frac{9}{2} + \frac{\sqrt{2}}{2}$. It is interesting that Cardano says that "this [$\sqrt{\frac{83}{4} + \frac{18}{4}\sqrt{2}} + \frac{3}{2} - \frac{\sqrt{2}}{2}$] is the true value of the thing [*[h]æc igitur est vera æstimatio rei*]", and not 6! Again, "this kind of value is general, because it can be made equal to a number or not, and [it can be] made equal to the *binomium* [*hoc genus æstimationis est generale, quia potest æquari numero, et non æquari, et posset æquari binomio*]".[83] On the one hand, this strengthens my interpretation of "general" solutions as solutions that go under certain codified shapes and that can encompass other shapes (see here Sections 3.1.4 and 4.2.4). On the other hand, this could be one of the earliest (chronologically speaking) passages where Cardano observes that x can be written as the sum of two parts (above called y, z), which in this case are the two radicals $\sqrt{\frac{83}{4} + \frac{18}{4}\sqrt{2}}$ and $\sqrt{\frac{11}{2} - \frac{6\sqrt{2}}{2}}$. Indeed, it is remarkable that Cardano exploits the "particular" rule (AM XXV.1) in order to study the shape of the solutions of $x^3 = a_1x + a_0$ instead of using the formula in (AM XII). It is true that, if one wants to amend the cubic formula, he may appeal to other results. Nevertheless, the fact that no mention of Ferrari's method for solving quartic equations is made makes us believe that this chapter has been written some time before the big discoveries of the *Ars Magna*.

Finally, at the very end of the *Aliza*, Cardano harks back on the topic of the possible shape for the solutions of $x^3 = a_1x + a_0$ with real coefficients. In **Chapter LVII** "On the treatment of the general value of the cube equal to some things and a number [*De tractatione*

[83] See [9, Chapter XXIV, pages 51-52].

æstimationis generalis capituli cubi æqualis rebus et numero]" Cardano states that

> the general value of the chapter of the cube equal to some things and a number is had neither by a general nor by a special rule, except by that [which is to] find a quantity that multiplied by a second [one] produces the number of the equality. That second quantity behaves like the gnomon, and is the first root or the side of the aggregate by the number of the things, and [is] discovered according to that quantity.[84]

The rule referred to is (AM XXV.2) (see above, page 84). The above quotation could also fall under the same pre-formula context as in Chapter XXIV.

In this chapter, Cardano massively appeals to the very end of **Chapter XL** "On three necessary [things] that is necessary to put before the discovery [*De tribus necessarijs quæ præmittere oportet ad inventionem*]". There, the last paragraph stands aside compared to the rest and deals with the equation $x^3 = a_1x + a_0$. Cardano states the following proposition.

A XL. *Consider the equation* $x^3 = a_1x + a_0$. *Then,*[85] $x = \sqrt{a_1 + \frac{a_0}{x}}$.

[84]"*Iam docui te quod æstimatio generalis capituli cubi æqualis rebus et numero non est habita, neque per regulam generalem neque specialem, nisi per illam, ut invenias quantitatem quæ ducta in secundam, producat numerum æquationis, et illa secunda quantitas gerit vicem gnomonis, et sit prima radix seu latus aggregati ex numero rerum, et secunda illa quantitate inventa*", see [9, Chapter LVII, page 104].

[85]We remark that this proposition could also provide an algorithm to calculate an approximate solution of the considered equation. For instance, let us consider $x^3 = 3x + 3$. We call $f(x) = \sqrt{3 + \frac{3}{x}}$ and we take $x_0 = 1$. Then, if we evaluate $f(x_0), f(f(x_0)), f(f(f(x_0))), \ldots$, and so on, up to the tenth iteration we obtain a very good approximation of the real value.

We also recall that, even though Cardano does not seem to use here the proposition in this way, there is a whole chapter of the *Ars Magna* that is devoted to calculate approximate solutions through a modified version f the method of the double false position, see [4, Chapter XXX, pages 52v-54r].

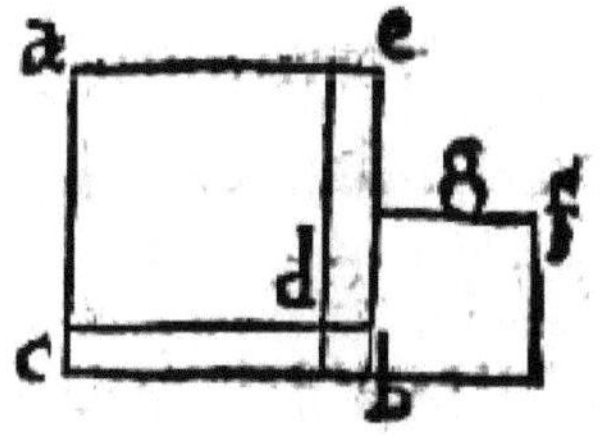

Figure 5.3. – *De Regula Aliza*, Chapter XL. Kind courtesy of Biblioteca Virtual del Patrimonio Bibliográfico and Universidad de Sevilla.

His justification trots out a so-called "orthogonal triangle [*triangulum orthogonius*]",[86] which is nowhere else mentioned in the *Aliza*.

A XL — *Proof*. As usual, Cardano interprets the equation in a geometrical environment. Considering the diagram in Figure 5.3, he takes $x = \overline{BC}$, $a_1 = (\overline{AD})^2$, $a_0 = g$. In this interpretation, where $a_1 x = (\overline{AD})^2 \overline{BC}$, it is natural to take $g = (\overline{C\,D\,E})\overline{BC}$ (and in turn Cardano makes it equal to $(\overline{BF})^2$ for the sake of brevity). Then $\overline{BC}^3 = \overline{BC}((\overline{AD})^2 + (\overline{BF})^2)$ and $\overline{BC}^2 = (\overline{AD})^2 + (\overline{BF})^2$, from where he gets the searched formula. □

Actually, the proposition as it is stated is wrong. Indeed, Cardano only considers the positive root (we have already remarked it on page 84, while dealing with (AM XXV.2), since indeed it is the same rule as there). By the way, this chapter sheds new light on how the "particular" rule in (AM XXV.2) was understood by Cardano. At the basis, the proposition describes a decomposition in two parts of the cube as a three-dimensional body. Anyway, and as before, no real positional property of the geometrical objects involved is singled out, since indeed it can be imitated by basic arithmetical operations.

86 "And its origin is from the orthogonal triangle" or "*est origo eius ex triangulo orthogonio*", see [9, Chapter XL, page 82].

Coming back to Chapter LVII, Cardano uses the above proposition[87] (or (AM XXV.2)) to test (again) the possible shape for an irrational solution of $x^3 = a_1 x + a_0$. He considers x respectively to be a *binomium* of the 2nd or 5th type $\sqrt{a} + b$, any *recisum*, and a *binomium* of the 1st or 4th type $a + \sqrt{b}$. He remarks that, substituting one of these values for x in $\sqrt{a_1 + \frac{a_0}{x}}$, he gets either a *binomium* or a *recisum*, or always a *binomium*, or a *recisum*. Then, only the shape $\sqrt{a} + b$ is admissible, since it is the only one that can lead to a true equality, which is 'a *binomium* equal to a *binomum*' (while the others respectively are 'a *recisum* equal to a *binomum*' and 'a *binomium* equal to a *recisum*'). Cardano's conclusion is then the following:

> [i]t is also agreed that this value [$\sqrt{a_1 + \frac{a_0}{x}}$] is common to the cubic *binomium* discovered in a part of the chapter and to the surface *binomium* here indicated and the common quantity is the general value.[88]

No explanation of the term "surface [*superficiale*]" *binomium* is given there or elsewhere. I am likely to compare it with the previously mentioned cubic *binomium*.

The subsequent **Chapter LVIII** "On the common quantity from two incommensurable [quantities], in how many ways it is said [*De communi quantitate duabus incommensis quot modis dicatur*]" seems to start where Chapter LVII ended. The topic is again the "common quantity [*quantitas communis*]". Cardano recalls once more that the

[87] Note that in Chapter LVII Cardano abruptly employs some new stenographic expressions. There '*c. p.*' means 'with what is produced from [*cum producente*]', '*d. n*' means '*n* divided by (something, which in our case is x) [*diviso*]', and '*f. n*' means 'fragment of *n* [*fragmento*]'. This kind of (to say the truth, not functional) formalism is employed only once elsewhere in the *Aliza* and with the very same examples, which are "20 *c. p.* 32", "20 p: *d.* 32", and "20 *f.* 32". They all stand for $20 + \frac{32}{x}$. See [9, Chapter LIII, page 98 and Chapter LVII, page 104].

[88] "*Constat etiam quod talis æstimatio est communis binomio cubico invento in parte capituli, et binomio superficiali hic declarato et communis quantitas est æstimatio generalis*", see [9, Chapter LVII, page 104].

possible shapes for the (irrational) solutions of $x^3 = a_1x + a_0$ (with a_1, a_0 rational) are the *binomium* of the 2nd or 5th type $\sqrt{a}+b$ and the cubic *binomium* $\sqrt[3]{a} + \sqrt[3]{b}$ (with $\sqrt[3]{ab}$ rational). His aim is to search for a "common quantity" or a "common value [*æstimatio communis*]" to them. Unfortunately (again), Cardano's start is not very plain.[89] But then the things get better. Cardano explicitly states that

> it is necessary that the common quantity is neither of the kind of AB nor of BC. And this can be, because the animal is common to the human being, to the donkey, to the ox, and to the horse, thus AB and BC are contained in a certain common quantity, which as long as it is common to all [AB and BC] has only this property, that, the simple number of the equality [being] divided by that same, it is the root of the things with what comes out.[90]

He takes $AB = \sqrt{a} + b$ a *binomium* of the 2nd or 5th type and $BC = \sqrt[3]{a} + \sqrt[3]{b}$ a cubic *binomium* (with $\sqrt[3]{ab}$ rational). We saw that in Cardano's opinion these are "general" solutions, meaning that they

[89] "Then two values of the cube equal to some things and to a number are already known, one in the biggest part of the number and is [the value] of the cubic *binomium*, and the other in the smallest part of the number [and is the value] of the second or fifth *binomium* by square roots, and the common value that cannot be incommensurable, indeed they would have been commensurable between them, and the fourth [value is known], namely [the one] that is realised in the smallest part of the number. Therefore it is necessary that the common fails, as it is said, by conjunction" or "*Sunt ergo iam notæ duæ æstimationes cubi æqualis rebus et numero, una parte in maiore numeri, et est binomii cubici, alia in parte minoris numeri binomii ex R quadratis secundi vel quinti, et communis æstimatio quæ non potest esse incommensis, essent enim inter se commensæ, et quarta scilicet quæ intelligitur in parte minoris numeri, deficere igitur commune oportet, ut dicatur per coniunctionem*", see [9, Chapter LVIII, page 106].

[90] "*Necesse est igitur ut sit quantitas communis genere non ab nec bc. Et hoc esse potest, nam animal est commune homini et asino et bovi et equo, ita ab et bc continentur sub communi aliqua quantitate, quæ donec communis est omnibus habet solam eam proprietatem, quod cum dividitur numerus simplex æquationis, per illam ipsam est R numeri rerum cum eo quo prodit*", see [9, Chapter LVIII, page 106].

are written under certain codified shapes and can encompass other more "particular" shapes (like a rational number). The "common value" that Cardano is looking for may then be of an even more encompassing shape, since it should also include those two irrational shapes. At first, Cardano hopes that the encompassing shape could be the sum of the two above irrational shapes. But then he realises that the only property of this "common value" is to be a solution of $x^3 = a_1x + a_0$. This implies that the equality in Chapters XL and LVII $x = \sqrt{a_1 + \frac{a_0}{x}}$ holds, which is clearly enough (even though a part of the text is missing) checked in the two examples $\sqrt{8} + 2$ and $\sqrt[3]{4} + \sqrt[3]{2}$. Cardano ends[91] by remarking – as he already did in Chapter LVII – that the "nature [*natura*]" of (one of) the solution(s) must be the same as in $\sqrt{a_1 + \frac{a_0}{x}}$.

Finally, I deal with an application. In **Chapter XIV** "On the discovery of the kind of the value [*De inventione generis æstimationis*]" Cardano considers not only the equations of the family $x^3 = a_1x + a_0$ with rational coefficients, but also of the family $x^3 + a_0 = a_1x$. Here he works the other way round and explains how to draw the numerical value of the coefficients, knowing that they must be rational and given a certain shape for a solution (namely, a *binomium* of the 2nd or 5th type for $x^3 = a_1x + a_0$ or a *binomium* of the 1st or 4th type and a *recisum* of the 2nd or 5th type for $x^3 + a_0 = a_1x$). Despite all the rambling discussions in the chapters of the *Aliza* that we have analysed above, here Cardano rather behaves in a very similar way to (AMA XXII.i) and (AMA XXII.ii), where he basically acted by substitution and comparison. My main point being only to shortly

[91] "Therefore, in both divisions, as you see, it is common that the *recisum* comes forth and, added the number of the things, it goes through a nature similar to the square of the thing. Therefore the number of the things changes its nature, which comes from the division of the number of the equality by the thing" or "*[c]ommune est ergo ut vides in utraque divisione prodire recisum, quod additum numero rerum, transeat in naturam similem quadrato rei. Numerus igitur rerum mutat naturam eius, quod provenit ex divisione numeri æquationis per rem*", see [9, Chapter LVIII, page 107].

echo the substitution techniques from the *Ars Magna Arithmeticæ*, I skip the details (which by the way are plainly understandable). It is interesting to remark that, as a sort of complement to the *Ars Magna Arithmeticæ*, Cardano also gives a proof of the formulae that he finds for the coefficients. As usual, he translates the hypotheses in a geometrical, non-positional environment. Moreover, in this case he only performs a sequence of steps completely parallel to the ones that he would have done if he had arithmetically drawn the coefficients.

Summing up, in the context of Cardano's surveys on $x = y + z$ in the equation $x^3 = a_1x + a_0$, it turns out that he lingers for more than a few moments on this substitution, or rather on a topic strongly tied to it. As we expected (since we are acquainted with the parallel results in the *Ars Magna Arithmeticæ*), the idea of the substitution could have been a side effect of the study of the possible algebraic shapes for the irrational solutions of the equation $x^3 = a_1x + a_0$, with a_1, a_0 rational. We remark that appreciating whether these algebraic shapes can or cannot express a solution of an equation is one of Cardano's most recurring (and probably ancient) strategies in dealing with equations. Indeed, concerning that equation, it comes to light that an irrational shape for one of its solutions must contain two parts (which in Section 5.2 we used to call y and z) that bear certain relationships between them. We recall that, in the end, these shapes are the *binomium* of the 2nd or 5th type $\sqrt{a} + b$ and the cubic *binomium* $\sqrt[3]{a} + \sqrt[3]{b}$ with $\sqrt[3]{ab}$ rational. Nowadays, it is clear that, for instance, if $\sqrt{a} + b$ is a solution of the equation $x^3 = a_1x + a_0$, the associated polynomial is reducible. Moreover, another solution is $-\sqrt{a} + b$ and the third solution is rational. Anyway, Cardano does not have the concept of irreducible polynomial available so that he must analyse the coefficients of the equation and their shapes. The same happens for $\sqrt{a} + \sqrt{b}$. We know that it cannot be a solution of $x^3 = a_1x + a_0$ (since the splitting field of the equation is of order 4, which does not divide 6), but Cardano needs to show that he gets an impossible system. The link between the substitution $x = y + z$ and the possible shapes for the irrational solutions is firstly made at the very beginning of Chapter I. Afterwards, we meticulously went and searched for the subsequent chapters in which the topic is developed. The messiness of the *Aliza*'s structure strongly appears.

The stages of Cardano's survey are the following. One stage, which explicitly recalls the substitution $x = y + z$ and is linked to Chapter I, is Chapter XI. There, Cardano's classification of y, z according to the Euclidean terminology for irrational numbers is employed in order to count how many parts there are in x^3. Afterwards, Cardano compares them to the rational and irrational parts in $a_1x + a_0$ and draws some information on how many parts there are in x. But Cardano's conclusions turn out to be sometimes inaccurate and the Euclidean terminology prevents him from having a clear view.

Another stage is Chapter X (which also recaps the result in Chapter V). There, by substitution and comparison, Cardano counts out many possible shapes for the irrational solutions of the considered equation.

An intermediary stage could be the surveys in Chapters XIII and XVI. We still have a few remainders of the Euclidean terminology (even if not as massively as in the classification in Chapter XI), but the more effective method by substitution and comparison is also employed to respectively count out all the *trinomia* and *quadrinomia* (and, accordingly, the *multinomia* with more than five terms). Chapter XIV attests that the techniques of the *Ars Magna Arithmeticæ* are still in use.

Furthermore, another stage is Chapter XXIV, where the guidelines are (AM XXV.1) and the third corollary to (AM XIII). There, Cardano wants to study the irrational part of a shape for an irrational solution.

Finally, but strictly related with the preceding one, there is the stage in which (AM XXV.2)) is employed in order to detect an even more encompassing shape that should account for both $\sqrt{a} + b$ and $\sqrt[3]{a} + \sqrt[3]{b}$. This is made in Chapters LVII and LVIII.

The study of the shapes for the irrational solutions of $x^3 = a_1x + a_0$, with a_1, a_0 rational, also helps in giving a (relative) estimate in time for the concerned chapters. Indeed, it seems that these surveys come before the discovery of the cubic formulae. This is at least true for Chapter XI, where Cardano strikingly counts out some shapes that actually are possible shapes (and maybe he would have not missed them, if he had had the cubic formula available), and for Chapter XXIV, where Cardano does not yet seem to be aware of Ferrari's method for solving quartic equations. But also the results in Chapters V, X, XIII, and XVI are prior, or at worst contemporary, to the cubic formulae. Indeed, they are fully comparable (when not akin) to some of the results in the *Ars Magna Arithmeticæ* that I have analysed in Section 4.2.4. We recall the interpretation of the corresponding part of the *Ars Magna Arithmeticæ* on page 212, where I have argued that the study of the shapes of the irrational solutions is an access point to the structure of the solutions themselves, and then to the formula. These (possibly previous) surveys are recovered in the *Aliza* with an additional aim: justify the fact that a solution x of $x^3 = a_1x + a_0$ is composed by two parts $x = y + z$. This, in turn, leads to the method of the splittings, which are intended to amend the cubic formula.

5.3.2. On the origin of the splittings: a hypothesis in relation with *Ars Magna*, Chapter XXV

As we have seen in Section 3.1.4, the "particular" rules in *Ars Magna*, Chapter XXV are very tightly associated with (AM VIII.2). We analyse the chapters of the *Aliza* that refer to (AM VIII.2) in Section 5.4.2. Nonetheless, I choose to deal here with the chapters of the *Aliza* that concern *Ars Magna*, Chapter XXV because they are in a stronger connection with *Aliza*, Chapter I. We moreover recall that this connection (or, at least, the connection between *Ars Magna*, Chapter XXV and the "*aliza* problem") already existed in *Ars Magna*, Chapter XII (see above, on page 114).

In **Chapter II** "On the special rules of Chapter XXV of the *Ars magna* of the cube equal to some things and a number [*De Regulis specialibus Capituli XXV Artis magnæ cubi æqualis rebus et numero*]" Cardano deals again with all the "particular" rules in *Ars Magna*, Chapter XXV that concern $x^3 = a_1x + a_0$ (see here, Section 3.1.4 on page 83). The chapter's start is disconcerting. The first paragraph looks like a proof, but I have not been able at all to find out what Cardano wants to prove. Indeed, on the one hand, he never says what his aim is. Worst, on the other hand, the proof itself is completely unclear. It has quite a lot of references to a certain diagram supposedly filled up with many letters (from A to O). Unfortunately, the diagram is displayed neither there (in both editions), nor elsewhere (namely, in *Ars Magna*, Chapter XXV or in *Aliza*, Chapter I, which are both quoted in the rest of this chapter, or in *Ars Magna*, Chapter VIII, which could have also been a reasonable option), and I have not been able to retrieve it from the text. Without the diagram it is impossible to understand the paragraph, which is massively based on it. This could be evidence of the fact that the *Aliza* was published in a hurry, since maybe the publisher had not realised that he also needed to reproduce the diagram.

Anyway, at the end of this paragraph, Cardano gets the equation $x^3 = a_1x + a_0$, and then a sequence of propositions follows.

A II.1. *The*[92] *system in the proposition (AM XXV.1) does not come from any splitting in* Aliza, *Chapter I.*

We remind indeed that, while commenting (AM XXV.1) on page 84, we did not bring it back to the substitution $x = y + z$.

Then, Cardano says that, from the demonstration at the beginning of Chapter II, it follows that $x^3 + a_0 = a_1x$ is "simpler [*simplicius*]" than $x^3 = a_1x + a_0$ and that one can pass from the knowledge of the first equation to the knowledge of the second.[93] He implicitly makes reference to the system common to (AM VIII.2) (or (A V.ii) for the particular case of $x^3 + a_0 = a_1x$, see below on page 358) and

[92]"This demonstration shows that this chapter [$x^3 = a_1x + a_0$] does not arise from those seven ways [in Chapter I]" or '*[H]demonstratio ostendit quod hoc capitulum non oritur ex illis septem modis*", see [9, Chapter II, page 5].

[93]"From this it also follows that the chapter of the cube and a number equal to some things is simpler and from it [it is] easier to [get to] the knowledge in the chapter of the cube equal to some things and a number" or "*[e]x hoc etiam sequitur quod capitulum cubi et numeri æqualium rebus est simplicius, et ex se magis obvium cognitioni capitulo cubi æqualium rebus et numero*", see [9, Chapter II, page 5].

to (AM XXV.1). The explanation is not very well detailed, but one can observe that the solution given in the first case, namely $\sqrt{g}$, is regarding its shape "simpler" than the one given in the second case, namely $\sqrt{f + \frac{g}{4}} + \frac{\sqrt{g}}{2}$. Cardano's statement is in any case evidence of the fact that he was aware of the connection between (AM VIII.2) and (AM XXV.1), as I have supposed on page 115.

Finally, Cardano remarks that "this rule is not general by itself to the whole chapter of the cube equal to some things and a number [...]. But it is general to the chapter of the cube and a number equal to some things [*Hæc etiam regula non est generalis per se toti capitulo cubi æquali rebus et numero, quia ubi numerus esset maior non satisfaceret, sed est generalis capitulo cubi et numeri æqualium rebus*]". I cannot account for this last remark, except assuming that the meaning of 'general' is not always well fixed. We recall, for instance, that we have already found a probable incongruity in the *Ars Magna*, where Cardano calls Chapter XII "general" (and this disagrees with his characterisation of the term in the rest of the book, see above on page 114). Maybe something similar is happening here in the *Aliza*, but I cannot provide any reasonable interpretation of "general" under which the above quotation makes sense.

A II.2. *The system in Proposition (AM XXV.2) does not come from any splitting in* Aliza, *Chapter I.*

Again, when commenting on (AM XXV.2) on page 84, we did not bring it back to the substitution $x = y + z$. Rather, Cardano explained that the idea behind (AM XXV.2) is to write $g^3 = (a_1 + f)g$. The same explanation is also recalled here in the *Aliza*.

A II.3. *The system in Proposition (AM XXV.3) comes from the splitting (A I.3).*

A II.4. *The system in Proposition (AM XXV.4) comes from the splitting (A I.4).*

A II.5. *The system in Proposition (AM XXV.5) comes from the splitting (A I.2).*

At page 85, we have already displayed a way to rephrase the three propositions of the *Ars Magna* that highlights the substitution $x = y + z$. Then, the three above propositions are clear, if one compares the way in which they have been rephrased with the propositions at issue in *Aliza*, Chapter I (respectively on pages 274, 276, and 277).

A II.6. *The system in Proposition (AM XXV.6) does not come from any splitting in* Aliza, *Chapter I.*

Cardano calls Proposition (AM XXV.6) "special [*specialis*]". Indeed – as we have already remarked in Section 3.1.4 on page 88, that proposition is mainly meant to perform polynomial division and, as such, stands aside from the others, since it does not depend on (AM VIII.2).

A II.7. *The system in Proposition (AM XXV.7) comes from the system (AM XXV.5)) [or from the splitting (A I.2)].*

Unfortunately, in Cardano's further commentary on the above proposition, his description[94] does not fit with (AM XXV.5), though he quotes the very same example $x^3 = 7x + 90$. Actually, the system in (AM XXV.7) seems rather to come from the system in (AM XXV.4), or in (A I.4), where the shape of the term of degree zero of the equation is $a_0 = y^2z + yz^2$.

A II.8. *The system in Proposition (AM XXV.8) comes from the splitting (AM XXV.3) [or (A I.3)].*

[94] "The seventh [rule] arises from the fifth [one]. But it is seen that [it is] different from that, because in that [fifth rule] the whole root, namely $\sqrt{28 - 3x^2}$, is supposed, in this [seventh rule] the half [of the root], [namely] $\sqrt{7 - \frac{3}{4}x^2}$, [is supposed]. And because in one [fifth rule] the parts are reciprocally multiplied by the squares, in the other [seventh rule] the aggregate [is multiplied] by the product of the parts. Nevertheless they are the same as it is demonstrated in the book on Proportions" or "*[s]eptima oritur ex quinta, sed videtur ab ea diversa, quia in illa supponitur R tota scilicet RV: 28 m: 3 quad., in hac dimidium R 7 m: $\frac{3}{4}$ quad.. Et quia in una ducuntur partes mutuo in quadrata, in alia aggregatum in productum partium; sunt tamen idem ut demonstratum est in Libro de proportionibus*", [9, Chapter II, page]. I could not find the exact reference to the *De Proportionibus*.

Cardano does not provide any detailed argument, but in this case the correspondence holds, since the term of degree zero of the equation has the shape $a_0 = 2y^2z + 2yz^2$.

Quite far from Chapter II, another whole chapter of the *Aliza* is devoted to *Ars Magna*, Chapter XXV. It is **Chapter XXV** "On the examination of the third rule of Chapter XXV of the *Ars magna* [*De examine tertiæ regulæ Capituli XXV Artis magnæ*]". More precisely, it is devoted to the comparison of the "special,[95] non-general rule [*regula specialis non generalis*]" (AM XXV.3) with the "general, non-special rule [*regula generalis non specialis*]" (AM XII). We remind that both rules are about $x^3 = a_1x + x_0$ and that the second one gives the cubic formula.

Note that here Cardano does not explicitly mention (AM XII). Anyway, in the examples, it is clear that he is using the cubic formula. Indeed, instead of immediately providing the integer solution (note that in this chapter all the equations, except the last one, have only one real solution, which is 6), Cardano writes it at first as the sum of two cubic radicals. This means that he should not have guessed the solution, but he is possibly using the cubic formula. Or rather, when Cardano details the calculations, we notice that he is not exactly using (AM XII), but instead a proposition that is directly implied by it. Maybe for the comparison's sake, this proposition is stated in the same style as (AM XXV.3). Here it follows.

A XXV.i. *Consider* $x^3 = a_1x + a_0$.
If

$$\begin{cases} a_0 = f + g \\ \left(\frac{a_1}{3}\right)^3 = fg \end{cases},$$

then $x = \sqrt[3]{f} + \sqrt[3]{g}$.

[95]See the remarks on pages 98 and 114 concerning the terms "particular" and "general" referring to these rules

If we take $f = \frac{a_0}{2} + \sqrt{\left(\frac{a_0}{2}\right)^2 - \left(\frac{a_1}{3}\right)^3}$ and $g = \frac{a_0}{2} - \sqrt{\left(\frac{a_0}{2}\right)^2 - \left(\frac{a_1}{3}\right)^3}$, the conditions in the system are fulfilled. Then, (A XXV.i) is implied by (AM XII).

Let us come back to Cardano's comparison between (AM XXV.3) and (AM XII) (or rather (A XXV.i)). He says that some equations of the family $x^3 = a_1x + a_0$ follow (AM XXV.3), others are associated with (AM XII), and finally there are certain equations that are associated with both.[96] He explains that in the first case $a_1 > a_0$, whereas in the second case $a_1 < a_0$. Cardano does not say that the third case is when $a_1 = a_0$, but we are led to suppose it. Indeed, the numerical equation that he mentions, namely $x^3 = 6x + 6$, pertains to this case, since 6 is not a solution of it, despite $a_1 = a_0$. Anyway, even dismissing the alleged condition on the third case, the remaining conditions are inaccurate.

Indeed, Cardano further specifies that (AM XII) (or rather (A XXV.i)) is not useful for $x^3 = 22x + 84$, since the equation has $\Delta_3 < 0$. Unfortunately, the example is very poorly chosen, because the equation actually has $\Delta_3 > 0$. Since Cardano's explanation is not ambiguous,[97] I consider that the condition in (AM XII) is the usual one on the discriminant, namely $\Delta_3 \geq 0$ or $\left(\frac{a_0}{2}\right)^2 \geq \left(\frac{a_1}{3}\right)^3$. On the other hand, Cardano says that (AM XXV.3) is not useful for $x^3 = 17x + 114$ (and for $x^3 = 15x + 126$) referring to a certain boundary given by a maximum problem. He knows that, if $f + g = a_1$, the maximum

[96] "Therefore there are three parts in this chapter, the first that covers the special, non-general rule when the number of the things is bigger in comparison to the number of the equality. The second that covers the general, non-special rule when the number of the equality is bigger in comparison to the number of the things. The third that covers both" or "*tres sunt partes in hoc capitulo, prima quæ servit regulæ speciali non generali, cum numerus rerum est magnus in comparatione numeri æquationis. Secunda quæ servit regulæ generali non speciali cum numerus æquationis est magnus comparatione numeri rerum. Tertia quæ servit utrique*", see [9, Chapter XXV, page 52].

[97] "[T]he general rule cannot reach $x^3 = 22x + 84$, because 21, the fourth part of 84, makes a square neither bigger nor equal to the cube of the third part of the things $\frac{22}{3}$" or "*non potest regula generalis attingere ad 1 cub. æqualem 22 rebus p: 84, quia 21 quarta pars 84 non facit quadratum, neque maius neque æquale cubo $7\frac{1}{3}$ tertiæ partis rerum*", see [9, Chapter XXV, page 52].

value of $\sqrt{f}g + f\sqrt{g}$ is attained when $f = g = \frac{a_1}{2}$ (and is $a_1\sqrt{\frac{a_1}{2}}$).[98] Finally, we remark that, even though there is a connection between

[98] Cardano justifies his assertion on the maximum value through a cross reference to *De Proportionibus*, Proposition 209: "If it appears [that] a rectangular surface [is] divided into two equal parts, which are both squared, likewise in two unequal [parts], the parallelepiped from the side of the middle part times the whole surface will be bigger than the aggregate of the parallelepipeds from the unequal parts times the sides of the other part by what is reciprocally made from the difference of the side of the smaller part from the side of the middle [part] times twice the difference of the bigger part of the surface from the middle [part] of the surface and from the difference both of the unequal sides joined to both sides and of the equal [sides] joined times the smaller part of the surface" or "*[s]i superficies rectangula in duas partes æquales divisa intelligatur, quæ ambę quadratæ sint, itemque in duas inæquales, erit parallelepipedum ex latere mediæ partis in totum [totam] superficiem maius aggregato parallelipedorum ex partibus inæqualibus, in latera alterius partis mutuo in eo, quod fit ex differentia lateris minoris partis a mediæ latere in differentiam maioris partis superficiei a media superficie bis, et ex differentia amborum laterum inæqualium iunctorum ad ambo latera, æqualia iuncta in minorem partem superficiei*", see [11, pages 241-242]. It says that, if we take three real numbers u, v, w such that $2u = v+w$, then

$$2u\sqrt{u} = v\sqrt{w} + w\sqrt{v} + \left(2(\sqrt{u} - \sqrt{v})(w - u) + (2\sqrt{u} - \sqrt{v} - \sqrt{w})v\right).$$

This implies that the maximum value of $\sqrt{f}g + f\sqrt{g}$ is $2u\sqrt{u}$, and is attained when $v = w = u$.

A few words on the proof. As usual, it employs the language of geometry, and mainly consists in drawing an effective diagram thanks to which the quantities at issue are identified. Then, readjusting the parts in the diagram, the proposition follows easily. The point is to make a good assignment of the quantities to the geometrical objects. More precisely, considering the diagram in Figure 5.4,

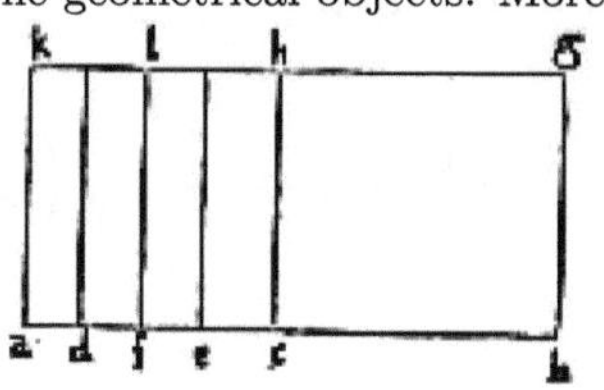

Figure 5.4. – *De Proportionibus*, Proposition 209. Kind courtesy of Biblioteca Virtual del Patrimonio Bibliográfico and Universidad de Sevilla.

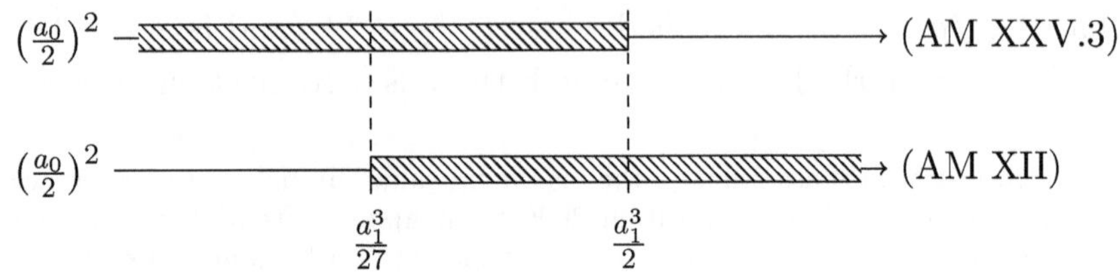

Figure 5.5. – Comparison between the conditions on (AM XXV.3) and (AM XII) (or rather (A XXV.i)) in *Aliza*, Chapter XXV.

(AM XXV.3) and (A I.3) (highlighted for instance in Chapter II), the condition in (AM XXV.3) is not comparable with the one stated in (A I.3), since it does not involve $a_1 x$. In the two above examples, this maximum is strictly smaller than $\frac{a_0}{2}$, so that one cannot find those (positive) f, g. Then, we conclude that (AM XXV.3) has the condition $a_1\sqrt{\frac{a_1}{2}} \geq \frac{a_0}{2}$, that is $\frac{a_1^3}{2} \geq (\frac{a_0}{2})^2$. Let us resume in the diagram in Figure 5.5 the conditions stated in (AM XXV.3) and (AM XII) (or rather (A XXV.i)).

The final aim of the comparison concerns the equations that fall under the third case, for which Cardano gives the following proposition.

A XXV.ii. *Consider* $x^3 = a_1 x + a_0$.

When both (AM XXV.3) and (AM XII) [or rather (A XXV.i)] apply, then f, g *in (AM XXV.3) can be obtained in the following way: find a solution* α *of the equation and take*

$$\sqrt{f} = \frac{\alpha}{2} + \sqrt{\left(\frac{\alpha}{2}\right)^2 - \frac{a_0}{2\alpha}} \quad \textit{and} \quad \sqrt{g} = \frac{\alpha}{2} - \sqrt{\left(\frac{\alpha}{2}\right)^2 - \frac{a_0}{2\alpha}}.$$

Cardano takes $(\overline{AH})^2 = (\overline{HB})^2 = u$, $\overline{AC} = \overline{CB} = \sqrt{u}$, $(\overline{KD}) = v$, $(\overline{DG}) = w$, $\overline{AF} = \sqrt{v}$, and $\overline{EB} = \sqrt{w}$. Then, he only needs to draw $\overline{FC} = \sqrt{u} - \sqrt{v}$, $\overline{DH} = w - u$, and $\overline{FE} = 2\sqrt{u} - \sqrt{v} - \sqrt{w}$, and this can be done by looking at the diagram.

Note that this proposition is actually (AM XXV.8), since $a_1 = \alpha^2 - \frac{a_0}{\alpha}$ holds. By the way, the link between Propositions (AM XXV.3) and (AM XXV.8) has already been highlighted in (A II.8).

We observe that the point of the proposition cannot obviously be to draw a solution of the equation (note that all the examples that Cardano provides have $\Delta_3 > 0$, so he perfectly knows how to draw a solution). Rather, his aim here is to study the shape of f, g in (AM XXV.3). If one wants to find a solution to the problem entailed by the *casus irreducibilis*, this could be of a certain importance, since, when (AM XII) does not apply (that is, when $\Delta_3 < 0$), (AM XXV.3) still applies.

Finally, in **Chapter LIX** "On the order and on the examples in the second and fifth *binomia* [*De ordine et exemplis in binomiis secundo et quinto*]" Cardano provides two methods to solve two particular classes of equations of the family $x^3 = a_1x + a_0$. He does not explicitly state the methods as such, but he provides 69 numerical equations with integer coefficients, divided into five lists, from which we gather two rules.

Firstly, from the first four lists we draw the following proposition.

A LIX.i. *Set $i = 1, 2, 3$ or 4 and consider the family of equations*

$$x^3 = (n + i^2)x + ni$$

for any integer $n \geq 0$. Then $x = \sqrt{\frac{4n+i^2}{4}} + \frac{i}{2}$.

In the examples, Cardano limits himself to consider $0 \leq n \leq 16$ when $i = 1, 2, 3$ and $0 \leq n \leq 17$ when $i = 4$. Note that the proposition can be trivially extended to any natural i. Moreover, we remark that all the equations $x^3 = (n + i^2)x + ni$ have the same (negative) rational solution, which is $-i$, and that thereby they have three distinct real solutions (otherwise, note that the term under the square root is $4n + i^2 \geq 0$). All the equations have then $\Delta_3 < 0$. Therefore, this proposition enables Cardano to devise a considerable number of equations that he is able to solve with $\Delta_3 < 0$.

It is very likely that Cardano comes to this proposition using (AM XXV.1). On the one hand, we can easily verify that one can use (AM XXV.1) to derive (A LIX.i), taking $n = f$ and $g = i^2$. On the other hand, Cardano himself recalls the algorithm of (AM XXV.1), even if he does not explicitly mention the proposition (he only says "in the previous [*in priore*]"). It is in the fourth corollary to this proposition, where he compares, through an example, the formulae given in (A LIX.i) and in (AM XXV.1), and remarks that they lead to the same result.[99]

Out of the other three corollaries, two of them describe the formula in (A LIX.i). The first is a commentary on the term '$\frac{i}{2}$' in the formula and the third[100] points out that all the equations have $\Delta_3 < 0$. In the second corollary, Cardano states that "the chapter cannot be general, because the first number is necessarily a square [*capitulum non potest*

[99]Concerning the equation $x^3 = 19x + 30$, Cardano says that "[i]n the following, we will take 1, the fourth part of 4, and we will add [it] to the difference 15, it is made 16, the root of which, which is 4 [is] added and constitutes the value 5. In the previous, we will add $\frac{9}{4}$, the fourth part of 9, to the difference 10, $\frac{49}{4}$ is made, the root of which is $\frac{7}{2}$, added $\frac{3}{2}$, the half of 3, [which is] $\sqrt{9}$, the value of the thing 5 is made as before" or "*[i]n posteriore accipiemus 1, quartam partem 4, et addemus ad 15, differentiam fit 16, cuius R, quæ est 4, addito constituit ęstimationem 5. In priore addemus $2\frac{1}{4}$, quartam partem 9, ad 10 differentiam, fit $12\frac{1}{4}$, cuius R, quæ est $3\frac{1}{2}$, addito $1\frac{1}{2}$, dimidio 3, R 9, fit 5, ut prius rei æstimatio*", see [9, Chapter LIX, corollary 4, page 109].

[100] Unfortunately, by a not completely clear argument: "the number of the equality can never be increased to such a point that the square of the half of that is bigger than the cube of the third part of the number of the things. Indeed, then by the first rule the value would be the cubic *binomium* and by that rule [(A LIX.i)] the squared *binomium*, and thus one would be equal to the other. Which can be allowed [. . .]. Nevertheless it cannot be extended and the value is loosened in an integer number" or "*nunquam numerus æquationis potest adeo augeri, ut quadratum dimidii eius sit maius cubo tertiæ partis numeri rerum; nam tunc per primam regulam fieret æstimatio binomium cubicum. Et per hanc regulam binomium quadratum, et ita unum æquale esset alteri. Quod licet esse possit, ut in hoc exemplo RV: cu.20 p: R 329 [392] p: RV: cu. 20 m: R 392, et est 2 p: R 2 et 2 m: R 2, quod est 4, non potest tamen continuari, et æstimatio resolvitur in numerum integrum*, see [9, Chapter LIX, corollary 4, page 109]. The "first rule" maybe refers to (AM XII), as it is stated in the *Ars Magna*, or to (A XXV.i).

esse generale, quia primus numerus necessario est quadratus]". It happens indeed that, when $n = 0$, then $a_1 = i^2$ (as Cardano has already remarked at the very beginning of this chapter). He spends some words explaining why a_1 is an (integer) square, whereas its connection with the non-"generality" of this chapter seems to be obvious for him. Then, we can guess that (A LIX.i) is not "general", since it concerns the class of equations $x^3 = (n + i^2)x + ni$, which is not "general", because the coefficients are not so.

From the fifth and last list of equations we gather the following proposition.

A LIX.ii. *A solution of the family of equations* $x^3 = nx + (\alpha^3 - \alpha n)$ *is* α *for any integer* $n \geq 0$.

In the examples, Cardano only considers $0 \leq n \leq 16$ and $\alpha = 6$ (but the proposition can be trivially adjusted in order to return any natural value for the solution α instead of 6). No corollary concerns this proposition. It likely comes from the rephrasing $a_0 = \alpha^3 - a_1\alpha$.

We remark that this chapter is the only passage (in all the texts that we have analysed) where Cardano writes that a coefficient can be '0'. Namely, it happens in the first line of each list. For instance, in the first list we find

$$\text{`1 cu. 0 p: 1 pos. R } \frac{1}{4} \text{ p: } \frac{1}{2}\text{'.}$$

In these lists, the equations of the family $x^3 = a_1x + a_0$ are usually written in a shortened form. In the first column there is x^3, while in the second and third columns there is $a_0 + a_1x$. In each line of the lists, between the first and the second columns, we should add '=' in order to correctly read the equation. Note then that the '0' in the second column of the above diagram means '$a_0 = 0$'.

Summing up, Chapters II, XXV, and LIX of the *Aliza* are characterised by the comparison with *Ars Magna*, Chapter XXV. Chapter II associates *Ars Magna*, Chapter XXV with *Aliza*, Chapter I, while Chapter XXV is on the shape of f, g in (AM XXV.3) (not asking this time for rational coefficients), and Chapter LIX is mainly meant to provide a list of equations with rational coefficients and $\Delta_3 < 0$ that Cardano is able to solve.

These chapters (together with the chapters considered in Section 5.4.2) have in common the interest for those "particular" rules that enable to solve an equation by finding two numbers f, g that fulfil certain requirements. Sometimes the *Ars Magna* is explicitly mentioned, sometimes not. Whether all these chapters of the *Aliza* had been written before or after the *Ars Magna* is – as said – a matter up to discussion.

It is likely that Cardano conceived the propositions in *Ars Magna*, Chapter XXV starting from the study of the irrational solutions of equations with rational coefficients. Indeed, on the one hand, for most of the time Cardano's equations only sporadically have irrational coefficients. On the other hand, the propositions in *Ars Magna*, Chapter XXV (and the chapters of the *Aliza* that link them with the splittings, or that analyse some of them in detail) have been refined after Cardano realised that the cubic formulae had a problem. Note that indeed there is nothing similar in the *Practica Arithmeticæ* or in the *Ars Magna Arithmeticæ*, and that in *Ars Magna*, Chapter XII Cardano himself says that the propositions of Chapter XXV are a temporary relief to the *casus irreducibilis*, while the *Aliza* should be the conclusive treatment. Then, I guess that some of the splittings, namely (A I.2)-(A I.4), could have originated from rethinking some propositions in *Ars Magna*, Chapter XXV, namely (AM XXV.3)-(AM XXV.5).

5.3.3. Cardano's last say on the origin of the splittings

Up to now, I have tried to gather from the *Aliza* as much information as possible concerning the origins of the substitution $x = y + z$ and of the splittings. All my hypotheses had a mainly arithmetical source. Nevertheless, there could also have been a loose geometrical inspiration at the origin of Cardano's conjectures.

We can track it down in **Chapter LX** "General demonstration of this chapter of the cube equal to some things and a number [*Demonstratio generalis capituli cubi æqualis rebus et numero*]". Just by reading the title, we have great expectations regarding this chapter. Unfortunately, it belongs to that hotchpotch of not-so-clear chapters of the *Aliza*. Its abstract language is recondite, even though it is not as poorly written as other chapters are. Luckily, there is also a part that contains some calculations, and this turns out to be of a certain importance. Cardano's start seems to be very promising.

> And, since there is this special rule that concerns the value, therefore it is also not remarkable if it is also special in the way of the discovery, supposing a squared number. Therefore, in order to be generally considered, we propose [...].[101]

It is not clear to which "special" (or "particular") rule Cardano is referring to. As we have seen on page 331, the foregoing Chapter LIX implicitly refers to (AM XXV.1) and the rules in *Ars Magna*, Chapter XXV are considered to be "particular". But this is a poor connection. I rather believe that a more suitable option could be a generic reference to the rules in *Aliza*, Chapter I (to which the term "particular" can also be applied, and some of which are indeed equivalent to some of the rules in *Ars Magna*, Chapter XXV, as Cardano himself remarked in the *Aliza*, see Section 5.3.2). Indeed, in the following – as we will see – Cardano assumes that $x = y + z$, which is closer to the hypotheses in *Aliza*, Chapter I, rather than to those in *Ars Magna*, Chapter XXV. Apart from identifying the exact reference, the rough idea that the very first lines of Chapter LX convey is that Cardano wants to explain the "general" way through which he got the relevant "rule" – whatever it may be.

---

[101]"*Et cum sit regula hæc quòd ad æstimationem attinet specialis, ideò etiam non mirum est si sit etiam specialis in modo inveniendi, cum supponat numerum quadratum. Ergo ut generaliter consideretur proponamus [...]*", see [9, Chapter LX, page 109].

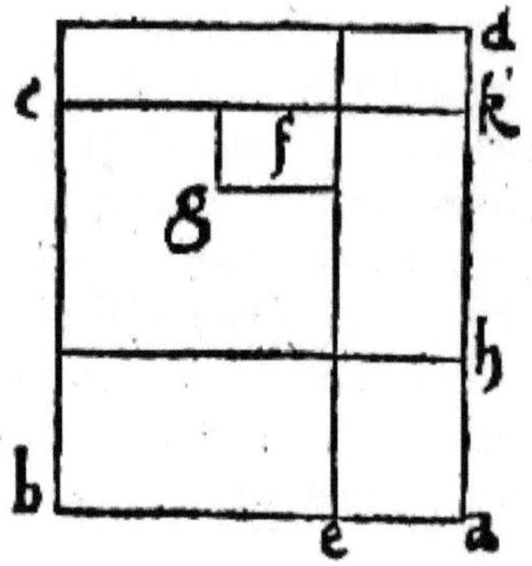

Figure 5.6. – *Aliza*, Chapter LX. Kind courtesy of Biblioteca Virtual del Patrimonio Bibliográfico and Universidad de Sevilla.

Cardano considers the equation $x^3 = a_1x + a_0$ and rewrites it as $x^2 = a_1 + \frac{a_0}{x}$. We have seen that it is one of his usual techniques. Then, he distinguishes three cases, depending on $a_0 = x^2$, $a_0 > x^2$, or $a_0 < x^2$. No commentary is given in the first case. In the second case Cardano provides a geometrical interpretation, where $\overline{AB} = x$ and $\overline{AB}\,\overline{AD} = a_0$ considering the diagram in Figure 5.6. Unfortunately, the description of this case is imprecise and sometimes simply borders on being inexplicable. For instance, Cardano takes $(\overline{BD}) > (\overline{AE})^2\ (\overline{DE})$, that is $\overline{AB}\,\overline{AD} > \overline{AE}^3\overline{AD}$, which is quite a bizarre statement regrding dimensional coherence, since it is a comparison between a rectangle and a four-dimensional body.[102] According to the diagram and to what he

[102] It is very likely that the text has been corrupted, since we find 'AE' instead of the expected 'AC' (and exchanging 'E' with 'C' is a common – maybe the most common – misprint in the *Aliza*). Anyway, it must have been corrupted in a massive way, since it is not only a matter of misprinting a type, but there is also the multiplication by $(\overline{DE})$ that makes no sense.

We remark that in the 1663 edition the letters in the description of the second (and of the third) cases are capitalised, as if someone revised the passage. This is a unique case in the 1663 edition of the *Aliza*. There, the above misprint has been emended. Indeed, we find $(\overline{BD}) > (\overline{AC})^2(\overline{DE})$ instead of $(\overline{BD}) > (\overline{AE})^2(\overline{DE})$. Nevertheless, the dimensional problem persists.

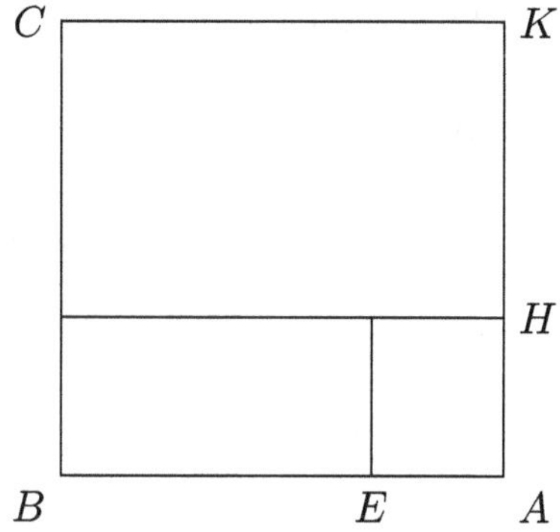

Figure 5.7. – *Aliza*, Chapter LX (partial).

makes in the third case, he should have taken instead $\overline{AB}\,\overline{AD} > \overline{AC}^2$. Then, we drop this case, since this is only the first of a very long list of incongruities. Let us turn to the third case, which is much more sensible. Indeed, we even happily manage to retrace the description of the second case by analogy with the description of this third case. Since the above diagram is indeed two diagrams in one (for the second and third cases), I only copy in Figure 5.7 the part that concerns the third case. Cardano takes $\overline{AB} = x$ and $\overline{AB}\,\overline{AH} = a_0$ (so that morally $\overline{AH} = \frac{a_0}{x}$). Anyway, since at the beginning he stated that he wanted $\frac{a_0}{x}$ to be a surface, he assumes $\overline{AE} = 1$ and $\overline{AH}\,\overline{AE} = \frac{a_0}{x}$. Since $x^2 = a_1 + \frac{a_0}{x}$, Cardano knows that $\overline{AB}^2 = a_1 + \overline{AH}\,\overline{AE}$, hence by the diagram $a_1 = (H\,C\,E)$. Then,

$$(H\,C\,E) - \overline{AB}\,\overline{AH} = (\overline{HK}\,\overline{AB}) - (\overline{AH}\,\overline{AE}).$$

Let us write $y = \overline{AH} = \frac{a_0}{x}$ and $z = \overline{HK}$ such that $x = y + z$. The last equality can be rewritten as

$$a_1 - a_0 = xz - y.$$

This could be another sign of the fact that the publisher did not proofread the draft of the *Aliza*, or that he proofread it very quickly.

Moreover, we have that $xy = x\frac{a_0}{x} = a_0$. To show that it is correct, Cardano repeats the whole procedure through an example in which he assumes $x = \sqrt[3]{12} + 2$ and $a_1 - a_0 = 3$, drawing the values of y, z and verifying that he falls back on the good values for $a_1 - a_0$ and a_0.

Afterwards, Cardano converts the above procedure into the following statement.[103]

A LX. *Consider the equation* $x^3 = a_1 x + a_0$.
If [there are two positive, real numbers y, z *such that]*

$$\begin{cases} xy = a_0 \\ xz - y = a_1 - a_0 \end{cases},$$

then $x = y + z$.

Cardano says that this problem can be settled "by the Rule *de modo* and [*de*] *positione* [*ex regula de modo et positione*]".[104] Then, the chapter (and the book) end with some cryptic remarks on huge fractions.

It is extremely interesting to have a closer look at the system in (A LX). The proposition is similar to the ones in *Aliza*, Chapter I, but it cannot be retrieved with the same method. It happens nevertheless that, once that one makes the substitution $a_0 = (y+z)y = y^2 + yz$ in the second line of the system, one gets $a_1 x = y^3 + 3y^2 z + 3yz^2 + z^3 - y^2 - yz$. Obviously, the sum $a_1 x + a_0$ gives $(y+z)^3$, but one can hardly argue that Cardano gets the system *via* the formula for the cube of a binomial, as it happened in Chapter I. Things should have

[103] "[F]ind the quantity that can be divided into two parts so that the product of the whole by one produces, for example, 3 and [the product of the whole] by the remaining [part] added the preceding [one] produces, for example, 8" or "*[i]nvenias quantitatem quę possit dividi in duas partes ut ductum totum in unam producat 3. Gratia exempli, et in reliquam partẽ addito priore producat 8 pro exemplo*", see [9, Chapter LX, page 111].

[104] The Rule *de modo* is a method to solve systems usually with two equations in two unknowns, see [64, pages 71-74]. The rule *de positione* is simply to assign the unknown to a certain quantity in a problem; this can be done in many ways, see for instance [4, Chapter XXXII].

gone the other way round, that is to say, that Cardano gets the idea of splitting $(y+z)^3$ from jiggering with the system in (A LX) (or with a similar one).

Here, it is difficult to determine where Proposition (A LX) comes from, and in this chapter Cardano himself gives an answer. The idea behind (A LX) – as we have said – is to rewrite the equation as $x^2 = a_1 + \frac{a_0}{x}$ and to interpret in the usual geometrical environment $\frac{a_0}{x}$ as a surface. Since Cardano knows very well how to sum two surfaces (but not how to sum a surface and a line), it is quite a natural choice to take $\frac{a_0}{x}$ as a surface. Since it is very easy to divide a square the side of which is x into two rectangles, the most obvious choice would have been to take a_1 one rectangle and $\frac{a_0}{x}$ the other. But it leads nowhere. Cardano has then the idea to consider the unity segment AE and in this way he manages to interpret $\frac{a_0}{x}$ as the surface $\overline{AHAE}$. Finally, an appropriate division of the square leads to the division of its side in $AH = y$ and $HK = z$.

This last chapter of the *Aliza* contains a hint on where the splittings at the beginning of the book could have been derived from. It is reasonable to expect that Cardano does not pull such a method out of the bag (for instance, considering a symbolic writing, which he did not have available). Instead, this chapter shows that also a kind of figurative (or geometrical) thinking plays a role. Still, the role of geometry in Chapter LX is very restricted. Or rather, the geometrical argument here involves reflections on the dimension of the geometrical objects at issue. It leads to assume that $\frac{a_0}{x}$ is a surface. Then, it is a very specific kind of geometry, since it does not essentially depend on the position of its objects.

5.3.4. Studying the splittings by themselves

Finally, I deal with a few chapters of the *Aliza* in which the splittings are considered from a different viewpoint, becoming a subject of inquiry by themselves.

In **Chapter VII** "On the examination of the values assumed by the second and third rules of the first chapter [*De examine æstima-*

tionum sumptarum ex regula secunda et tertia primi capituli]" Cardano analyses one of the splittings of Chapter I.

First of all, a remark on the title of this chapter. In the 1570 and 1663 editions it originally is "On the examination of the values assumed by the second and third rules of the second chapter [*De examine æstimationum sumptarum ex regula secunda et tertia secundi capituli*]". The title would supposedly make reference to certain "values assumed" in (A II.2) and (A II.3). Actually, there are no values assumed in those propositions. I rather think that the reference is to *Aliza*, Chapter I, since Cardano is indeed using (A I.2) in the rest of Chapter VII. Moreover, in all the splittings considered in Chapter I Cardano really assumes a value (even if not numerical) for a solution of the equation, namely he assumes that a solution is written as $x = y + z$. Nevertheless, (A I.2) is the only splitting explicitly mentioned in Chapter VII, so that the reference to the "third rule" is empty. Note that Cossali[105] not only replaced the "second chapter" in the title by a reference to *Aliza*, Chapter I (as we did), but also changed the numbers of the propositions (then, according to Cossali, Cardano would have quoted (A I.1) and (A I.2)). But, as a matter of fact, no true reference to (A I.1) can be found in the text.

Let us now address the rest of this chapter.

A VII. *Let the equation $x^3 = a_1 x + a_0$ be given as well as one of its solutions α. Consider Y, Z such as in (A I.2).*
Then, $YZ = \frac{a_0}{3\alpha}$,

$$Y = \frac{\alpha}{2} + \sqrt{\left(\frac{\alpha}{2}\right)^2 - YZ} \quad \textit{and} \quad Z = \frac{\alpha}{2} - \sqrt{\left(\frac{\alpha}{2}\right)^2 - YZ}$$

No proof is given, but Cardano shows as usual the procedure in detail on some examples. The only justification that he quotes is

[105] "*[N]el Capo VII De Regula Aliza erroneamente intitolato:* De examine æstimationum sumptarum ex regula secunda et tertia secundi capituli *in cambio di:* ex regula prima et secunda capituli primi", see [60, Chapter I, Paragraph 6, page 42].

Elements II.5. In the examples we remark that, given an equation $x^3 = a_1 x + a_0$ and one of its solutions α, Cardano takes $\alpha = Y + Z$ as in (A I.2). The product YZ can be immediately found from the first line $a_0 = 3Y^2 Z + 3YZ^2$ of the system in the splitting. Knowing the product and the sum of Y, Z, *Elements* II.5 provides an equality[106] that enables Cardano to immediately draw the values of Y and Z.

This chapter is short and overall clear. Then, let us look for its real place in the *Aliza*'s framework. Since only numerical equations with $\Delta_3 > 0$ are considered, I guess that here Cardano is not considering the splittings as a method to possibly overwhelm the problem entailed by the *casus irreducibilis*. Rather, he considers the splittings by themselves. More precisely, Cardano seems to be interested in a kind of algebra of the splittings (meant as a collection of calculation rules), which could possibly turn out to be useful when he wants to solve an equation, but which, as such, is independent from solving equations.

Cossali went one step further and guessed that Cardano was explaining how to pass from one splitting to another, and *vice versa*. Cossali took firstly into account the splittings (A I.1) and (A I.2), and then also (A I.3) and (A I.4),[107] as he finds a compact way to

---

[106] According to Campano "*[s]i linea recta per duo æqualia duoque inæqualia secetur: quod sub inæqualibus totius sectionis rectangulum continetur cum eo quadrato quod ab ea quæ inter ustrasque est sectiones describitur æquum est ei quadrato quod a dimidio totius lineæ in se ducto describitur*" and, according to Zamberti, "*[s]i recta linea secetur in æqualia et non æqualia: rectangulum compræhensum ab inæqualibus sectionibus totius una cum quadrato quod a medio sectionum æquum est ei quod a dimidia fit quadrato*", see [32, page 25v]. This means that $yz + \left(\frac{y+z}{2} - y\right)^2 = \left(\frac{y+z}{2}\right)^2$.

Obviously, we remark that giving $Y + Z = \alpha$ and $YZ = \frac{a_0}{3\alpha}$ is equivalent to give the equation $x^2 - \alpha x + \frac{a_0}{3\alpha} = 0$. Then, if one allows a loose interpretation of *Elements*, Book II, it would give the quadratic formula. Anyway, we do not need to go that far to account for Cardano's calculations.

[107] "*[N]el Capo VII De Regula Aliza erroneamente intitolato:* De examine æstimationum sumptarum ex regula secunda et tertia secundi capituli *in cambio di:* ex regula prima et secunda capituli primi *Cardano espone la via di uno spezzamento all'altro. Io la estenderò agli spezzamenti 3º e 4º*", see [60, Chapter I, Paragraph 6, page 42].

write them all in one.[108] Cossali took the solution of the equation $x^3 = a_1x + a_0$ to be $\alpha = y + z$, where y, z are such as in (A I.1). Then, by (A VII) he assumes that $\alpha = Y + Z$, then infers the product YZ, and eventually Y, Z depending on $y + z = Y + Z$. The remark is non-trivial at all and his interpretation is very appealing. However, it is borderline on overinterpreting. Indeed, let us consider Cardano's first example $x^3 = 18x + 30$, the (real) solution of which is $\sqrt[3]{18} + \sqrt[3]{12}$. It is true that $y = \sqrt[3]{18}$ and $z = \sqrt[3]{12}$ verify the condition in the splitting (A I.1). Cardano's justification of how he found the solution is the following:

> the value of the thing $\sqrt[3]{18} + \sqrt[3]{12}$ is discovered according to the part of the chapter, and it is extended to infinite increasing beyond the number[109]

(and no justification at all is given for the subsequent examples). Then, the validity of Cossali's interpretation entirely depends on how one is likely to interpret 'the part of the chapter', namely if it is (very implicitly) referring to (A I.1) or rather to (AM XII).

Anyway, Cossali's interpretation is not completely unrelated to Cardano's intents. Cossali could have had the idea through another chapter

---

[108] The following proposition recaps Cossali's statements.

A I.2-4 (Cossali). *Consider* $x^3 = a_1x + a_0$. *Write* $x = Y + Z$ *with* Y, Z *two [real, positive] numbers.*
If Y, Z *can be chosen such that the splitting*

$$\begin{cases} Y^3 + Z^3 + bY^2Z + bYZ^2 - a_1(Y + Z) = 0 \\ (3 - b)Y^2Z + (3 - b)YZ^2 - a_0 = 0 \end{cases},$$

holds with $b = 0, 1, 2$, *then the following condition is entailed*

$$\begin{cases} a_0 \leq \frac{3-b}{4}(Y + Z)^3 \\ [a_1(Y + Z) \geq \frac{1+b}{4}(Y + Z)^3] \end{cases}.$$

We have (A I.2) for $b = 0$, (A I.3) for $b = 1$, and (A I.4) for $b = 2$.

[109] "*[R]ei æstimatio iuxta partem capituli inventam sit* Rcu: *18 p:* Rcu: *12 et supra augendo numerum extenditur in infinitum*", see [9, Chapter VII, page 17].

of the *Aliza*, which is tightly linked to Chapter VII, namely **Chapter LIII** "On the accurate consideration of [the things] above said in Chapter VII [*De diligenti consideratione quorundam superius dictorum Capitulo 7*]". There, Cardano explicitly mentions (A I.1) and moreover he compares two splittings.

At first, Cardano considers the equation $x^3 = 12x + 20$, the (real) solution of which is $x = \sqrt[3]{16} + \sqrt[3]{4}$. He says that the solution

> can be assigned giving similarly the number 20 to the cubes, and the number can also be given to both cubes and to two mutual [parallelepipeds], and [the number can] also [be given] to both cubes and to four mutual parallelepipeds, and thus in three ways.[110]

He is indeed describing the first line of the splitting (A I.1) and of the splittings parallel to (A I.3) and (A I.4). Then, Cardano evaluates $2\left(\frac{x}{2}\right)^3 = \frac{1}{4}x^3$ when $x = \sqrt[3]{16} + \sqrt[3]{4}$ (which is smaller than 20, as it should be in the condition of (A I.1)). He remarks that, when a_0 becomes smaller, the splitting can no longer be applied (since the condition is not fulfilled).[111]

Afterwards, Cardano considers the equation $x^3 = 12x + 34$, the (real) solution of which is $x = \sqrt[3]{32} + \sqrt[3]{2}$, and the splitting parallel[112]

---

110 "*Et iam dicamus quod cubus æqualis sit 12 rebus p: 20, et rei æstimatio est R cub.16 p: Rcu: 4, et hæc potest tribui dando 20 numerum cubis similiter, et potest idem numerus dari ambobus cubis et duobus mutuis, et etiam ambobus cubis et quatuor mutuis parallelipedis, et ita trifariam*", see [9, Chapter LIII, page 96].

111 "Therefore it is clear that that root cannot be divided because of the smallness of the number" or "*liquet igitur non posse dividi sic hanc R propter numeri parvitatem*", see [9, Chapter LIII, page 96].

112 It is the following.

A I.3 parallel. *Consider* $x^3 = a_1x + a_0$. *Write* $x = y + z$ *with* y, z *two [real, positive] numbers.*
If y, z *can be chosen such that the splitting*

$$\begin{cases} a_0 = y^3 + y^2z + yz^2 + z^3 \\ a_1x = 2y^2z + 2yz^2 \end{cases}, \quad \textit{that is} \quad \begin{cases} a_0 = (y+z)(y^2+z^2) \\ a_1 = 2yz \end{cases},$$

to (A I.3). It happens that the coefficients of this equation match the condition of the splitting parallel to (A I.3), but not the one of (A I.1) (as well as the coefficients in the preceding example $x^3 = 12x + 20$ match the condition of (A I.1), but not the one of the splitting parallel to (A I.3)). Then, as in (A VII), Cardano draws the values of Y, Z in the splitting parallel to (A I.3), the sums $Y + Z = x$ and $Y^2 + Z^2$ being given.

A LIII.i. *Let the equation $x^3 = a_1x + a_0$ be given as well as one of its solutions α. Consider Y, Z such as in the splitting parallel to (A I.3).*
Then, $Y^2 + Z^2 = \frac{a_0}{\alpha}$, and

$$Y = \frac{\alpha}{2} + \sqrt{\frac{Y^2 + Z^2}{2} - \left(\frac{\alpha}{2}\right)^2}, \quad \text{and} \quad Z = \frac{\alpha}{2} - \sqrt{\frac{Y^2 + Z^2}{2} - \left(\frac{\alpha}{2}\right)^2}.$$

As before, once that one assumes that $\alpha = Y + Z$, the value of $Y^2 + Z^2$ immediately comes from the splitting. Eventually, one infers Y, Z, this time using *Elements* II.9 instead of *Elements* II.5 (none of the two propositions is however mentioned here).

But then Cardano oversteps the ending point of Chapter VII and seems to suggest a comparison between some of the splittings, such as the one proposed by Cossali[113] for Chapter VII. Cardano devises a geometrical environment in which he interprets the splitting (A I.1) and that parallel to (A I.3), both applied to the example $x^3 = 6x + 40$. Namely, considering the diagram in Figure 5.8, he takes $\overline{AB} = \overline{CD} = 4$, which is the (real) solution of the considered equation. Moreover, in the first square, he takes $\overline{AE} = y$ and $\overline{EB} = z$ as in (A I.1) and, in

holds, then the condition

$$\begin{cases} a_0 \geq \frac{1}{2}x^3 \\ a_1x \leq \frac{1}{2}x^3 \end{cases}$$

follows.

[113]By the way, Cossali's interpretation of Chapter LIII is centred on the passage from (A I.1) to one of the splittings parallel to (A I.3) and (A I.4), and *vice versa*, see [60, Chapter I, Paragraph 6, pages 44-46].

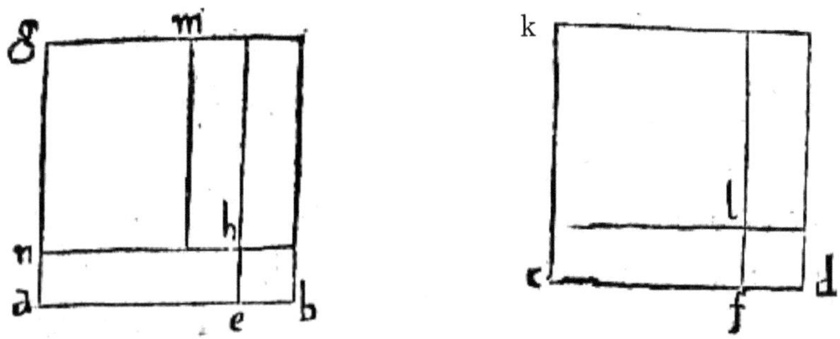

Figure 5.8. – *Aliza*, Chapter LIII. Kind courtesy of Biblioteca Virtual del Patrimonio Bibliográfico and Universidad de Sevilla.

the second square, he takes $\overline{CF} = Y$ and $\overline{FD} = Z$ as in the splitting parallel to (A I.3).[114] Then, he deduces y, z solving the system in (A I.1) (or using the formula, which is the same). He could have also deduced Y, Z using (A LIII.i), but – here the problems begin – he says that

> that rule [the splitting parallel to (A I.3)] does not help to that equation so understood. Therefore it is necessary to discover another [equation] typical of that [rule].[115]

[114] Cardano explicitly describes the splitting (A I.1) and that parallel to (A I.3), without explicitly mentioning them. In the comparison between the two, he shows that $(y + z)(y^2 - yz + z^2) = 40 = (Y + Z)(Y^2 + Z^2)$. We remark that, in order to prove the first part in the above equality, he needs to identify in the first square the geometrical object the measure of which is $y^2 - yz$ (he already knows that $z^2 = \overline{EB}^2$), since he assumed that $y^3 + z^3 = 40$ because of (A I.1) and he knows that $y^3 + z^3 = (y+z)(y^2 - yz + z^2)$. Passing from the (imaginary) three-dimensional diagram (with height AB) to the two-dimensional diagram as it is printed in the text, he clears out the term $(y + z)$. Then, he only needs to build the rectangle (MH), the measure of which is equal to $(\overline{AH}) = yz$, and to take it away from GH, the measure of which is y^2.

We remark that here Cardano is accurately imitating in a geometrical environment the calculations that he makes with those y, z, or $\overline{AE}, \overline{EB}$.

[115] "*[N]ec regula illa servit huic æquationi sic intellectæ, ergo oporteret invenire aliam ei propriam*", see [9, Chapter LIII, page 97].

He suggests then an appropriate example, which is $x^3 = 12x + 34$, that – as we have already seen – matches the condition in (A I.3). I can only suppose that the equation $x^3 = 6x + 40$ was a poorly chosen example. Then, the things get worst, since we expect that Cardano applies, as above, the same splittings to $x^3 = 12x + 34$, but he rather considers the ones parallel to them, which are (A I.2) and (A I.3) (so that our nice talk about the conditions of the splittings is completely disrupted). However, Cardano says that one cannot find such a Y, Z by the means of one splitting, but only the y, z of the other splitting.[116] The splitting parallel to (A I.4) is no longer mentioned. Then, despite all the incongruities that there may be in the text, the idea of the comparison between the splittings is really there, thus legitimising Cossali's interpretation.

Cardano ends this chapter with a note, in which he reconsiders his opinion on the fact that the propositions in *Ars Magna*, Chapter XXV were "particular". Indeed, he says[117] that they are actually "general", as we have already observed (see above, Section 3.1.4 on page 83).

We recall (see above on page XI) that in this chapter Cardano introduces a new kind of notation to talk about a solution of an equation depending on the conditions given in some propositions in *Ars Magna*, Chapter XXV. For instance, referring to $x^3 = 20x + 32$,

'20 d. p. R p. 32'

[116]"And with the equality $x^3 = 12x + 34$ one may not discover CD so that it is composed by CF and FD, but by the other rule. But we will find AB so that it is divided into the parts AE, EB" or "*nec licebit cum æquatione 1 cu. æqualis 12 rebus p: 34, invenire cd ut composita est ex cf et fd, sed ex alia regula. Sed inveniemus ab ut est divisa in partes ae, eb*", see [9, Chapter LIII, page 97].

[117]"From [the things] seen here and above it appears clearly that all the rules of Chapter twenty-five of the Ars magna that they call special are general and they are said [to be] special for the only reason of the kind of the value" or "*[e]x visis hic et superius apparet liquido, quod omnes regulæ vigesimiquinti capituli Artis magnæ, quas vocant speciales, sunt generales, et dicuntur speciales solum ratione generis æstimationis*", see [9, Chapter LIII, note, page 97].

means '20 divided into the part and the root that produce 32 [*divisum in partem et radicem producentes 32*]' and refers in a very convoluted way to the solution of the considered equation depending on the f, g in (AM XXV.1). I could not retrieve for sure the proposition to which

"32 p. 20 cum p. 32",

or "what produces 20 with what produces 32 [*producentis 20 cum producente 32*]", refers, since it is a very generic description. On the contrary,

'Ag. R p: 20 p: n: 16'

means 'the aggregate of the roots of the parts of 20 that reciprocally multiplied produce 16 [*aggregatum radicum partium 20, quæ mutuo ductæ producunt 16*]' (or the half of 32), and makes reference to the f, g in (AM XXV.3) (or (A I.3)). This kind of stenography is also used in **Chapter LVII**, where Cardano explicitly mentions Chapter LIII and the very same example $x^3 = 20x + 32$. We recall (see here on page 315) that at the beginning of this chapter the "particular" rule (AM XXV.2) is mentioned as an introduction to the study of the shape of a solution. We also remind that another access point to the topic was the writing $x = \sqrt{a_1 + \frac{a_0}{x}}$ in (A XL). The stenography used by Cardano is employed with the intent of describing this last writing. Indeed, Cardano correlates the three following stenographic writings:

'32 p: 20 c. p. 32',

which means 'what produces 20 with what produces 32 [*producens 20 cum producente 32*]',

"R 20 p: d. 32",

which means 'the root of 20 plus 32 divided by the same root [*R 20 p: diviso 32 per ipsam radicem*]', and

"R 20 f. 32",

which means 'the root of 20 with a fragment of 32 [*R 20 cum fragmento 32*]' (a "fragment [*fragmentum*]" is "what comes forth from a division [*quod ex divisione prodit*]"), as to have the same meaning. Despite the obscurities of the above writings, at least the last one points to $\sqrt{20+\frac{32}{x}}$, considering that Cardano says that "$\sqrt{20}$ f. 32 is the value of $x^3 = a_1x + a_0$ [*R 20 f. 32 est æstimatio cubi ęqualis 20 rebus p: 32 numero*]" (even though we expected a universal root instead of 'R').

Chapters VII and LIII deal with the study of the splittings by themselves. The fact that all the examples have $\Delta_3 > 0$ could indeed point at the little interest showed here for the *casus irreducibilis*. This study eventually leads to establish some calculation rules on the splittings, which could possibly in turn be useful when one wants to solve an equation, also (and especially) when it falls into the *casus irreducibilis*. In Chapter VII, given the coefficients of an equation of the family $x^3 = a_1x + a_0$ and one of its solutions, Cardano explains how to find Y, Z in the splitting (A I.2) (depending on the sum and on the product of Y, Z). There is a similar argument in Chapter LIII, where, under the same assumptions, Cardano explains how to find Y, Z in the splitting parallel to (A I.3) (depending this time on the sum of Y, Z and on the sum of their squares).

Nevertheless, in Chapter LIII Cardano goes further on – and this makes us review our interpretation of Chapter VII in the wake of Cossali's guidance. Despite all the incongruities, in Cardano's text there is indeed the idea of passing from one of the first four splittings of Chapter I to another, possibly also including the splittings parallel to (A I.3) and to (A I.4). Indeed, if we write y, z according to the splitting of departure and Y, Z according to the splitting of arrival, and if we assume that $y + z = Y + Z$ (since they are both equal to a solution of the equation), the second condition on Y, Z will be given by the splitting of arrival itself. What goal could allegedly have this method? Regarding to my hypothesis on the final aim of the splittings, namely – in short – that Cardano studies (A I.2)-(A I.7) in order to find another cubic formula, the following observation strikes me. If one manages to find a way to pass from one splitting to another when the conditions of two splittings are satisfied, then they could pass from that same splitting to the other also when the conditions of one of the two splittings are not satisfied. This would be extremely useful, in particular, when the conditions on the splitting (A I.1) do not hold. Then, in this case, one would not need to embrace the drastic relief to the *casus irreducibilis* and completely discard the cubic formula for $x^3 = a_1x + a_0$. On a small scale, one could also be contented with keeping the usual cubic formula and using another one (non-"general", as the one in (A I.1) was not either) when the first one cannot be applied. This entails that one should relax their requirements on what a cubic formula is, namely it will no longer be unique. This seems to be in any case no problem for Cardano, since we have seen that also in Chapter I the possibility of more than one cubic formula is foreseen.

5.3.5. Summing up

There is a sizeable number of chapters in the *Aliza* that can be gathered around the thread given by the splittings in Chapter I for the pivotal equation $x^3 = a_1x + a_0$. In the above sections, we have analysed seventeen chapters out of sixty. All together, these could have formed the original core, or part of the original core, of that "*aliza* problem" mentioned in the *Ars Magna*. In short, we have identified two main topics linked to the splittings. One concerns their alleged origin, while the other is a possibly unconventional way in which the splittings could have been employed in order to face the problem entailed by the *casus irreducibilis.*

The topic on which most of evidences can be found is the first one. Basically, we have conceived three different hypotheses on the alleged origin of the splittings. Firstly, in Section 5.3.1, we have questioned the substitution $x = y + z$. Whether he got the idea from Tartaglia's poem or not, the *Aliza* attests the deep interest that Cardano had in this shape for a solution of $x^3 = a_1x + a_0$. It is, by the way, a long standing interest, since similar surveys were also in the *Ars Magna Arithmeticæ*. Regarding these early beginnings, we have made the hypothesis that the study of the shapes of the solutions started before the discovery of the cubic formulae. According to this viewpoint, it becomes interesting to consider the *Aliza* as prelude to the *Ars Magna*, and not only the other way round. In Cardano's mathematical writings this is indeed a well recognisable issue, which more precisely aims to describe the irrational solutions of cubic equations with rational coefficients depending on some codified shapes. It is likely that Cardano's dexterity in calculations with irrational numbers (see, for instance, Section 5.4.1), opened the door to the development of his researches towards the shapes of the irrational solutions. Anyway, in the *Aliza*, unlike in the *Ars Magna Arithmeticæ*, these researches are not only scattered all along the book, but also very heterogeneous. There are indeed great dissimilarities among the methods employed in the different chapters (which either refer to Book X of the *Elements*, or proceed by substitution and comparison, or use propositions from the *Ars Magna*). These could maybe testify to the different developing stages of Cardano's surveys. But, in the end, Cardano comes to the same conclusions as in the *Ars Magna Arithmeticæ*, namely that an irrational solution of $x^3 = a_1x + a_0$, with a_1, a_0 rational, must be a *binomium* of the 2nd or 5th type $\sqrt{a} + b$ or a cubic *binomium* $\sqrt[3]{a} + \sqrt[3]{b}$ such that $\sqrt[3]{ab}$ is rational. In this way, a very important remark surfaces: the shape of the solutions of the considered equation must contain two parts. This, in turn, possibly leads to write $x = y + z$. Hence, the idea of substituting $x = y + z$ in $x^3 = a_1x + a_0$ could have been a side effect of the study of the possible shapes for its irrational solutions.

Secondly, in Section 5.3.2, we have analysed the connection with *Ars Magna*, Chapter XXV, which is quite an obvious link, as *Ars Magna*, Chapter XII already suggested. Moreover, when analysing *Ars Magna*, Chapter XXV, we have already remarked that three propositions could be written in a way that recalls three splittings – and now we clearly see it. But, even though both the three splittings and the corresponding propositions in *Ars Magna*, Chapter XXV concern the problem entailed by the *casus irreducibilis*, they are intended to achieve different purposes. The splittings are meant to provide a new formula, while the rules in *Ars Magna*, Chapter XXV only give a case-by-case method. We have suggested that the splittings could have also come from a further improvement of some of the propositions in that chapter of the *Ars Magna*, and hence that they respond to the need of a conclusive relief to the problem entailed by the *casus irreducibilis*. Nonetheless, it happens that in the considered chapters of the *Aliza* Cardano limits himself to rational coefficients. This led me to conjecture that, if Cardano did not conceive the propositions in *Ars Magna*, Chapter XXV (and thus some of the splittings) starting from the study of equations with rational coefficients, he at least planned to apply these propositions to equations with rational coefficients. This is indeed what happens in the great majority of the cases. Moreover, these rules must have been refined after Cardano had realised the problem that the cubic formulae entail, since no mention of such a method is made either in the *Practica Arithmeticæ* or in the *Ars Magna Arithmeticæ*.

Thirdly, in Section 5.3.3 we have accounted for Cardano's last say on the topic of the splittings. As it was indeed very reasonable to expect, a geometrical inspiration can also be found at the origin of the mainly arithmetical methods that Cardano devised. Anyway, it is a very specific kind of geometry, where the position of the geometrical objects is not essential and the main constraint is due to dimensional coherence.

5.4. Some technical supplements

In the present section, I deal with some chapters that, on the one hand, complete my analysis in the preceding sections by providing some technical details, even though they do not display a real uniformity with respect to their contents. On the other hand, their variety helps in giving a flavour of the diversity of the contents in the *Aliza*.

The variety of the technical methods indeed contributes to the *Aliza*'s usefulness and they find their *raison d'être* if we consider the

Aliza not only as a failed attempt to emend the cubic formula, but also as a real treasury of algorithmic results.

5.4.1. Preliminary study of irrational numbers

Many chapters of the *Aliza* seem to provide the starting point for the subsequent deeper survey on the shapes of the irrational solutions of $x^3 = a_1x + a_0$ with rational coefficients (see Section 5.3.1) and consequently on the substitution $x = y + z$ (see Section 5.2). These chapters deal essentially with different kinds of irrational numbers.

Pretty soon we come across a classification of the six types of *binomia* and *recisa* by square roots,[118] and we find some calculation rules and properties. As such, these chapters could maybe be paired to the arithmetisation of Book X of the *Elements* in the first part of the *Ars Magna Arithmeticæ* (even though they do not constitute such a compact block as there). Indeed, these chapters are scattered all along the *Aliza* and I could only gather them with regard to (the main consistent part of) their contents.

In **Chapter IV** "On the way to reduce all the quantities that are called 'first sides' by the tenth [Book] of Euclid in short [*De modo redigendi quantitates omnes, quæ dicuntur latera prima ex decimo Euclidis in compendium*]" Cardano suggests a classification for irrational numbers. The treatment is essentially analogous to the one in the *Ars Magna Arithmeticæ* that I have recalled in Section 3.1.4 on page 80. Note that Euclid is explicitly mentioned at the beginning of this chapter and his Book X is a steady reference. Cardano longerover quotes a "third book [*tertio libro*]" – very likely referring to the *Opus Arithmeticæ Perfectum* (as he often does when he does not specify the title of the referred writing), the third book of which should have been devoted to irrational numbers.[119] Then, he teaches case-by-case how to take the square root of all the six *binomia* and *recisa*,

[118] According to the definitions in Book X, see Footnote 61 or [32, Book X, page 133r].
[119] See above, on page 13.

interspersing his explanation by some rules.[120] These deal with the (in)commensurability and order relationships between the different types of *binomia* and *recisa*, teach how to multiply radicals, and show how to rationalise fractions where square roots appear in the denominator and a rational number in the numerator.

This last topic, together with the rationalisation of fractions where cubic roots appear, is the starting point of **Chapter XVII** "In how many ways the number can be produced by the non-number [*Quot modis numerus possit produci ex non numero*]". It helps to achieve the final aim of this chapter, which is to show how a rational number can be obtained by performing operations on irrationals. In particular, taking a, b rational, Cardano considers the products of the form $(a+\sqrt{b})(a-\sqrt{b})$ or $(a+\sqrt{b})(ka-k\sqrt{b})$, with k rational, the differences of the form A^2-B^2, with $A=\frac{b-a^2}{2a}+a$ and $B=\frac{b-a^2}{2a}$, and the powers of the form $(\sqrt[n]{a})^n$, with n natural, which all give a rational result.

In **Chapter XXX** "What [is] the equality of the cubes of the parts of a divided line [*Qualis æqualitas cuborum partium lineæ divisæ*]" Cardano shows the following proposition.

A XXX. *If a line AB is divided into C, then*

$$\overline{AC}^3+\overline{CB}^3=(\overline{AC}+\overline{CB})(\overline{AC}^2-\overline{AC}\,\overline{CB}+\overline{CB}^2).$$

This proposition is useful in the context of the splittings. Indeed, if we take $\overline{AC}=y$ and $\overline{CB}=z$ (as we did on page 269) and rewrite (A XXX) as $y^3+z^3=(y+z)(y^2-yz+z^2)$, we remark that the proposition justifies the passage from the system

$$\begin{cases} a_0=3y^2z+3yz^2 \\ a_1x=y^3+z^3 \end{cases}$$

[120] We remark that a similar topic had already been presented in *Ars Magna Arithmeticæ*, Chapters XIII, XIV, and XVII, see here on page 186.

to the system

$$\begin{cases} a_0 = 3y^2z + 3y\,z^2 \\ a_1 = y^2 - yz + z^2 \end{cases},$$

which was left implicit in the splitting (A I.2) (see above, on page 273).

Since this chapter is quite a detailed explanation of a simple identity, it is unusually easy to detect in the proof (that I do not reproduce[121]) the common pattern of (AM VI.6). As before, the geometrical environment in which the proposition is interpreted fixes the reference of the objects involved. Then, the relative position of the geometrical objects identified in this way is no longer essential and the proof strikingly reduced to an arithmetical calculation.

Summing up, these few chapters do not contain new, revealing results on irrational numbers. They rather contain some technical propositions that complete the picture of the *Aliza* in the above sections. Collaterally, they also give an insight of the extreme dexterity and acquaintance with calculation methods that Cardano shared with the mathematicians of his time.

5.4.2. Again on Proposition (AM VIII.2)

At the very beginning of the *Aliza*, there are two chapters that both implicitly refer to the same proposition of the *Ars Magna*, namely (AM VIII.2). Since in the framework of the *Ars Magna* it was proved that this proposition had many connections, I deal with its echoes in the *Aliza*.

In **Chapter III** "On the way to discover the quantities that help to the chapters by the product of one part by the other and by the square of the difference of the parts [*De modo inveniendi quantitates quæ serviant capitulis per producta unius partis in aliam, et quadratum differentiæ partium*]" Cardano uses (without explicitly giving the reference) a particularised version of (AM VIII.2). We remind that in

[121] See [9, Chapter XXX, pages 59-60].

the end this proposition is nothing more than a help to the intuition to guess a solution of the considered equation. What is remarkable in the *Ars Magna* is that Cardano considers equations of any degree, provided that they belong to the large class of $x^q + a_0 = a_p x^p$ with $0 < p < q$ (see here, on page 73). In the *Aliza*, instead, Cardano considers only a family of cubic equations (more precisely, the ones lacking in the first degree term), but we find a more detailed insight. This is not surprising at all in a book devoted to the *casus irreducibilis*. The final aim of this chapter is to try to better determine the two numbers f, g that enable him to find a solution of the considered equation. Then, Cardano takes into account the shape of f, g and searches for some constraints on it. In particular, he forces the coefficients of the equation to be rational and exploits (AM VIII.2).

A III.i. *Let $x^3 + a_0 = a_2 x^2$ be given [with a_2, a_0 rational] and assume that two positive, real numbers f, g exist such that*

$$\begin{cases} a_2 = f + g \\ a_0 = fg^2 \end{cases}.$$

Then [$x = g$ and], if f is a binomium *of any type, g is a* recisum *of the same type, and* vice versa.

Note that Cardano does not specify that the coefficients of the equation are rational. We can only observe that Cardano's subsequent calculations require them to be so (see below, Footnote 125). Moreover, we recall that we have already run into similar considerations in the *Ars Magna* and *Ars Magna Arithmeticæ*, and there the coefficients were rational.

We also remark that, if Cardano is fully aware that $x = g$ is a solution, he does not seem to have an understanding of the fact that the *binomia* and *recisa* of the 3rd or 6th type cannot be a solution of a (non-complete) cubic equation[122] (see for instance *Aliza*, Chapters

[122]"Then the first part [g] can be the first, the second, and the third *binomium* or *recisum* and can also be the fourth, the fifth, and the sixth [*binomium* or *recisum*], nevertheless not the *recisum*, because, being the first [part] a

V and X, in Section 5.3.1). Still, in the examples that follow the proposition, he obviously only considers some *binomia* and *recisa* of the 1st or 4th and of the 2nd or 5th type.

We observe that we have already run into a similar (even if more detailed) proposition in (AM XXV.15). We remind that (AM XXV.15) is linked to (AM VIII.2) and gives, according to certain conditions, two solutions of $x^3 + a_0 = a_2x^2$ under the shape of a *binomium* and a *recisum* (see above, on page 95). Indeed, we recall that $\frac{f}{2} \pm \sqrt{\frac{f^2}{4} + fg}$ are the two other solutions (see above, on page 77).

Cardano does not provide any justification for (A III.i). One may – and maybe Cardano did – consider that the examples that follow the next proposition display a clear enough explanation also for this proposition. For instance, Cardano considers the equation $x^3+8 = 7x^2$ and suggests the couples $f = 3 - \sqrt{8}$, $g = 4 + \sqrt{8}$ and $f = 3 + \sqrt{8}$, $g = 4 - \sqrt{8}$.

A III.ii. *Let the equation $x^3 + a_0 = a_2x^2$, where only a_2 is given positive, rational. We write $a_2 = F + G$ with F, G positive rational. Then, referring to the proposition (A III.i), we have*

$$f = F \pm \sqrt{H} \quad \textit{and} \quad g = G \mp \sqrt{H},$$

where $H = 2FG - G^2$. Moreover,

$$a_0 = FG^2 + FH - 2GH.$$

We remark that in this case the rational solution is $F - G$.

Moreover, note that Cardano considers $g = G - \sqrt{H}$ only when $g > 0$, since g is also a solution, and obviously $H = 2FG - G^2$ only

root, the second part is necessarily a *binomium*, because the first is a *recisum*, therefore in both there is a positive root, therefore that number that is divided at the beginning cannot be" or "*[p]rima ergo pars potest esse binomium vel recisum primum, secundum, et tertium et potest etiam esse binomium quartum, quintum, et sextum, non tamen recisum, quia cum prima sit R et secunda pars necessario sit binomium, quia prima est recisum, igitur in utraque esse R p:, igitur non potest esse numerus ille qui ab initio divisus est*", see [9, Chapter III, page 6].

when $H > 0$. His abstract explanation of this proposition could seem confusing,[123] but his examples (all with $\Delta_3 < 0$) are indeed very clear.[124] As it frequently happens, Cardano justifies his statement only through these examples, showing that the algorithms that he gave work. It is nevertheless not difficult to recover the formulae of the above proposition by a straight calculation.[125]

On the one hand, as already observed, the *Ars Magna* gives a more comprehensive result (for more equations with any coefficients). This result is nothing but a rephrasing of the original equation, thanks to which one switches from searching for a solution of the original equation to searching for the f, g in the rephrasing. On the other

---

[123]"Then I say that, divided the number of the squares in two parts, that we call principal, and we call principal [the part] that is multiplied by itself, multiply the first by the double of the second to find the other two parts. And you will deduce the square of the first from the product and the root of the *residuum* is the part to add or to subtract to the principal [parts] with the agreed conditions. And similarly, to produce the positive number, multiply the difference of the principal [parts] between them and the product by the double of the first principal, and what is produced is the searched number" or "*[d]ico ergo quod diviso numero quadratorum in duas partes, quas vocabimus principales, et eam quæin se ducitur, vocabimus principales, et pro alijs duabus partibus inveniendis, duc primam in duplum secundæ et a producto deducito quadratum primæ, et R residui est pars addenda principalibus, aut detrahenda cum conditionibus condictis. Et similiter pro numero producendo duc differentiam principalium in se, et productum in duplum primæ principalis, et quod producitur, est quæsitus numerus*", see [9, Chapter III, page 6].

[124]See [9, Chapter III, pages 6-7]. The examples concern the equations $x^3+8 = 7x^2$, $x^3 + 6 = 7x^2$, $x^3 + 48 = 7x^2$, $x^3 + 24 = 8x^2$, $x^3 + 40 = 8x^2$, $x^3 + 45 = 8x^2$, and $x^3 + 75 = 8x^2$.

[125]Indeed, we take two positive, rational numbers F, G such that $a_2 = F + G$ and a positive, rational number H. We write $f = F \pm \sqrt{H}$ and $g = G \mp \sqrt{H}$. We want to draw a value for H and a_0 depending on F, G.

Then $g^2 = G^2 + H \mp 2G\sqrt{H}$. We have that

$$fg^2 = FG^2 + FH - 2GH + \sqrt{H}(\mp 2FG \pm G^2 \pm H).$$

By (A III.i) we know that $a_0 = fg^2$ must be rational. Therefore, we need that $\sqrt{H}(\mp 2FG \pm G^2 \pm H) = 0$. Then, either $H = 0$ (but Cardano does not consider this case), or $\mp 2FG \pm G^2 \pm H = 0$, from where we infer the value of H. Moreover, $a_0 = FG^2 + FH - 2GH$.

hand, (A III.ii) particularises this result to a certain family of equation with rational coefficients[126] (but note that this does not make it easier to find F, G).

In **Chapter V** "On the consideration of the *binomia* and *recisa* that contain a rational figure, whence on the value of the chapters [*De consideratione binomiorum et recisorum continentium figuram rheten, ubi de æstimatione capitulorum*]" Cardano uses again a particular version of (AM VIII.2), which now concerns $x^3 + a_0 = a_1 x$ (again without any explicit reference). Consider f, g as in (AM VIII.2) and assume that the solution x is a certain type of *binomium* or *recisum*. One of the aims of this chapter is then to specify which type of *binomia* and *recisa* f, g are.

A V.ii. *Let*[127] $x^3 + a_0 = a_1 x$ *[with a_0, a_1 positive, rational] and one of its solutions x be given.*
[We consider (AM VIII.2). In particular, if two positive, real numbers f, g exist such that

$$\begin{cases} a_1 = f + g \\ a_0 = f\sqrt{g} \end{cases},$$

then $x_1 = \sqrt{g}$, $x_2 = -\frac{\sqrt{g}}{2} + \sqrt{\frac{g}{4} + f}$, and $x_3 = -\frac{\sqrt{g}}{2} - \sqrt{\frac{g}{4} + f}$ (see above on page 75).]
Let x be a solution of the equation. Then, the following relations between x, f, and g hold:

- *if x is* binomium *of the 1st type, then g is a* binomium *of the 1st type and f is a* recisum *of the 1st type;*
- *if x is* recisum *of the 1st type, then g is a* recisum *of the 1st type and f is a* binomium *of the 1st type;*

[126] Note that – as we have already remarked, in the vast majority of Cardano's examples the equations have rational coefficients.

[127] My commentary on Proposition (A V.i) is on page 296. I use this numbering since it reproduces the order of the results in Cardano's text.

- *if x is* binomium *of the 4th type, then g is a* binomium *of the 1st type and f is a* recisum *of the 4th type;*
- *if x is* recisum *of the 4th type, then g is a* recisum *of the 1st type and f is a* binomium *of the 4th type;*
- *if x is* recisum *of the 2nd type, then g is a* recisum *of the 1st type and f is a* binomium *of the 2nd type;*
- *if x is* recisum *of the 5th type, then g is a* recisum *of the 1st type and f is a* binomium *of the 5th type.*

The proposition is expounded by some examples (again, all with $\Delta_3 < 0$). We recall that Cardano is well acquainted with the operations of rising to the power and taking the root of this kind of numbers (for instance in *Aliza*, Chapter IV, see above on page 352, and in the *Ars Magna Arithmeticæ*, see above on page 186), on which all the calculations that are at the bottom of this statement depend.[128] Note that – as Cardano himself remarks – the *binomia* of the 2nd or 5th type are not listed among the possible solutions. Indeed, when a *recisum* of the 2nd or 5th type $\sqrt{a} - b$ is one of the irrational solutions

[128]Note that the relation between x and $g = x^2$ is always true, namely that the square of a *binomium* of the 1st or 4th type is a *binomium* of the 1st type and that the square of a *recisum* of the 1st, 2nd, 4th, or 5th type is a *recisum* of the 1st type. Indeed, let us firstly consider a *binomium* or *recisum* of the 1st or 4th type $a \pm \sqrt{b}$ (that is, with $a^2 - b > 0$). Then, $(a \pm \sqrt{b})^2 = (a^2 + b) \pm 2a\sqrt{b}$. This is a *binomium* or *recisum* of the 1st type, since $(a^2 + b)^2 - 4a^2b > 0$ and $\frac{\sqrt{(a^2+b)^2-4a^2b}}{a^2+b} = \frac{a^2-b}{a^2+b}$ is always rational (in particular, it does not depend on the type of $a \pm \sqrt{b}$, that is in fact that $\frac{\sqrt{a^2-b}}{a}$ is rational or not). Secondly, let us consider a *recisum* of the 2nd or 5th type $\sqrt{a} - b$ (that is, with $a - b^2 > 0$). Then, $(\sqrt{a} - b)^2 = (a + b^2) - 2b\sqrt{a}$. This is a *recisum* of the 1st type, since $(a+b^2)^2 - 4ab^2 > 0$ and $\frac{\sqrt{(a+b^2)^2-4ab^2}}{a+b^2} = \frac{a-b^2}{a+b^2}$ is always rational (in particular, it does not depend on the type of $\sqrt{a} - b$, that is in fact that $\frac{\sqrt{a-b^2}}{\sqrt{a}}$ is rational or not).

On the other hand, the conclusions on f depend on the intrinsic nature of the coefficients of the equation.

(and the equation has more than one real solution, that is $\Delta_3 < 0$), then the other irrational solution is $-(\sqrt{a}+b) < 0$.

Finally, Cardano makes some remarks that describe the proposition and recalls the statement in (AM XIII bis). This abrupt reference to (AM XIII bis) stroke me. He simply states the formula for the second solution of $x^3 + a_0 = a_1 x$. But the fact that he chooses to recall it right in the same chapter in which he is speaking of the connection between the shapes of the *binomia* and *recisa* and a solution of the considered equation suggests that the idea of the formula in (AM XIII bis) (and also the one in (AM XIII)) could have been somehow originated in relation to the environment of the *binomia* and *recisa*. Unfortunately, I cannot provide any evidence, except maybe the fact that in this way we can account for some of the carelessness in those propositions of the *Ars Magna*, as I have already remarked on page 116.

Summing up, Chapters III and V of the *Aliza* (even though they do not display any uniformity) both recall the same result from the *Ars Magna*, which is no longer proposed as such in the *Aliza*. On the one hand, one may believe that there is no need to reprint it, since in the 1570 edition the *Aliza* is appended to the *Ars Magna*. On the other hand, it is useless to recall a too general result when one is interested only in cubic equations (or rather – I would say – in cubic equations with $\Delta_3 < 0$, since all the examples are of this kind). Cardano considers the two families of equations $x^3 + a_0 = a_1 x$ and $x^3 + a_0 = a_1 x$, and takes f, g as in (AM VIII.2). His main aim in the *Aliza* is to try to better specify f, g, possibly in order to get a solution of the equations. In the *Ars Magna*, the whole difficulty has been displaced from directly solving the considered equation to find such a f, g – and we recall that finding those f, g is not trivial at all. There, fixing f, g depended on intuition. Now, Cardano assumes to have rational coefficients and suggests another rephrasing of (AM VIII.2), which is maybe supposed to make it easier to find those f, g. As we have seen in Section 5.3.2 while dealing with the chapters of the *Aliza* that refer to *Ars Magna*, Chapter XXV (which is tightly linked to (AM VIII.2)), limiting to rational coefficients is quite a common strategy in the *Aliza*, also considering that most of Cardano's examples have only sporadically irrational coefficients.

5.5. How far goes geometry in the *Aliza*?

In Section 2.3, we have seen that the book that we call nowadays '*Aliza*' possibly originates from an "*aliza* problem" that Cardano planned to include in a certain "book of the geometrical problems", very likely referring to the last book of the (never written) *Opus Arithmeticæ Perfectum*. Therefore, we have been led to expect a certain amount of geometry in this treatise. Considered that – as we have seen – the *Ars Magna* contains many proofs in a loose geometrical fashion, the interesting question is 'is there any geometry in the *Aliza* and, if yes, which kind of geometry is it?'

Since we are now more acquainted with this book, we are able to remark that in the *Aliza* there is not as much geometry as we were expecting to be. We only came across a few examples, for instance while dealing with Chapter XXX (on page 353), Chapter XL (on page 316), or Chapter LX (on page 336). In all these cases we found that, in the geometrical environment in which Cardano translates his hypotheses, the relative positions of the involved geometrical objects do not play an essential role. We recall that this is exactly what happened in the *Ars Magna*. Moreover, we have seen in Section 5.3.3 that the splittings themselves could have had a loose geometrical inspiration.

Given these premises, it is time to have a look at another chapter of the *Aliza* in which geometry – and an essentially different kind of geometry – is massively employed.

5.5.1. Cardano's proof of the existence of a solution for $\mathbf{x^3 + a_0 = a_2x^2}$

In the *Aliza*, **Chapter XII** "On the way to show geometrically the value of a cube and a number equal to some squares [*De modo demon-*

strandi geometrice æstimationem cubi et numeri æqualium quadratis]" we find for the very first (and unique) time a different geometrical approach.

Cardano is mainly interested in the equation $x^3 + a_0 = a_2x^2$ when it falls into the *casus irreducibilis.* Indeed, he says that, if a condition equivalent to $\Delta_3 > 0$ holds, "the purpose is false [*propositum esset falsum*]."[129] If, instead, a condition equivalent to $\Delta_3 = 0$ holds, Cardano is able to solve the equation (since in the cubic formula he does not come across square roots of negative numbers) and spends a very few words to explain how to find a solution.[130] We resume his reasoning in the following proposition.

A XII.i. *Consider* $x^3 + a_0 = a_2x^2$. *If* $a_0 = \frac{4}{27}a_2^3$, *then* $x = \frac{2}{3}a_2$.

Let us then assume $x^3 + a_0 = a_2x^2$ with $\Delta_3 < 0$. In this case, the cubic formula contains some square roots of negative numbers and Cardano needs to face this problem to get a (real) solution. Yet, there is a particular case in which Cardano is able to find a real solution through the formula. This happens when a condition equivalent to $q = 0$ (see above, Section 2.5.3) holds. Then, Cardano is able to draw a solution using the fact that $\sqrt[3]{-y} = -\sqrt[3]{y}$.[131]

A XII.ii. *Consider* $x^3 + a_0 = a_2x^2$. *If* $a_0 = \frac{2}{27}a_2^3$, *then* $x = \frac{a_2}{3}$.

[129] Note that Cardano makes reference to "the things said above [*ex supradictis*]" (more specifically, to a certain "Proposition 27") to justify the fact that, if $a_2 = 12$ and $a_0 > 256$, then "the purpose is false". We remark that the condition on the coefficients implies that $\Delta_3 > 0$. Unluckily I did not find such a "Proposition 27" neither in the preceding chapters of the *Aliza*, nor in the *Ars Magna*, or in the *De Proportionibus*.

[130] Indeed, taking $\Delta_3 = 0$, that is $a_0 = \frac{4}{27}a_2^3$, we have $q = \frac{2}{27}a_2^3$ (see above, Section 2.5.3). Then, $x = \sqrt[3]{q} + \frac{a_2}{3} = \frac{2}{3}a_2 > 0$ is the (positive) double root. Moreover, $x = -2\sqrt[3]{q} + \frac{a_2}{3} = -\frac{1}{3}a_2 < 0$ is the (negative) remaining root.

[131] Indeed, taking $q = 0$, we have $\Delta_3 < 0$ and $x = \sqrt[3]{\sqrt{\Delta_3}} + \sqrt[3]{-\sqrt{\Delta_3}} + \frac{a_2}{3}$. Since the cubic root is an odd function, $x = \sqrt[6]{\Delta_3} - \sqrt[6]{\Delta_3} + \frac{a_2}{3} = \frac{a_2}{3}$ gives a real, positive solution.

Remark that in this case we need to use the properties of imaginary numbers to obtain the other solutions. Indeed, $x = \pm\frac{i\sqrt{3}}{3}\sqrt[6]{-a_2^6}$.

Once the limit cases have been cleared out, we can focus on Cardano's main issue in this chapter: the equation $x^3 + a_0 = a_2x^2$, when square roots of negative numbers appear in the cubic formula and cannot be deleted (that is, when $\Delta_3 < 0$ and $q \neq 0$). In particular, this is the case of $x^3 + 192 = 12x^2$. Through this example Cardano shows that this family of equations (under the above conditions) always has a (real) solution. As he reminds, he will use the very same method that Eutocius of Ascalon employed in his commentary on Proposition II.4 of *On the sphere and cylinder* by Archimedes.[132]

A XII.iii. *Consider the equation $x^3 + 192 = 12x^2$. Then, it has a [real] solution.*

A XII.iii — *Proof.* Considering the diagram in Figure 5.9, Cardano takes $\overline{AB} = 12$ and E on AB such that $\overline{EB} = 2\overline{AE}$. He considers $\overline{DQ}^2\overline{DZ} = 192$ [with $\overline{DZ} = \overline{AE}$].[133] He takes u such that $\overline{EB} : \overline{DQ} = \overline{DQ} : \overline{u}$ and AC perpendicular to AB such that $\overline{DZ} : \overline{AC} = \overline{EB} : \overline{u}$. From these hypotheses it follows that $\overline{DZ} : \overline{AC} = \overline{EB}^2 : \overline{DQ}^2$, that is $\overline{AC}\,\overline{EB}^2 = \overline{DQ}^2\overline{DZ}$. Cardano wants to find a point O on AB such that $\overline{AO}\,\overline{OB}^2 = \overline{AC}\,\overline{EB}^2$.

He performs the following construction. He takes the line passing by C and E in order to draw the rectangle of vertexes C, F, G, H. Then, he takes M on GH such that $\overline{GM} = \overline{DQ}$ and considers $\overline{FN}$ such that $\overline{AB} : \overline{EB} = \overline{EB} : \overline{FN}$. Referring to Book II of Apollonius' *Conics*,[134]

[132] Cardano says that Eutocius gave two solutions, but that he will quote only the first. He also says that he will not quote instead the propositions by Euclid, see [9, Chapter XII, page 25].

In his commentary, Eutocius reports in details on three solutions to the same problem: one found in an old book, one by Dionysodorus, and one by Diocles. As we will see, Cardano follows Eutocius's first solution almost step by step. He does not mention the other two solutions.

[133] Cardano does not explicitly assume that $\overline{DZ} = \overline{AE}$, but it is clear that he must have done so. Indeed, immediately after the construction, he states that $\overline{AE} : \overline{AC} = \overline{EB}^2 : \overline{GM}^2$, having drawn $\overline{GM} = \overline{DQ}$.

[134] In particular, *Conics* II.4 says that "[g]iven the asymptotes and a point P on a hyperbola, to find the curve", see [28, page 56].

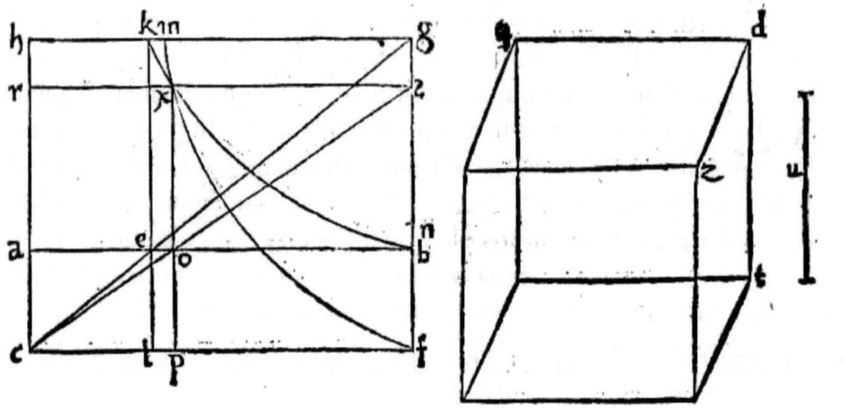

Figure 5.9. – *Aliza*, Chapter XII. Kind courtesy of Biblioteca Virtual del Patrimonio Bibliográfico and Universidad de Sevilla.

he draws the parabola with axis FG, parameter $\overline{FN}$, passing through F and the hyperbola with asymptotes HC, CF, passing through B.

[Then, Cardano shows by a continuity argument on the sly that the parabola and the hyperbola intersect.] On the one hand, since $\overline{DZ} : \overline{AC} = \overline{EB}^2 : \overline{DQ}^2$, [$\overline{AE} = \overline{DZ}$], [$\overline{EB}^2 = \overline{AB}\,\overline{FN}$ by the hypothesis on $\overline{FN}$], [$\overline{AB} = \overline{CF}$], and $\overline{GM} = \overline{DQ}$, it holds that $\overline{AE} : \overline{AC} = \overline{CF}\,\overline{FN} : \overline{GM}^2$. [By similar triangles,] $\overline{AE} : \overline{AC} = \overline{CF} : \overline{FG}$ [and then $\overline{AE} : \overline{AC} = \overline{CF}^2 : \overline{FG}\,\overline{CF}$]. Therefore, $\overline{CF}\,\overline{FN} : \overline{GM}^2 = \overline{CF}^2 : \overline{FG}\,\overline{CF}$. By *Elements* VIII.9, Cardano gets $\overline{FG} : \overline{GM} = \overline{GM} : \overline{FM}$, [that is $\overline{GM}^2 = \overline{FN}\,\overline{FG}$,] which means by Apollonius' *Conics* I[.11][135] that M is on the parabola. On the other hand, since $\overline{(HE)} = \overline{(EF)}$, adding $\overline{(AL)}$ to both, it holds that $\overline{FB} : \overline{HK} = \overline{FG} : \overline{GH}$. This means that K is on the hyperbola. [Since the parabola passes through

[135]"First let the diameter PM of the section be parallel to one of the sides of the axial triangle as AC, and let QV be any ordinate to the diameter PM. Then, if a straight line PL (supposed to be drawn perpendicular to PM in the plane of the section) be taken of such a length that $PL : PA = BC^2 : BA\,AC$, it is to be proved that $QV^2 = PL\,PV$", see [28, page 8].

F and M, the hyperbola passes through B and K, and both are continuously traced, they must intersect.] Cardano calls X the point of intersection.

Then, he performs the last construction. He draws the parallel and perpendicular lines to AB passing through X. The point O is where the perpendicular cuts AB.

Cardano needs now to show that the intersection point S between FG and the parallel to AB is aligned with C, O, so that he can exploit the properties of similar triangles. By the hyperbola definition, $\overline{(XP)} = \overline{(AF)}$ and, taking away the common $\overline{(AP)}$, he gets $\overline{(RO)} = \overline{(OF)}$. [By the converse of *Elements* I.43,[136]] they are the complements of two parallelograms around the diameter that pass through C, O, S. Then, $\overline{CF} : \overline{FS} = \overline{AO} : \overline{AC}$. Since $\overline{EB}^2 = \overline{CF}\,\overline{FN}$, Cardano gets $\overline{FS}\,\overline{FN} : \overline{EB}^2 = \overline{AO} : \overline{AC}$. By the definition of the parabola and since $\overline{SX} = \overline{OB}$, he finds that $\overline{OB}^2 : \overline{EB}^2 = \overline{AO} : \overline{AC}$, that is $\overline{AO}\,\overline{OB}^2 = \overline{AC}\,\overline{EB}^2$ by *Elements* XI.34. □

In short, Cardano provides as usual the good assumptions under which the concerned equations can be translated in an appropriate geometrical environment. In this case, this means that the coefficients of the equation $x^3 + 192 = 12x^2$ are represented by the measures of some geometrical quantities – the segment AB and the parallelepiped DQ^2DZ with a squared basis. Note that Cardano introduces the point E, the segments AC and u, and later the point M in order to interpret DQ^2DZ in the same diagram as AB. Cardano wants then to show that it exists O on AB such that $\overline{AO}\,\overline{OB}^2 = \overline{DQ}^2\overline{DZ}$ holds. This means in turn that it exists $\overline{OB} = x$ that is a solution of the equation $x^3 + 192 = 12x^2$, since $\overline{AO} = 12 - x$ and $\overline{DQ}^2\overline{DZ} = 192$. Cardano does what is needed in order to draw the parabola and the hyperbola, shows (partially implicitly by a continuity argument) that

[136] According to Campano, "*[o]mnis parallelogrammi spacii eorum quæ circa diametrum sunt parallelogrammorum supplementa æqua sibi invicem esse necesse est*" and, according to Zamberti, "*[o]mnis parallelogrammi eorum quæ circa dimentientem sunt prallelogrammorum supplementa sibi invicem sunt æqualia*", see [32, page 20v].

they intersect, and finally verifies that the intersection point identifies the segment searched for. We remark indeed that the proof does not provide any actual instructions on how to calculate the measure of the segment OB. The segment exists, but one can explicitly find its measure only if they know how to calculate the abscissa of the intersection point of the parabola and the hyperbola, which is indeed equivalent to solve the cubic equation.[137] But clearly enough – as Cardano recalls in the close of this chapter[138] – this is far beyond his skills and the aim of the proposition. We cannot say for sure whether Cardano wrote this chapter before or after having found the cubic formulae. The close could fit both alternatives, since "the arithmetical operation that does not satisfy" could refer either to the fact that Cardano has not yet found the cubic formula or that he has found a cubic formula that does not always work. In any case, the geometrical environment in which the proposition at issue is translated is not suitable to draw the numerical value of the solution or an algorithm that gives it.

At the end of the proof, Cardano adds a few words as an explanation. He firstly observes how "the reasoning of the construction [*ratio constructionis*]" relies on the definitions of the parabola (for instance, $\overline{GM}^2 = \overline{FN}\,\overline{FG}$) and of the hyperbola (for instance, $\overline{(XP)} = \overline{(AF)}$). Then, he remarks that the parameter of the parabola depends only on the coefficient of the term of degree two, while the hyperbola depends on both the coefficients of the considered equation. Finally, Cardano

[137]Indeed, if we take C to be the origin of a Cartesian coordinate system, FC on the x-axis (right to left oriented, such that $\overline{OB} = x > 0$), and FG on the y-axis, the equations of the parabola and hyperbola respectively are $x^2 = \frac{4}{9}a_2 y$ (or $x^2 = \frac{16}{3}y$) and $(a_2 - x)y = \frac{9}{4}\frac{a_0}{a_2}$ (or $(12 - x)y = 36$). Then, their intersection points are given by the solutions of $x^3 + 192 = 12x^2$.

We moreover remark that the parabola and hyperbola have three intersection points (since the equation has three real solutions). They are one on the left and one on the right of the intersection point shown by Cardano in the diagram.

[138]"And the easy geometrical operation is the most difficult in arithmetic and it does not even satisfy" or "*Et ideo facilis operatio Geometrica difficillima est arithmetice, nec etiam satisfacit*", see [9, Chapter XII, page 27].

seems to suggest that one can draw $\overline{FS}$, and consequently $\overline{OB}$, thanks to the parabola.[139] Unfortunately, this last part is quite corrupted,[140] and in any case contains a serious miscalculation.[141]

We remark that, in this proof, some geometry is undeniably employed and is of a different kind than usual, beginning with the diagram itself, which is not the one that Cardano usually draws. This, for sure, also depends on the fact that the aim of the proof is not to calculate a solution, but only to show its existence.

In order to seize more clearly the implications of Cardano's proof, let us have a look at where Proposition (A XII.iii) comes from.

5.5.2. Comparing Cardano's proof with his forerunners' one

Cardano explicitly quotes Eutocius' commentary on Proposition II.4 of Archimedes' *On the sphere and cylinder* to be the source of his proof. Therefore, let us go and check this precursor text.

Being a basic treatise with possibly a good deal of applications, *On the sphere and cylinder* had a relatively lucky history compared to other Archimedes' books. Anyway, it experienced, as it was common, some alterations. For instance, already during the 2nd century BC, Diocles and Dionysodorus no longer disposed of the proof that Archimedes maintained to have written at the end of Proposition II.4. During the 4th century AD, Eutocius confirmed that this proof is missing from all the codices that he has consulted. Nevertheless, in a passage from an old book, he bumped into a theorem that he ascribed to Archimedes and that he believed to be the missing proof. It must be said that nowadays Eutocius' opinion is no longer trusted, since

[139] This would correspond to draw y in the system given by the equations of the parabola and hyperbola, but the calculations do not correspond to Cardano's ones.

[140] See for example "*Posita ergo fs quad. c*", [9, Chapter XII, page 27].

[141] While "reducing to one the cube [*reduxerimus ad unum cubum*]", Cardano finds 8 for the coefficient of the term of degree two, whereas the correct equation – according to all his hints up to now – should have been $3\sqrt{3}y^2 - y^3 = 9\sqrt{3}$. Moreover, not even the formulae provided by Cardano to calculate these coefficients are consistent with his previous procedures.

his arguments (amounting to the use of Doric dialect and of some ancient terminology for the conic sections) are not decisive.

Proposition II.4 is about cutting a sphere into two parts such that they have a given ratio. This result relies on a so-called auxiliary problem,[142] the proof of which was promised by Archimedes and got lost.

Archimedes' auxiliary problem to *On the sphere and cylinder*, II.4. *Let*[143] *the segment* AB *and a point* E *on* AB *such that* $\overline{EB} = 2\overline{AB}$, *and let* $\overline{AC} < \overline{AB}$.
To find a point O *on* AB *such that* $\overline{AO} : \overline{AC} = \overline{EB}^2 : \overline{OB}^2$.

For comparison's sake, I have used the same lettering as Cardano's. Then, we can make reference to the diagram 5.9 on page 364. We remark that this is exactly what Cardano needs to prove in (A XII.iii) in order to show that the family of equations $x^3 + a_0 = a_2x^2$, with $\Delta_3 < 0$ and $q \neq 0$, has a (real) solution.

Eutocius proves a slightly more general version in his commentary.

Eutocius' auxiliary problem to *On the sphere and cylinder*, II.4. *Let*[144] AB, AC *be given segments and* Δ *be a given surface.*
To find a point O *on* AB *such that* $\overline{AO} : \overline{AC} = \overline{\Delta} : \overline{OB}^2$.

This version requires a condition of existence ("*diorismos*") on O. Indeed, $\overline{AO} : \overline{AC} = \overline{\Delta} : \overline{OB}^2$ implies that $\overline{AO}\,\overline{OB}^2 = \overline{AC}\,\overline{\Delta}$, where $\overline{AC}\,\overline{\Delta}$ is given by hypothesis. But, if the segment AB is cut in any point E, the maximum value of $\overline{AE}\,\overline{EB}^2$ is attained when E is such that $\overline{EB} = 2\overline{AE}$ (as Eutocius shows later on[145]). Then, it must be $\overline{AC}\,\overline{\Delta} \leq \overline{AE}\,\overline{EB}^2$. Note that the condition is automatically matched by the particular $\Delta = EB^2$ chosen by Cardano. Actually,

[142] See [29, pages 62-64] or [41, pages 101-105].
[143] See [29, page 64] or [41, pages 103 and 635].
[144] See [29, page 66] or [41, page 636].
[145] I skip the proof, since it would lead us too far from our interests.

the condition holds when the problem is associated (in the same way as Cardano does) to any equation in which[146] $\Delta_3 \leq 0$.

Eutocius gives an analysis and a synthesis for this problem. I only sketch the idea behind the analysis,[147] since it was the part that, according to the common practice, Cardano did not make explicit.

Eutocius' auxiliary problem to *On the sphere and cylinder*, II.4 — *Analysis.* [Assume that O is given.]

Eutocius considers the lines that pass by C and O in order to draw the rectangle of vertexes C, F, S, R and the line perpendicular to AB that passes through O. He takes $\overline{FN}$ such that $\overline{\Delta} = \overline{AB}\,\overline{FN}$.

[By hypothesis, by some trivial identities, and by the properties of similar triangles,] it follows that $\overline{XS}^2 = \overline{FN}\,\overline{FS}$, that is, that the parabola of axis FS, parameter $\overline{FN}$, that passes through F also passes through X. Similarly, it follows that $\overline{AB}\,\overline{FB} = \overline{RX}\,\overline{XL}$, that is, that the hyperbola with asymptotes RC, CF, that passes through B also passes through X. Since the parabola and the hyperbola are given, the intersection point X is also given, and then the projection point O is given.

Note that this analysis explains how the intersection point X of the parabola and hyperbola is linked to the point O, which identifies the segment sought for, but gives no hint on how the parabola and hyperbola were discovered.

The synthesis[148] is identically retraced by Cardano's proof of (A XII.iii), so I omit it. From now on I speak of 'Cardano's proof of the auxiliary problem' to highlight the peculiar rephrasing of the statement that links the auxiliary problem to a cubic equation *via* certain assumptions – but keep in mind that it is a particular case of Eutocius' synthesis. As Cossali remarked,[149] Cardano's idea is to

---

[146]Indeed, it is enough to show that $\overline{AC} \leq \overline{AE}$ and this is true because $\overline{AC} = \frac{9}{4}\frac{a_0}{a_2^2}$, $\overline{AE} = \frac{a_2}{3}$, and $a_0 \leq \frac{4}{27}a_2^3$ since $\Delta_3 \leq 0$.

[147]See [29, pages 67-68] or [41, pages 636-639].

[148]See [29, pages 68-69] or [41, pages 639-641].

[149]"*Di Cardano fu dunque propriamente solo l'avere a tale equazione condotta la geometrica ricerca della radice dell'equazione* $x^3 + q = nx^2$ *nel caso, che*

apply the solution of the auxiliary problem to the case in which its statement hides a cubic equation that falls into the *casus irreducibilis.*

We have pointed out that Proposition (A XII.iii) and its proof follow Eutocius' general version of the auxiliary problem and his synthesis. We know for sure that Cardano was overall interested in Archimedes' works. Indeed, in the 1544 edition of his autobiography,[150] Cardano explains that he was planning to achieve a revision of the *Elements*, which should have also collected the whole geometrical knowledge from Euclid on and some original propositions by Cardano himself. Cardano reveals that he left aside Archimedes' works, but that he had already agreed with Filippo Archinto (to which the *Ars Magna Arithmeticæ* was dedicated and who helped Cardano to get a position at the Scuole Piattine in 1534) to add them.

Then, it is natural to wonder what version of *On the sphere and cylinder* Cardano could have at hand.[151] When Charles of Anjou, called by Pope Clement IV, took possession of the kingdom of Frederick II, some Greek manuscripts went from Naples to the Papal Court. Among them, there was almost the whole of Archimedes' works (including *On the sphere and cylinder*) and Eutocius' commentary. In 1269 the literal Latin translation by Willem van Moerbeke, which was based on these manuscripts, was prepared. Anyway, this was not enough to boost the study of Archimedes' texts. Indeed, their mathematics lingered fairly unknown through the Middle Ages, since, on the one hand, it did not match the prevailing interest for calculation of that time. On the other hand, it was difficult mathematics that required a very good knowledge of the Greek *corpus*, such as the proportion theory, the conic sections, and Book XII of the *Elements*, and all of them were missing in the Middle Ages. In 1455 *On the*

nella formola di algebraica risoluzione presentasi da immaginario implicamento svisata", see [57, page 434].

[150]"*Euclides omnia collegit, quæ in Geometria ab antiquis scripta essent, nos quæ invenimus, suis autoribus adscriptis, nostris inventis adiunximus, uno Archimede relicto, quem cum Archinto illi adiungere decrevimus*", see [14, page 57]. See also [63, page 127].

[151]For the following, I mainly refer to [77].

sphere and cylinder was translated again into Latin by Iacopo da Cremona, but his translation was even less popular than Moerbeke's one.

Later on, around the first half of the 16th century, Archimedes' works came to be better known again. Johannes Müller von Königsberg, also known as Regiomontanus, reexamined Iacopo's translation for the first time from a completely different viewpoint. He was a real scientist and an expert mathematician whose aim was to give a new birth to the Greek science. But Regiomontanus' ambitious project was prevented by his death. His papers were partially used by Thomas Geschauff, also known as Venatorius, to publish in 1544 in Basel the Greek *editio princeps* of Archimedes' works (containing also Eutocius' commentary). This is the first printed (and consequently truly widespread) version.

Note that I have only mentioned Archimedes' editions and manuscripts that are previous to 1570 and in which *On the sphere and cylinder* appears. By the way, also Tartaglia made an edition of Archimedes' works in 1543, but it does not contain the book in which we are interested. We do not know which edition of Archimedes (and of Eutocius) Cardano could have read, but it is extremely unlikely that he had available one of the handwritten translations by Moerbeke or Iacopo.[152] Then, only one plausible possibility remains, that is to say, that Cardano had the *editio princeps* available with Eutocius', Dionysodorus', and Diocles' solutions to the auxiliary problem.

At this point we ought to spend a few words on the solution to the very same problem[153] given by 'Umar al-Khayyām. Once and for all, it must be clearly said that we are not privy to any relationships with Cardano's proof. Nevertheless, I deal with it, since Khayyām's strikingly similar use of the intersection of a parabola and a hyperbola

[152] Nevertheless, one of the manuscripts by Iacopo circulated in the same Milanese milieu frequented by Cardano some years before his birth. Indeed, Giorgio Valla attests that his mathematics teacher Giovanni Marliani, medicine professor in Pavia and member of the Collegio dei fisici in Milan, had a copy of one of Iacopo's translations. Marliani died in 1483.

[153] For a detailed history of the auxiliary problem, see [80].

to get the searched point helps to shed a new light on Cardano's usage.

Both Archimedes' *On the sphere and cylinder* and Eutocius' commentary were translated into Arabic, but, in most of the cases, separately. The translation of Archimedes' works was made around the 9th century and was mostly unabridged, which was not the case for Eutocius' commentary. Not later than in the 10th century, fragments of the commentary were translated and very often did not go together with Archimedes' book.[154] Then, it is not surprising that many Arabic mathematicians, probably unaware of Eutocius' efforts, tried to complete Archimedes' text.

We find Khayyām's solution in the *Treatise of algebra and almuqābala* (probably written in the first part of the 12th century). Apart from the fact that Khayyām explicitly mentions Proposition II.4 of *On the sphere and cylinder*, we know virtually nothing on his relationship with Eutocius' text. As it is well known,[155] Khayyām's aim is not merely to solve the auxiliary problem. Rather, the problem is embedded in a family of equations, and its solution comes as a drawback of the proof of the existence of (at least) a (real) solution of this family of equations. Khayyām starts with a complete classification for equations up to degree three – following some criteria (namely, to have positive coefficients and positive solutions, never to equate something to zero, and to group the equations according to the number of terms) that will be readopted by Cardano. Whenever Khayyām is able to, he solves the equations under an arithmetical interpretation to provide the value of the solution(s). This is obviously not the case for the equation $x^3 + a_0 = a_2x^2$ at issue.

This equation fills up the fifth position in the group of six equations in three terms. Modern editors used to number the equations one after the other, so that this equation became 'Equation 17'. Khayyām's treatment is highly articulated. On the one hand, assuming that the solutions must be positive entails a condition on the coefficients

154 See [80, pages 139-131].

155 See for example [90] and for an interpretation [87, pages 125-142].

(namely, that $\sqrt[3]{a_0} < a_2$). Khayyām argues by the absurd in order to count out the other cases. On the other hand, the geometrical environment in which he interprets his hypotheses leads him to further discuss three cases. Since we are only interested in the comparison with Cardano's text, I directly make the good assumptions in order to be in the same case as $x^3 + 192 = 12x^2$.

Khayyām's Equation 17. *Consider*[156] *the equation* $x^3 + a_0 = a_2x^2$. *Assume*[157] *that* $\sqrt[3]{a_0} < \frac{a_2}{2}$.
Then, it has a [real] solution.

Khayyām's Equation 17 — *Proof*. Considering the diagram in Figure 5.10, Khayyām takes $\overline{AC} = a_2$. We consider a_0 as a parallelepiped with a squared basis. By a previous lemma,[158] Khayyām knows that it exists a segment H such that $\overline{H}^3 = a_0$. He takes B on AC such that $\overline{BC} = \overline{H}$. By hypothesis it follows that $\overline{BC} < \overline{AB}$.

Then, Khayyām performs the following construction. He draws the square of BC, the vertexes of which are C, B, D, E. He draws the parabola of axis AC, parameter $\overline{BC}$, passing through A and the hyperbola with asymptotes AC, CE, passing through D. [Since $\overline{BC} < \overline{AB}$, then $\overline{BC}^2 < \overline{AB}\,\overline{BC}$ and] the point D is inside the parabola. Therefore, the parabola and the hyperbola intersect in two points. Khayyām calls T one of them and Z its projection on AC. He wants to prove that ZC is a solution of the equation.

[Since T, D belong to the hyperbola,] $(\overline{TC}) = \overline{BC}^2$, and then $\overline{ZC} : \overline{BC} = \overline{BC} : \overline{TZ}$. Since T belongs to the parabola, $\overline{TZ}^2 = \overline{BC}\,\overline{AZ}$, and then $\overline{BC} : \overline{TZ} = \overline{TZ} : \overline{AZ}$. Therefore, $\overline{ZC} : \overline{BC} = \overline{BC} : \overline{TZ} = \overline{TZ} : \overline{AZ}$, then $\overline{ZC}^2 : \overline{BC}^2 = \overline{BC} : \overline{AZ}$, and then $\overline{BC}^3 = \overline{ZC}^2\overline{AZ}$.

[156] See [43, pages 40-44] and [90, pages 46-52, 174-180].
[157] Since $\frac{1}{8} < \frac{4}{27}$, the assumed condition implies that $a_0 < \frac{4}{27}a_2^3$, that is $\Delta_3 < 0$.
[158] "*Étant donnés le carré ABCD, base du parlallélépipède rectangle ABCDE, et le carré MH, construire sur la base MH un parallélépipède rectangle égal au solide donné ABCDE*", see [43, pages 30-31], or also [90, pages 156-158].

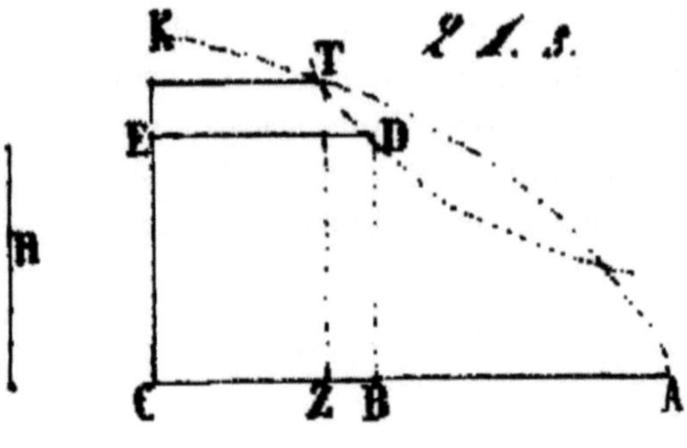

Figure 5.10. – Woepke's diagram in Khayyām's *Treatise of algebra* for the proof concerning Equation 17. Kind courtesy of University of Chicago Library.

Finally, $\overline{ZC}^3 + a_0 = \overline{ZC}^3 + \overline{ZC}^2\overline{AZ} = \overline{ZC}^2(\overline{ZC} + \overline{AZ}) = \overline{ZC}^2\overline{AC}$, and then $\overline{ZC}^3 + a_0 = a_2\overline{ZC}^2$. □

Note that the structure of Khayyām's proof does not substantially differ from Eutocius' and Cardano's ones. Indeed, once chosen the appropriate conic sections, the core of the proof is to use their properties to show that (one of) their intersection point(s) defines the searched segment.

Then, we shall search elsewhere for the difference. The choice of the conic sections[159] is a good option to inquire into. We are able to guess a sensible analysis for Khayyām's proof[160] and, contrary to Eutocius' one, it lets us understand how Khayyām chose the conic sections.

---

[159] Note that Khayyām's choice of the conic sections differs from Cardano's one. Let us suppose $a_0 = 192$ and $a_2 = 12$. If we take C to be the origin of a Cartesian coordinate system, AC on the x-axis (right to left oriented), and EC on the y-axis, the equation of the parabola and hyperbola respectively are $y^2 = 4\sqrt[3]{3}(x+12)$ (or $y^2 = \sqrt[3]{a_0}(x+a_2)$) and $xy = -16\sqrt[3]{9}$ (or $xy = -\sqrt[3]{a_0^2}$). Then, their intersection points are given by the solutions of $x^3 + 192 = 12x^2$. We recall that Cardano's conics are $x^2 = \frac{4}{9}a_2y$ and $(a_2 - x)y = \frac{9}{4}\frac{a_0}{a_2}$, see above, page 366.

[160] I was inspired by [87, pages 136-137] and I thank Marco Panza for his notes.

Indeed, it is enough to retrace backwards the proof of Khayyām's Equation 17 on page 373.

Khayyām's Equation 17 — *Analysis*. Consider $x^3 + a_0 = a_2x^2$. It follows that [$a_0 = a_2x^2 - x^3$ and] $a_0 = x^2(a_2 - x)$, [that is, the coefficient of the term of degree zero can be written under the shape of a parallelepiped with squared basis.[161]] Thanks to a lemma,[162] we know that it exists a (real) number h such that $h^3 = x^2(a_2 - x)$. Then, $x^2 : h^2 = h : a_2 - x$. It follows that it exists a (real) number y such that $x : h = h : y = y : a_2 - x$. [This gives the equations of the conics sections, if we take $h = \sqrt[3]{a_0}$. The parabola is $y^2 = h(a_2 - x)$ and the hyperbola is $xy = h^2$.[163]]

If we try to gather some clues on how Cardano chose his conic sections simply by retracing backwards the last steps of the proof of (A XII.iii) (that is, the analysis of (Eutocius)), we do not get the conic sections straight away. Rather, we do not get them at all, since indeed we need to already know how they are defined. Nevertheless, it is not a major problem to imitate Khayyām and wildly derive Cardano's (or Eutocius') supposed conics directly from the equation.[164] The point is that in Cardano's text you do not find anything similar. This suggests that Cardano draws the conic sections in a way that does not directly depend on the equation. I do not have any guess on that, but we remark that Cardano's conic sections are tightly related to $\frac{4}{9}a_2$,

[161] We remark that Khayyām employs the common language for geometrical objects and numbers. See [87, pages 127-128].

[162] See above Footnote 158

[163] Note that, up to the sign, these are the same conic sections of the Footnote 366, on page 366. This sign depends on the orientation of the x-axis that I chose in order to have the same sign for the solutions compared to Cardano's one.

[164] Consider $x^3 + a_0 = a_2x^2$. Then, it follows that $x^2(a_2 - x) = a_0$. We write $x^2(a_2 - x) = a_0\left(\frac{9}{4a_2^2}\right)\left(\frac{4}{9}a_2^2\right)$, or better $x^2(a_2 - x) = \left(\frac{9}{4}\frac{a_0}{a_2^2}\right)\left(\frac{4}{9}a_2^2\right)$.

Up to now, this is simply Cardano's retraced. But then he uses the definition of the parabola, so that we cannot hope to derive it. Let us leave Cardano's way. We write $x^2 = \left(\frac{4}{9}a_2\right)\left(\frac{9}{4a_2}\frac{a_0}{(a_2-x)}\right)$. Then, the parabola is $x^2 = \frac{4}{9}a_2y$ and the hyperbola is $y = \frac{9}{4a_2}\frac{a_0}{(a_2-x)}$.

which is the parameter FN of the parabola and which appears in the hyperbola together with a_0. In turn – as Cardano says at the end of the proof – $\frac{4}{9}a_2$ depends only on $AB = a_2$ and on E, which is taken at two-thirds of AB. The point E is needed when the problem is shaped under the terms of the auxiliary problem, since one needs to have the point in which the maximum value of $\overline{AE}\,\overline{EB}^2$ is attained in order to state the *diorismos*. Then, Cardano's conic sections are tightly related to the point E, which arises because of the geometrical (positional) properties of the diagram. But this does not happen in Khayyām's text, where the conic sections are chosen starting from the equation itself. As Netz says, "[s]ince Khayyam's study of cases is logically prior to his study of geometrical properties, he is not interested in the geometrical properties of the points that define cases".[165]

[165]See [80, page 168]. His point is that

> [t]he author faced raw geometrical reality and transformed it into a statement in words, and the words still have impressed on them this fresh stamp of reality. The same is no longer true for Khayyam. While we do not know exactly which works Khayyam was aware of, we know – from his own words [while dealing with $x^3 + a_0 = a_2x^2$, Khayyām quotes Abū al-Jūd and al-Māhānī] – that he was acquainted with several treatments, some successful, some not, of the very same problem. Thus, when Khayyam sets out to produce his new version, he faces not geometrical reality in the raw, but geometry already transformed into verbal forms. He did not transform reality into words, but words into words,

see [80, page 160]. Netz provides some linguistic arguments to support his view, of which I am not fully convinced. Anyway, I choose to pick up the ones that can also be applied to Cardano so that the reader can appreciate them.

First, there is the line called 'H'. Since it does not intersect with other geometrical objects, it cannot be distinguished by any of his points. Then, "[t]his heterogeneous way of naming lines makes it somewhat less natural to see the expressions 'H', 'AC' as mere symbols. As mere symbols, they are homogeneous; their heterogeneity is a function of the geometrical configuration".

5.5.3. Summing up

Summing up, while analysing Chapter XII, we have discussed all the geometrical arguments that are in the *Aliza*. As said, Cardano usually employs a kind of geometry with no positional features, which, as such, imitates arithmetic. This chapter of the *Aliza* turns out to be the exception that proves the rule.

It is indeed true that the comparison with Khayyām's choice of the conic sections highlights that, on the contrary, in Cardano's case, this choice is tightly related with a certain point that depends in turn on the positional properties of the diagram. But we have also seen that here Cardano is substantially reproducing Eutocius' commentary on Proposition II.4 of *On the sphere and cylinder.* No surprise then that he is using his language of proportions together with his conception of geometry. But Cardano does not add any significant feature to Eutocius' proof. Actually, Chapter XII is one of those scattered notes that has been collected and included in the *Aliza* because it is focused on a cubic equation. The chapter stands, indeed, aside from the others and, in particular, I do not see any true connection with the splittings or with some method linked to them.

Therefore, if one wants to search for the genuine geometrical inspiration in Cardano's reasoning – which is a completely reasonable aim – it is better to refer, for instance, to *Aliza*, Chapter LX rather than to Chapter XII.

Still, there is the "permutability of names", as AZ and ZA, which destroys the identity as symbols of two-lettered objects.

Finally, Netz ends: "[i]n short, we see that Khayyam opens up the possibility of considering his objects symbolically, as elements manipulated by the rules of calculation; yet essentially conceives them as components in a geometrical configuration". For more details, see [80, page 163].

5.6. Trying for an arithmetic of nasty numbers

In Sections 5.2 and 5.3, we have focused on Cardano's main strategy to overcome the problem of the *casus irreducibilis*. In short, the idea was to try to replace the cubic formula for $x^3 = a_1x + a_0$ with another one that had no conditions on the coefficients. This should have been achieved through the splittings. Then, we gathered as many chapters as possible around this topic. But, if one wants to ride out the problem entailed by the *casus irreducibilis*, there is at least one other possibility to consider. One could simply accept the cubic formula as it is and try to live with the square roots of negative numbers that appear there. These ways are not mutually exclusive, and indeed we also find evidences for the second one in the *Aliza*, namely in Chapter XXII.

If one chooses to preserve the cubic formula, they must show how to deal with these nasty numbers. First of all, they need to look for their arithmetic. Eventually, they might hope that in the formula these numbers are going to vanish somehow. Indeed, it is possible that one has experienced some cases – and Cardano for sure had – in which a nice (that is, natural) solution of a cubic equation was known independently from the formula, but some nasty numbers were returned when he tried to calculate this solution through the formula. Apart from speculations, Cardano must have foreseen this direction. Indeed, if we believe Tartaglia's account,[166] Cardano asked him how to find the rule to derive the equalities $\sqrt[3]{\sqrt{108}+10} = \sqrt{3}+1$ and $\sqrt[3]{\sqrt{108}-10} = \sqrt{3}-1$. Note that those cubic roots are exactly the ones that the formula returns[167] for $x^3 + 6x = 20$. It is not unlikely that he could have searched for a similar rule when there are negative numbers under the square roots.

If then you choose this second way, you would probably try to define how to perform operations with square roots of negative numbers. One of the very first natural questions that arises is which sign these numbers should have. As we will see, Cardano tries for the easy way: to make the square roots of negative numbers disappear by imitating what happens with the cubic roots, where $\sqrt[3]{-a} = -\sqrt[3]{a}$. For that, he suggests an alternative sign rule, which, as such, is wrong. But the point is that Cardano's attempt displays (again) his will to overpass the problem entailed by the *casus irreducibilis*.

In the *Aliza*, **Chapter VI** "On the operations plus and minus according to the common usage [*De operationibus p: et m: secundum communem usum*]" Cardano recalls the usual sign rules, in particular that "minus multiplied by minus [...] produces always plus".[168] More precisely, he firstly does that for (rational) numbers and then for the *binomia* and *recisa*.

He moreover says that, according to the common usage, the square root of a positive number is positive, but the square root of a negative number does not exist.[169] We observe that this partially contradicts Cardano's statement in the *Ars Magna*, Chapter I and in the *Ars Magna Arithmeticæ*, Question 3, both on the fact that $\sqrt{9} = \pm 3$.

---

[166] See [40, Libro IX, Quesito XL, pages 125-6].

[167] We remark that this equation is one of the examples in *Ars Magna*, Chapter XI. There, it is clear that Cardano does not know that $\sqrt{108}+10$ and $\sqrt{108}-10$ are two cubes, since he leaves the (real) solution written with the two cubic radicals.

[168] "*In multiplicatione et divisione p: fit semper ex similibus, m: ex contrariis, unde p: ductum in p: et divisum per p:, et m: ductum in m: et divisum per m: producunt semper p:. Et ita p: in m: vel m: in p: vel p: divisum per m: vel m: per p: producit m:*", see [9, Chapter VI, Paragraph 1, page 15].

[169] "*R p: est p: R m: quadrata nulla est iuxta usum communem, sed de hoc inferius agemus*", see [9, Chapter VI, Paragraph 3, page 15].

But in **Chapter XXII** "On the contemplation of plus and minus, and that minus times minus makes minus, and on the causes of these according to the truth [*De contemplatione p: et m: et quod m: in m: facit m: et da causis horum iuxta veritatem*]", Cardano provides a radically different account. I do not try to dictate an interpretation of this text, since it is one of those chapters where unluckily there is very little consistency. I content myself with highlighting the relevance of Cardano's doubts about the usual sign rule, namely concerning the part that states that a negative number times a negative number makes a positive number. Eventually, we interpret these doubts as another attempt to overpass the problem entailed by the *casus irreducibilis.*

Cardano softly starts by quite an orthodox beginning. When one sums two numbers – he says – it happens that they can write them as a unique term (for instance, $6+2=8$) or not (for instance, $6+\sqrt{2}$ or $6+x$). He incidentally adds – and we keep it in mind for later – that "a multiple quantity is nevertheless equivalent to a simple [one], and this mostly happens with universal and abstruse roots, as it is declared in the *Ars magna* (Chapter 11) that $\sqrt[3]{\sqrt{108}+10}-\sqrt[3]{\sqrt{108}-10}$ is the same as 2".[170] In any case, when one wants to calculate the square of the sum of simple or "composite" numbers, like 8 or $6+\sqrt{2}$ and $6+x$, Cardano suggests to apply *Elements* II, 4.

Afterwards, he addresses the square of the difference of two numbers. Here we find his argument to maintain that the usual sign rule is wrong and that it should be replaced by the following one.

[170] "*[A]liquando tamen quantitas multiplex æquivalet simplici, et hoc maxime accidit in RV: et abstrusis, velut declaratum est a nobis in Arte magna, quod RV: cu.R 10, 8 p: 10 m: RV: cu.R 10, 8 m: 18, idem est quod 2*", see [9, Chapter XXII, page 43].

A XII. *A negative number times a negative number makes a negative number.*

Cardano's argument is false, and this makes even more interesting the justification that he gives. Let us have a look at it in detail.

A XXII — *Proof.* Considering the diagram in Figure 5.11, Cardano[171] takes $\overline{AB} = 10$ and $\overline{CB} = -2$. Then, $[\overline{AB}^2 = 100,]$ $\overline{AC} = 8$, and $\overline{AC}^2 = 64$. He considers now the gnomon (GCE). On the one hand, by the same *Elements* II, 4, $\overline{(GCE)} = 2\overline{AC}\,\overline{CB} + \overline{CB}^2 = \overline{(AD)} + \overline{(DE)} + \overline{(BD)}^2$. On the other hand, $\overline{(GCE)} = [\overline{AB}^2 - \overline{AC}^2 = 100 - 64 =]36$. But, since $\overline{(AD)}, \overline{(DE)} < 0$ [as they both are the product of the positive number $\overline{AC}$ and of the negative number $\overline{CB}$, and] $\overline{(AD)} + \overline{(DE)} = -32$, then it must be that $\overline{(BD^2)} = \overline{CB}^2 = -4$, otherwise one would have had that $\overline{(GCE)} = [-32 + 4 =]28$ and

[171]The full quotation of this controversial passage follows. "Then it is clear that AC is truly 8 and its square DF will be 64, but the gnomon is the whole remainder, as I have similarly said, and if BC is of someone else, therefore the whole gnomon also [belongs] to him, as I will show. And it is agreed that that gnomon is made by AC times CB twice, and they are the rectangles AD, DE, with the square of BC by the same proposition (fourth [Proposition] of the second [Book] of the *Elements*). But that whole gnomon is 36, because the square of AB is 100 and FD [is] 64, therefore the remaining gnomon GCE is 36, and AD and DE are minus and are 32, and the gnomon is -36. Therefore the square of BC, which is 4, is also minus. Indeed, if it had been plus, the gnomon would not be minus, if not 28, and DF 72, and AC $\sqrt{72}$ and not $\sqrt{64}$, which is 8. Therefore the square of BC is minus and is made by minus multiplied by itself. Therefore minus multiplied by itself produces minus" or "*liquet ergo quod ac vere est 8, et eius quadratum df erit 64, sed totus residuus gnomo est, ut dixi perinde, ac si bc esset alterius, ideoque totus gnomo etiam illius, ut ostendam, et constat quod ille gnomo per eandem propositionem fiet ex ac in cb bis, et sunt rectangula ad de cum quadrato bc iste autem gnomo totus est 36, quia ab quadratum est 100 et fd 64, igitur gce gnomo residuus est 36, et ad et de sunt m: et sunt 32, et gnomo est 36 m: igitur quadratum bc, quod est 4, est etiam m:, nam si esset p: non esset gnomo m: nisi 28 et df 72 et ac R 72, et non R 64, quod est 8. Igitur quadratum bc est m: et fit ex m: in se ducto, igitur m: in se ductum, producit m:*", see [9, Chapter XXII, page 44].

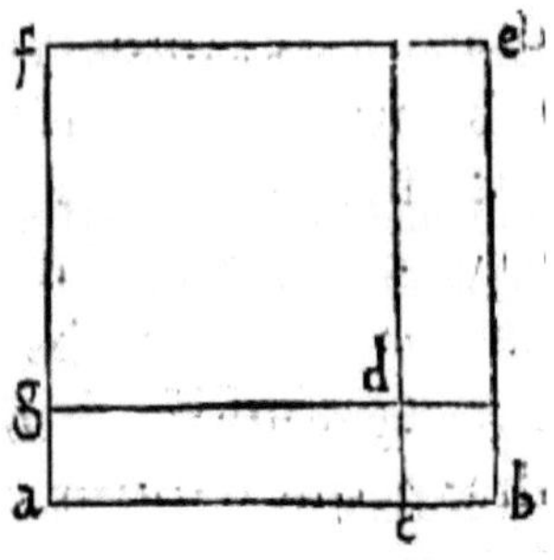

Figure 5.11. – *Aliza*, Chapter XXII. Kind courtesy of Biblioteca Virtual del Patrimonio Bibliográfico and Universidad de Sevilla.

$\overline{AC}^2 = [\overline{AB}^2 - \overline{(GCE)} = 100 - 28 =]72$. This means that "the square BC is minus, and is made by a minus multiplied by itself, then a minus multiplied by itself is a minus". □

As said, Cardano is mistaken. First of all, we remark that he assumes that the segment CB has a negative measure – which he never did before and will never do later, since he does not define what a segment with a negative measure is. The trick of his argument flows from this. Indeed, Cardano repeatedly shifts between considering $\overline{CB} = -2$ and $\overline{CB} = 2$. Namely, he takes at first $\overline{CB} = -2$, but then, deriving $\overline{AC} = 8$, he acts as if $\overline{CB} = 2$. This makes him infer a value for the measure of the gnomon (GCE). After that, he comes back to his first assumption, that is $\overline{CB} = -2$, to deduce that the measures of AD, DE (both equal to $AC\,CB$) are negative, and therefore that also the value of the measure of the gnomon must be negative. This argument is followed by a second one, which is completely likely, except that the diagram is a rectangle instead of a square.

Hereafter, Cardano explains why it is instead commonly believed (as he has said in Chapter VI) that the product of two negative numbers is positive. Referring to the above diagram [and[172] by

[172]In the very same paragraph Cardano states that one must employ *Elements* II, 4 for the square of a *binomium* and *Elements* II, 7 for the square of a *recisum*:

Elements II, 7], he gets that $\overline{AB}^2 + \overline{CB}^2 = 2\overline{AB\,CB} + \overline{AC}^2$, that is $\overline{AB}^2 + \overline{CB}^2 - 2\overline{AB\,CB} = \overline{AC}^2$. Since $2\overline{AB\,CB} = \overline{(GCE)} + \overline{CB}^2$, then

$$\overline{AB}^2 + \overline{CB}^2 - \overline{(GCE)} - \overline{CB}^2 = \overline{AC}^2.$$

Once cleared the ground, Cardano argues that

> only the gnomon is really a minus. Then we subtract as much as the square *BC* plus what we had [to subtract] from the square *AB* as if [it were] plus. Therefore, in order to rebuild that minus that we subtracted beyond measure, it is necessary to add as much as the square *BC* plus. Therefore, as *BC* is minus, we say that −2 is the inverted square in plus, then that minus times minus produces plus. But it is not true. We added as much as the square *BC* is plus. [It is] not that the square *BC* is plus, but that another arbitrarily assumed quantity equal to *BC* is added and made plus.[173]

Finally, Cardano needs to justify the remaining sign rules. At this point, he abruptly starts to call the minus 'alien [*alienum*]'.

"as well as with the *binomia* we perform operations by the fourth proposition and according to the substance of the composite quantity, so also with the *recisa* we perform operations by the same [to know] the substance, but by the seventh of the same to know the terms" or "*in binomiis operamur per quartam propositionem, et secundum substantiam quantitatis compositae , ita etiam in recisis quo ad substantiam et vere operamur cum eadem. Sed ad nominum cognitionem operamur in virtute septimæ eiusdem*", see [9, Chapter XXII, page 45]. This partially contradicts his foregoing practice (when he employed *Elements* II, 4 to calculate the square of a difference), but inconsistencies are the overall course in the *Aliza*.

173 "*[S]solus gnomo vere sit m:, quia ergo detrahimus quantum est quadratum bc, plusquam deberemus a quadrato ab, tamquam p:, ideo ad restitutionem illius m: quod detrahimus præter rationem oporteret addere, quantum est quadratum bc p:, et ideo cum bc sit m:, dicemus quod 2 m: quadratum conversum est in p: ideo quod m: in m: produxit p:. Sed non est verum; sed nos addidimus quantum est quadratum bc p: non quod quadratum bc sit p:, sed alia assumpta quantitas pro arbitrio nostro æqualis bc addita est, et facta est p:*", see [9, Chapter XXII, page 44].

Tautologically, he argues that "nothing is able beyond its powers, then plus is able of as much as itself"[174] and that the product of two positive numbers is positive. The same argument holds for the product of a positive number and a negative one, which makes a negative number.

Starting from these rules for the multiplication, Cardano says that one can infer the corresponding rules for the division and for taking the square root. He explicitly suggests only the rules for the division, and here come the drawbacks of his brand-new sign rules. Indeed, in this way it happens that the division of a positive number by a negative one is no longer well-defined. Moreover, a negative number divided by another negative number gives both a positive and a negative number (though Cardano does not mention it here, but rather in the *Sermo de plus et minus*, see below on page 386). Cardano does not go on detailing the consequences for taking the square root. If he had done so, he would have found that the square root of a negative quantity is negative, hence contradicting again his earlier claims in Chapter VI and both the ones in the *Ars Magna* and *Ars Magna Arithmeticæ*. Let us schematically resume Cardano's new rules and their consequences in Table 5.3. There I write for short '+' for a positive number and '−' for a negative one.

All along this chapter we followed Cardano attempting to alternative sign rules for numbers – and maybe also for nasty numbers, that is for square roots of negative numbers. He employs an overstretched argument (which also involves *Elements* II, 4) to justify that the product of two negative numbers is negative. Anyway, we should not focus on the potential (and wrong) result, but rather on the reason that pushes Cardano towards rethinking such a topic. I suggest to frame the whole problem in connection with the remark on simple and "composite" numbers at the very beginning of this chapter. This also has the advantage to provide the chapter with a little more internal cohesion. In particular, I consider the example $\sqrt[3]{\sqrt{108}+10}-$

174 "*[N]ihil potest ultra vires suas, ergo p: potest quantum est ipsum*", see [9, Chapter XXII, page 45].

Table 5.3. – New sign rules for the multiplication and division in *Aliza*, Chapter XXII, from which I derive the consequences concerning the square root.

Multiplication	Division	Square root
$+ \cdot + = +$	$\frac{+}{+} = +$	$\sqrt{+} = +$
$+ \cdot - = -$	$\frac{-}{+} = -$ and $\frac{-}{-} = +$	
$- \cdot + = -$	$\frac{-}{-} = +$ and $\frac{-}{+} = -$	
$- \cdot - = -$	$\frac{-}{-} = -$	$\sqrt{-} = -$

$\sqrt[3]{\sqrt{108} - 10} = 2$, which was quoted from the *Ars Magna*. My point is that, if it had been true that the product of two negative numbers is negative, then $\sqrt{-a} = -\sqrt{a}$ for any number a. Thus, Cardano would have been enabled to deal with all the expressions, such as the ones coming from a cubic formula that contains square roots of negative numbers.

At this point, a mention of the *Sermo de plus et minus* is due. This is a very short pamphlet, never published during Cardano's lifespan (and never mentioned in his autobiography), but included in the fourth volume of his *Opera Omnia*. It is a reaction[175] to Rafael Bombelli's *Algebra*, which was published in 1572: this means that the *Sermo* is a very late work. Due to its cryptic character, it is far from my purpose to enter all the details of this text. I limit myself to briefly take into account its beginning,

The *Sermo* starts where the *Aliza*, Chapter XXII stopped. There is even an explicit cross-reference to it.[176] Here, the survey on the consequences of the sign rules for division is fully achieved, since

[175] See [94].

[176] Tanner remarks the misprint '*Li. Aliza cap. 2*' instead of '22', see [94, page 168].

Cardano lists the cases $\frac{+}{+} = +$, $\frac{-}{+} = -$, $\frac{-}{-} = +$, $\frac{-}{-} = -$, and "nothing [*nihil*]" for $\frac{+}{-}$. The second paragraph starts in a very rambling way, but it is here that Proposition II, 7 of the *Elements*, Rafael Bombelli, and the equation $x^3 = a_1x + a_0$ are mentioned all together. In the middle, we read (in Tanner's translation) that

> [o]n the other hand, Raphael Bombellus of Bologna confined this to $\sqrt[3]{\ }$ of a binomial and its conjugate, because this minus appeared useful exclusively for the complete treatment of cube equal to unknown and number. But there the $\sqrt[3]{\ }$ *l.* is two-fold, namely that of the binomial and that of the conjugate of the binomial. So he was right to reserve this minus solely for these two cases, namely the $\sqrt[3]{\ }$ *l.* of a binomial and of its conjugate.[177]

Cardano carries on by remarking that what Bombelli called '*p: m:*' and '*m: m:*' are to be intended as '*p: di m:*' and '*m: di m:*' (which mean $+\sqrt{-}$ and $-\sqrt{-}$). This yields to Tanner's "first impression that Bombelli's notation was the result of an intentional misprint, whether condoned by him or not, and misguidedly perpetuated in its modern reprinting".[178] Then, Cardano recalls the sign rules given by Bombelli for these numbers ("*p: di m:*" times "*m: di m:*" and "*m: di m:*" times "*p: di m:*" make "*p:*", "*p: di m:*" times "*p: di m:*" makes "*m:*", and "*m: di m:*" times "*m: di m:*" makes "*m:*") and says that Bombelli has found (a) cubic root of $2 \pm \sqrt{-121}$. This enables Bombelli to draw a real solution of the equation $x^3 = 15x + 4$, which is

$$x = \sqrt[3]{2 + \sqrt{-121}} + \sqrt[3]{2 - \sqrt{-121}} = 2 + \sqrt{-1} + 2 - \sqrt{-1} = 4,$$

[177] See [94, pages 169-170]. "*Raphael autem Bombellus Bononiensis contraxit hanc ad R cub. Binomij et recisi, quia non videbatur m: hoc utile nisi pro perfectione cubi æqualis rebus et numero. Sed ibi est R cub. l. duplex binomii scilicet et sui recisi. Ideo recte contraxit hoc m: ad illas duas conditiones R cub. scilicet l. et binomii cum suo reciso*", see [20, page 435].

[178] See [94, page 171].

by the cubic formula. I have not been able to retrieve other useful information from the very obscure remaining three pages. Therefore, I am dropping here the *Sermo*.

I would only suggest a last, short remark on the *Sermo*. It is true that there is no explicit relation between Cardano's sign rule on the product of negative numbers and Bombelli's one on "*m: di m:*" times "*m: di m:*". But nevertheless it remains the fact that Cardano himself found a link between the two, since he started the *Sermo*, which is devoted to Bombelli's "*p: di m:*" and "*p: di m:*", by recalling the *Aliza*, Chapter XXII. This strengthens my hypothesis that the grounding reason of Cardano's rethinking of the sign rules lies in his willing to handle the square roots of negative numbers.

Summing up, we have good reasons to believe that Cardano has seriously been questioning about the square roots of negative numbers, and in particular about their sign. Cardano tries out various possibilities: they exist but behave differently, they do not exist, they are negative. We recall that in both the *Ars Magna Arithmeticæ* and the *Ars Magna* Cardano keeps the safest road, since he only suggests that the square roots of negative numbers do not possibly behave like the others. Indeed, he says – they cannot be neither a positive or a negative number, since we have, for instance, that $\sqrt{9} = \pm 3$. He assumes that $(\sqrt{-a})^2 = -a$ and that, once the square root has vanished, they behave as usual. Thus, the common sign rules hold – everything is fine. This low-profile handling of the square roots of negative numbers is in the same style as *Aliza*, Chapter VI, even if it is not completely consistent with it (in the *Aliza* Cardano seems rather to believe that the square roots of negative numbers do not exist).

Then, we are bewildered by Chapter XXII's rift, which is carried on in the *Sermo de plus et minus*. We do not have to be too surprised by them. Inconsistency being the general feature of the *Aliza*, we only have to pay attention not to throw the baby out with the bathwater. And the baby here is the aim that Cardano wanted to achieve, no matter if he failed. Rethinking the common sign rules, indeed, has an intrinsic value by itself, since it is a symptom of another attempt to overpass the problems caused by the presence of those nasty numbers in the cubic formula.

Conclusions

The book on which the present work has been focused, the *De Regula Aliza*, does not have the features of a masterpiece – but it is definitely the treatise of one of the most brilliant mathematicians in the 16th century. Reading it is a constant fight against the messiness of its structure. Apart from some extrinsic causes, like misprints and calculation mistakes, there is a profound reason behind this that origins from its composition process. The present work is not thought to be an exhaustive analysis of the *Aliza* – on the chance that such a kind of work may ever be written – but rather one stage in decoding it. We recall that an early step in this direction, however controversial, has already been made by the mathematician Pietro Cossali at the end of the 18th century.

Even though the *Aliza* was published twenty-five years after the latest of Cardano's treatises on equations, it easily finds its own place among his other mathematical treatises. We have followed the development of Cardano's treatment of equations in the *Practica Arithmeticæ*, *Ars Magna Arithmeticæ*, and *Ars Magna*. Equations pass from being a side-effect of proportions in the *Practica Arithmeticæ*, which is still an *abaco*-oriented treatise, to become more and more a proper subject of inquiry. This transition become obvious in the *Ars Magna Arithmeticæ*, where the first part on irrational numbers, expounded with the Euclidean terminology of Book X, is juxtaposed to the second part where the study of some cubic equations is the central subject matter. Then, Cardano's masterpiece, the *Ars Magna*, comes. All the cubic equations are solved: algorithmic rules or sometimes even out-and-out formulae are given together with their geometrical proofs. Unluckily, Cardano cannot provide a completely general method, since it happens that under certain conditions his formulae are nonsensical.

Indeed, when a cubic equation has three real distinct solutions, the formula (necessarily) involves imaginary numbers – this is the *casus irreducibilis*, as it has been lately called.

In this flowing, the *Aliza*, or at least a part of it, however inconsistent and obscure, is meant to overcome the problem. A very few pages have been written on this writing and – as said – only Pietro Cossali's work is a detailed account. Nevertheless, we should again have a look at the *Aliza*, mainly aiming to integrate its mathematics with the general framework of Cardano's solving methods for cubic equations.

There are two main strategies to overcome the *casus irreducibilis* in this treatise. In short, one (the most represented one) is to reject it, the other is to try to live with it. Rejecting the *casus irreducibilis* basically means to reject the cubic formulae of the *Ars Magna*. In particular, since the problem entailed by the *casus irreducibilis* is originated in the cubic formula for $x^3 = a_1x + a_0$, one would reject the cubic formula for that family of equations, and try to search for another one that has no conditions. This position can be softened by rejecting the *casus irreducibilis* only partially, which means to accept the possibility of the existence of more than one cubic formula. This is the main idea that leads to the splittings. Since Cardano has managed to derive the cubic formula from the first splitting for $x^3 = a_1x + a_0$, why not try to derive another formula from some other splitting either to replace or to complement this first one? Unluckily, this is as much unattainable as avoiding the square roots of negative numbers in the formula.

From the viewpoint of nowadays history of mathematics, nevertheless, the splittings have a fundamental role, since they are the main thread that we have managed to trace in order to give coherence to the *Aliza*. Indeed, there is a sizeable number of chapters of the *Aliza* that can be grouped around them. This has enabled us to advance three hypotheses on their origins. The first one put in relation the choice of the substitution $x = y + z$ with the study of the algebraic shapes for the irrational solutions of $x^3 = a_1x + a_0$, with a_1, a_0 rational. Thus, the splittings are also connected to the very early surveys on irrational numbers, like the ones in the *Ars Magna Arithmeticæ*, which possibly comes before the discovery of the cubic formulae. As such, the *Aliza* should be also considered as prelude to the *Ars Magna*, and not only the other way round. We recall that these surveys, scattered all along the *Aliza*, are very heterogeneous, and could attest the development of Cardano's enquiries. We remark that appreciating whether these algebraic shapes can or cannot express a solution of an equation is one of Cardano's most recurring strategies in dealing with equations. The fact that, in the end, it turns out that there are only two possible irrational shapes, both formed by the sum of two terms, could then have been at the origin of Cardano's substitution $x = y + z$.

The second hypothesis suggests that the splittings come from a further development of some rules in *Ars Magna*, Chapter XXV, which are intended to rewrite the equation at issue in a more comfortable way so that it is easier to guess a solution. We recall that the link between the *Aliza* and *Ars Magna*, Chapter XXV had already been suggested in *Ars Magna*, Chapter XII while dealing for the first time with the *casus irreducibilis*. Nevertheless, even though the splittings in the *Aliza* and the rules in *Ars Magna*, Chapter XXV respond to the same need, which is to overcome the problem entailed by the *casus irreducibilis*, they are intended to achieve different purposes. Indeed, the splittings are meant to provide a new formula, while the rules in *Ars Magna*, Chapter XXV only to give a case-by-case method.

The third hypothesis concerns a loose geometrical inspiration. It involves a very specific kind of geometry, where the focus is not on the relative position of the geometrical objects, but rather on their dimension. Hence, Cardano's treatment of the splittings in the *Aliza* shows how he possibly used to deal with equations. The solving methods in the *Ars Magna* are a sort of black box that make a clean sweep of all the previous inquiries, which are probably seen from that point on as worthless. It is therefore very reasonable to believe that, when Cardano wanted to overcome the problems of these formulae, he came back to his previous researches.

Moreover, there is the strategy that is a softened branching of the main one: one rejects the *casus irreducibilis*, at the same time trying to relax the constraints on the cubic formulae. For instance, one could accept to have two (or more) formulae or solving methods to deal with the same family of equations, each one with a condition, provided that these conditions are complementary. Then, when the condition of a formula is not matched, one can resort to the other and *vice versa*. In the *Aliza*, the splittings also become a subject of inquiry by themselves, when Cardano establishes some calculation rules on them. In turn, these could be allegedly used to complement the conditions on the usual cubic formula for $x^3 = a_1x + a_0$.

The second strategy is, on the contrary, to live with the *casus irreducibilis*. In fact, this will be the way followed in studying equations, and will eventually end up with the birth of the field of complex numbers. But it is by far too early for Cardano's time. In the *Aliza*, he tries to give an alternative sign rule, which could possibly be of help in transforming the square roots of negative numbers into square roots of positive numbers, but this account is not satisfactory at all.

During our survey we have also gathered some hints on the kind of geometry that Cardano employs. It would be very astonishing if Cardano had not been driven by a geometrical way of thinking. But it is a very special kind of geometrical intuition. In the vast majority of the cases (and, when it is not so, some ancient sources, like Eutocius, are rephrased) Cardano does not need any positional features, which means that the relative position of the involved geometrical objects is unessential. As such, this kind of geometry is the most suitable to imitate arithmetic. More precisely, the geometrical feature to which Cardano aims is the dimensional coherence, since he makes operations only with geometrical objects of the same dimension never overshooting the third dimension. Nevertheless, when equations are involved, this is not as plain as it could seem, and Cardano sometimes has to find tricks to respect the dimensional constraint.

In our surveys, we have also taken into account in detail the results of the *Practica Arithmeticæ*, *Ars Magna Arithmeticæ*, and *Ars Magna* concerning equations. This was a fundamental step that could not be skipped: the threads that we have managed to highlight in the *Aliza* are so well hidden that one has to try to make his way as plain as possible by appealing to all the other sources.

At this point, one last question still needs to be discussed: why did Cardano choose to publish the *Aliza*? Of course answers can only be tentative. From a contemporary, mere mathematical viewpoint, this treatise is inconclusive regarding its declared aim, that is, to overcome the problem entailed by the *casus irreducibilis*. For sure, Cardano was a man of his time, when sharing knowledge and letting others progress was not a priority. But we should also consider that Cardano's main focus are procedures and algorithms, rather than formulae. Under this perspective, the *Aliza* is not a failure since it contains many interesting results worth publication.

A. A short history of Cardano's life

Cardano's life is known fairly in detail, thanks to several editions of his autobiography, which he wrote when he was an old man.[1] In the following I will try to resume the most outstanding episodes of his life, with particular focus on the mathematical side.

Girolamo Cardano was born in Pavia, Italy, in 1501, probably of illegitimate birth. His father Fazio was a lawyer in Milan, but he also had wide interests in mathematics, medicine, and occult sciences. Cardano himself inherited this typical attitude of his time. Aside from the lawyer practice, his father lectured on geometry at the University of Pavia and at the Scuole Piattine in Milan. His proficiency was so deep that Leonardo da Vinci himself asked him an advice on some geometrical problems.

Cardano got his first education at home. His father taught him writing and reading in Vernacular and Latin, some basic arithmetic, a smattering of occult sciences, astrology. When Cardano was twelve years old, he began to study the first six books of Euclid's *Elements*. He became his father's flunky more than his assistant, so he soon aspired to get an education at the university. Cardano studied medicine in Pavia and in Padua, where both his capacities of brilliant student and his strong character were noticed. In Pavia, he distinguished himself in such a way that he was called to lecture on Euclid's geometry when he was only twenty-one years old. Nevertheless, he had to leave the city a few years later, because of the war between France and Spain to conquer Milan. Then, he moved to Padua, where he graduated

[1] See [3], [23], [6],[7], [17], [18]. See also [21] for an English translation, and [84] for an accurate resume.

in 1524. But his repeated applications for the Collegio dei fisici in Milan were all refused because of his illegitimate birth. His stubborn and unkind character did not earn him many friends, so that he had trouble in finding a work. Cardano never had great resources and gambling became for him not only a pastime, but a source of income. Eventually, he moved to a village near Padua where he managed to practise medicine and met his wife – in his words, the happiest period of his life. They had three children.

Cardano obtained in 1534 a teaching position at the Scuole Piattine[2] in Milan, thanks to a personal relationship with Filippo Archinto, a diplomat in Milan who knew his father's reputation. He should have lectured on arithmetic, geometry, and astrology (or astronomy) on non-working days, but, in order to attract a larger audience, he taught architecture instead of arithmetic and geography instead of geometry. Thanks to the reputation earned in this way, the German printer Petreius offered him to publish whatever manuscript he had to submit. It is in this period, and more precisely after failing to gain with his astrological works the patronage of Pope Paul III, that Cardano devoted himself more and more to mathematics.

In the meanwhile, he illegally practised medicine to supplement his exiguous salary, since he had not the Collegio's authorisation. He gradually acquired some reputation as physician. When he published his first book on bad practice in medicine, Cardano became abusive towards the Collegio. In this way, he finally managed to draw the attention of the Collegio, which had to accept him as a member. Within a few years, he became the most prominent physician in Milan. He began travelling. This was the highest point in his career.

Afterwards, some sad happenings troubled Cardano's private life. In 1546 his wife passed away. In 1560 one of his sons was sentenced to death for having poisoned his unfaithful wife, despite of all the

[2]The Scuole Piattine was founded in 1503 and financed by the nobleman Tommaso Piatti. There, a student could be instructed in Greek, dialectics, arithmetic, geometry, and astronomy. In 1503 Fazio Cardano, Girolamo's father, was teaching geometry, see [64, page 66].

support that Cardano gave him in every possible way. After the loss of his son, he retired from his public engagements.

In 1562 Cardano managed to get a position at the University of Bologna through the intervention of the future cardinal Francesco Alciati and of Federico Borromeo. He moved there with his youngest son, who at that time had already begun a criminal career. In Bologna, Cardano was recognised as the noteworthy scholar he was and he spent another good period of his life. He created his own network of friendships, which helped him in obtaining the freedom of the city and tax exemption. But he was more and more on strained terms with his youngest son. In 1569 the latter stole a large amount of his father's properties and Cardano had to report the theft to the authorities, who banished his son from the city.

The last years of Cardano's life were marked by misfortune. In October 1570 he was arrested and put in jail without warning. The causes have never been completely clarified. Maybe his son revenged himself denouncing him, but the charges are not known. More likely, Cardano's arrest was connected with the renewal impulse given by the Counter-Reformation. Among the varied works he wrote, it was not difficult to find statements that could be interpreted as impious. The most frequently quoted example is a horoscope of Jesus Christ, but even Emperor Nero's eulogy or the comparison between Christianity and other religions suited well. At that time Cardano was an old man. He was kept in jail for a few months, apparently without torture. Public opinion was favourable to him and he never directly opposed the church, so that the inquisitors were merciful. Cardano does not provide a more detailed account, probably because they imposed the condition to him to keep a secret. He had to abjure, but not publicly, was excluded from the university, was prevented from lecturing in public, and forbidden to publish any other work – he was deprived of all the worthwhile things in his life. Furthermore, he was obliged to pay a certain amount of money to the ecclesiastical authorities. They finally accorded him a modest annual income.

In 1571 Cardano moved to Rome, where Rodolfo Silvestri, one of his good friends who had also been a former pupil of him, lived. When the new pope Gregory XIII succeeded Pius V, Cardano's modest annual income was commuted into an even more modest pension, which nevertheless carried greater prestige as a sign of papal favour. Cardano still continued writing books until the very end of his life, hoping to get the restrictions on publishing and teaching lifted. At last, in 1572 he received the licence to publish his existing medical works and in 1574 the right to publish again new works. But at that time he already had destroyed 130 books of his treatises. In 1574 he was accepted at the Collegio dei fisici. In 1576 Cardano finally got the right to return to teaching in Bologna.

Cardano died on September 20, 1576.

A.1. Chronicle of a controversy

During Cardano's lifespan, a typical way of testing and developing knowledge was controversies between scholars. Cardano himself was often involved in and likes to take part in this kind of debates.

Usually, scholars challenged each others on professional opinions, but it was not infrequent that a controversy degenerated into personal attacks. In mathematical contests, they used to send letters to the adversaries and to a notary, specifying a time interval within which certain problems had to be solved. An implicit rule was not to submit to the adversary problems that one did not know how to solve by himself. Traditionally, a controversy ended with a public debate. Depending on the standing of the adversaries, the debate could take place before the authorities of the university and some judges. The winner could be awarded prize money, but in any case he gained prestige.

One can fully grasp how important it was to win controversies only if you appreciate that custom in the context of the Italian educational system of the time. As said, since the 13th century, the *abaco* schools spread from Italy throughout Europe. In the 15th century, after an elementary level of instruction, a child could enter a so-called *scuola di grammatica*, and then a university, or an *abaco* school, which prepared for a profession or an apprenticeship. In particular, in the *abaco* schools, students learnt how to solve problems that came from mercantile life and practical mathematics were taught. It often happened that the handbooks coming from teaching, the so-called *trattati d'abaco*, were published. In the main, they were based on simplified revisions of the *Liber Abaci* by Leonardo Pisano. But the universities and the *abaco* schools were not always parallel paths. In a controversy, the high reputation gained by the winner could even get him students who would have to pay to be taught. Moreover, the system of controversies explains the tendency of the time to keep secret new results and discoveries. In the history of the solution of cubic equations, controversies played a relevant role.

Cardano became renowned for having published for the first time in 1545 the cubic formulae in the *Ars Magna*. But the formulae's paternity is highly disputed. Allegedly, Scipione del Ferro first discovered some cubic formulae to solve the depressed equations, or at least one of them. We know very little of del Ferro except that he was professor of arithmetic and geometry for a long time in Bologna. He kept the lid on his results, but it is said that just before his death in 1526 he revealed his discoveries to his pupil Antonio Maria Fiore. Furthermore, at Scipione's death his papers came into possession of his son-in-law Annibale della Nave.

In the meanwhile Niccolò Fontana, also known as Tartaglia, *maestro d'abaco* in Verona, applied himself to the problem of solving cubic equations. Because of his poor economic conditions, he was very used to entering controversies. In 1530 he was challenged by Zuanne de Tonini da Coi, who taught mathematics in Brescia. Zuanne proposed to him two problems, which were equivalent to solve the cubic equations $x^3 + 3x^2 = 5$ and $x^3 + 6x^2 + 8x = 1000$. Tartaglia, recalling the authority of Luca Pacioli who considered it impossible to solve cubic equations, doubted that the weak mathematician that Zuanne was could have found by himself the solutions. So, he threatened Zuanne to report him to the authorities for boasting. In a subsequent letter to Zuanne, Tartaglia claimed to know how to solve the first equation, but not the second. Then, Antonio Maria Fiore challenged Tartaglia five years later in Venice with thirty problems. Exploiting his teacher del Ferro's results, Fiore prepared thirty questions all equivalent to solve a depressed cubic equation of the family $x^3 + a_1x = a_0$ with rational coefficients. Tartaglia, however, also submitted a variety of problems. During one sleepless night, shortly before the end of the time interval granted to solve the problems, Tartaglia found out the cubic formula he needed. The day after – it was February 13th according to Tartaglia's account – he came to the solution of the depressed equations of the family $x^3 = a_1x + a_0$. Fiore, who was strong in calculation but weak in theory, did not manage to solve his problems and was declared the loser.

Zuanne was informed about the controversy. He deduced that Tartaglia found out how to deal with (at least one of) the problems of their past controversy and asked Tartaglia to explain to him his method. At that time, Tartaglia was not inclined to spread his secret. This caused the relationship between Zuanne and Tartaglia to be broken off. Anyway, Zuanne's role is not negligible, since he also acted as a connection between Tartaglia and Cardano. In fact, Zuanne informed Cardano of the contest and reported to him Tartaglia's denial to release the formulae. Cardano was greatly impressed, all the more so because he was preparing the oncoming *Practica Arithmeticæ*. The years passed by, and even though Cardano attempted by his own to discover the formulae, he was unsuccessful.

Then, Tartaglia was approached in Venice by the bookseller Zuanantonio da Bassano in 1539. On Cardano's behalf, Zuanantonio asked Tartaglia for the cubic formulae. Neither flattery nor the promise to publish Tartaglia's results under his own name (or not to publish them at all) convinced Tartaglia to release the formulae. He continued to sidestep Cardano's requests until he realised that a connection with Cardano could turn out to be advantageous. Cardano had some good relationships in the court of Milan, especially with the Spanish governor of Lombardia, Alfonso d'Avalos, who was renowned to be an unusually generous patron. Tartaglia hoped to be introduced by him to the court of Milan and maybe to be engaged as a technical expert. To cut a long story short, after Cardano's repeated solicitations, Tartaglia imparted in the end to Cardano the cubic formulae for the depressed equations in verse. I will explain the poem in detail in the next section. Cardano, who was just about to pass for press the *Practica Arithmeticæ*, had to promise to Tartaglia not to publish his results. Eventually, Tartaglia never entered the court in Milan and returned to Venice, from where he continued the correspondence with Cardano, clarifying his further questioning. But he was anxious to verify that Cardano kept his promise and asked him a copy of his *Practica Arithmeticæ*. So, Cardano sent him a copy, but an unbound one.

Just after having received the poem, Cardano obviously began to work on the cubic equations lacking in the term of second degree and on the complete ones. Quite soon he seems to have come about the so-called *casus irreducibilis*, where a cubic equation has three real distinct solutions but the cubic formula involves imaginary numbers. So he knew that in some cases the formulae fail (or seem to fail) to return the solutions. On August 4th, 1539 Cardano wrote to Tartaglia, asking about the *casus irreducibilis*, but he got no answer.

Then, other results concerning equations popped up. Ludovico Ferrari, Cardano's talented pupil and his best friend, solved quartic equations in 1540. The two had met four years before, when Ferrari was a young boy. He soon became Cardano's secretary and was provided with an excellent education. But despite his popularity as mathematician, Ferrari chose a more remunerative position as tax assessor in Bologna. He always remained faithful to Cardano and was his occasional assistant. He was one of the supporters for Cardano's moving to the university of Bologna. So, Cardano and his pupil were step by step accumulating a large body of knowledge on third and fourth degree equations and Cardano's promise to Tartaglia remained an obstacle to make it public. It is still not clear why Tartaglia did not hurry to publish the results that he discovered.

But then something changed. In 1542 Cardano and Ferrari found Scipione del Ferro's papers in the house of Annibale della Nave. Thus, Cardano felt freed from his promise, since it was clear that Tartaglia could no longer be credited as the first discoverer of the cubic formulae.

He hurried to polish his results and only three years later the *Ars Magna*, which contained all the cubic formulae and Ferrari's method for quartic equations, went into press. There, Cardano mentioned both Scipione del Ferro and Tartaglia twice,[3] and Tartaglia alone

[3]In *Ars Magna*, Chapter I Cardano says that

> [i]n emulation of him [Scipione del Ferro], my friend Niccolò Tartaglia of Brescia, wanting not to be outdone, solved the same case [$x^3 + a_1x = a_0$] when he got into a contest with his [Scipione's] pupil, Antonio Maria Fior, and, moved by my entreaties, gave it to me. [...] Then, however, having received Tartaglia's solution and seeking for the proof of it, I came to understand that there were a great many other things that could also be had. [...] Hereinafter those things which have been discovered by others have their names attached to them; those to which no name is attached are mine. The demonstrations, except for the three by Mahomet and the two by Ludovico, are all mine

or

> *Huius æmulatio Nicolaus Tartalea Brixellensis, amicus noster, cum in certamen cum illius discipulo Antonio Maria Florido venisset, capitulum idem, ne vinceretur, invenit, qui mihi ipsum multis precibus exoratus tradidit. [...] Inde autem, illo habito, demonstrationẽ venatus, intellexi complura alia posse haberi. [...] Porro quæ ab his inventa sunt, illorum nominibus decorabuntur, cætera, quæ nomine carent, nostra sunt. At etiam demonstrationes, præter tres Mahometis et duas Lodovici, omnes nostræ sunt,*

see [22, pages 8-9] or [4, Chapter I, paragraph 1, page 3r]. The second reference is in Chapter XI

> Scipio Ferro of Bologna well-nigh thirty years ago discovered this rule [to solve $x^3 + a_1x = a_0$] and handed it on to Antonio Maria Fior of Venice, whose contest with Niccolò Tartaglia of Brescia gave Niccolò occasion to discover it. He [Tartaglia] gave it to me in response to my entreaties, though withholding the demonstration. Armed with this assistance, I sought out its demonstration in [various] forms. This was very difficult

or

> *Scipio Ferreus Bononiensis iam annis ab hic triginta ferme capitulum hoc invenit, tradidit vero Anthonio Mariæ Florido Veneto, qui cum in certamen cum Nicolao Tartalea Brixellense aliquando venisset, occasionem dedit, ut Nicolaus invenerit et ipse, qui cum nobis*

once.[4] The *Ars Magna* was immediately recognised as a masterpiece by the mathematicians of that time, and had a strong influence on the development of 16th century mathematics.

And then, the controversy began. One year later, the *Quesiti et inventioni diverse* by Tartaglia appeared. The last part is devoted to the exchange with Cardano, reconstructing the making of his promise and its breaking.

Cardano never directly replied, but then Ferrari came to his master's defence. He firstly attacked Tartaglia with one *cartello* and the exchange lasted six *cartelli* over one year and a half, until Tartaglia and Ferrari met in Milan. It was a gala occasion where the controversy took place. Ferrante Gonzaga himself, the governor of Milan, was named arbiter. Many distinguished personalities attended. The topics on which Tartaglia and Ferrari challenged were the problems contained in the *cartelli.* We have no records of the controversy and only a brief, one-sided account in Tartaglia's literary remains. In the late afternoon of August 18, 1548 in the Church of the Garden of the Frati Zoccolanti, Ferrari turned out to be the winner and Tartaglia was ingloriously defeated. As Oystein Ore has pertinently observed:

> Cardano and Ferrari represent by far the greater mathematical penetration and wealth of novel ideas. Tartaglia

> *rogantibus tradidisset, suppressa demonstratione, freti hoc auxilio, demonstrationem quæsivimus, eamque in modos, quod difficillimum fuit,*

see [22, page 96] or [4, Chapter XI, page 29v].

[4] It is in Chapter VI

> When, moreover, I understand that the rule that Niccolò Tartaglia handed to me had been discovered by him through a geometrical demonstration, I thought that this would be the royal road to pursue in all cases

or

> *Cum autem intellexissem capitulum quod Nicolaus Tartalea mihi tradiderat, ab eo fuisse demonstratione inventum geometrica, cogitavi eam viam esse regiam ad omnia capitula venanda,*

See [22, page 52] or [4, Chapter VI, paragraph 5, page 16v].

> was also, doubtless, an excellent mathematician, but his great tragedy was the head-on collision with the only two opponents in the world who could be ranked above him. [...] But on one point a deep injustice has persisted. Some of the early Italian algebraists referred to the formula for the solution of the cube and the cosa equal to a number proposition as "del Ferro's formula", but the influence of the *Ars Magna* was so great that it has forever since been known as "Cardano'a formula". Cardano never made any claim to this invention. In present-day mathematical texts it should in no way be too late to begin to pay homage again to the original discoverer, del Ferro, and thus give credit where credit is due.[5]

A.2. Tartaglia's poem

On March 25, 1539 Cardano receives from Tartaglia a poem containing the cubic formulae for the depressed equations. Here it follows.

Quando che'l cubo con le cose appresso
Se agguaglia à qualche numero discreto
Trovan dui altri differenti in esso.
Da poi terrai questo per consueto
Che 'l lor produtto sempre sia eguale
Al terzo cubo delle cose neto
El residuo poi suo generale
Delli lor lati cubi ben sottratti
Varra la tua cosa principale.

In el secondo de codesti atti
Quandi che 'l cubo restasse lui solo
Tu osservarai quest'altri contratti,
Del numero farai due tal part'àl volo
Che l'una in l'altra si produca schietto

[5] [84, page 106].

El terzo cubo delle cose in stolo
Delle qual poi, per comun precetto
Torrai li lati cubi insieme gionti
E tal somma sara il tuo concetto.

El terzo poi de questi nostri conti
Se solve col secondo se ben guardi
Che per natura son quasi congionti.
Questo trovai, et non con passi tardi
Nel mille cinquecentè, quatro e trenta
Con fondamenti ben sald'è gagliardi
Nella città dal mar'intorno centa.[6]

Firstly, considering the equation $x^3 + a_1 x = a_0$, Tartaglia suggests to solve the linear system

$$\begin{cases} y - z = a_0 \\ yz = \left(\frac{a_1}{3}\right)^3 \end{cases}$$

in two unknowns. Solving this system is equivalent to solve the quadratic equation $t^2 + a_0 t - \left(\frac{a_1}{3}\right)^3 = 0$, which was an easy exercise by that time. Then, Tartaglia says that a solution of the equation can be found thanks to the difference of the sides of certain cubes, that is $x = \sqrt[3]{y} - \sqrt[3]{z}$. If we want to draw an explicit formula, we would obtain

$$x = \sqrt[3]{\frac{a_0}{2} + \sqrt{\left(\frac{a_0}{2}\right)^2 + \left(\frac{a_1}{3}\right)^3}} - \sqrt[3]{-\frac{a_0}{2} + \sqrt{\left(\frac{a_0}{2}\right)^2 + \left(\frac{a_1}{3}\right)^3}}.$$

Analogously, considering the equation $x^3 = a_1 x + a_0$, Tartaglia suggests to solve the system

$$\begin{cases} y + z = a_0 \\ yz = \left(\frac{a_1}{3}\right)^3 \end{cases},$$

[6]See [40, Quesito XXXIII fatto con una lettera dalla eccellentia de Messer Hieronimo Cardano l'Anno 1539 A di 19 Marzo, page 119].

which is equivalent to solve $t^2 + a_0 t + \left(\frac{a_1}{3}\right)^3 = 0$. Then, Tartaglia says that a solution of the equation is a certain $x = \sqrt[3]{y} + \sqrt[3]{z}$. If we want again to draw an explicit formula, we would obtain

$$x = \sqrt[3]{\frac{a_0}{2} + \sqrt{\left(\frac{a_0}{2}\right)^2 - \left(\frac{a_1}{3}\right)^3}} + \sqrt[3]{\frac{a_0}{2} - \sqrt{\left(\frac{a_0}{2}\right)^2 - \left(\frac{a_1}{3}\right)^3}}.$$

Finally, Tartaglia says that the solution of the equation $x^3 + a_0 = a_1 x$ can be easily reduced to the solution of the equation $x^3 = a_1 x + a_0$.

We remark that none of the above calculations and formulae (except the systems) are conveyed by Tartaglia's poem. But Tartaglia writes to Cardano the key idea which leads to the cubic formulae: that the sum or the difference of the sides of two certain cubes has to be considered. Anyway, no hints on how Tartaglia had this idea and on how he got the systems are given.

B. List of internal and external references in and to the *De Regula Aliza*

Here follows the list of the internal and external references made by Cardano in and to the *De Regula Aliza*. Where the reference is between square brackets, it is implicit in the text.

With regard to what edition of the *Elements* Cardano could have read, we remark that in the short preface to the manuscript *Commentaria in Euclidis Elementa*, he explicitly mentions[1] the edition by Jacques Lefèvre d'Étaples or Jacopus Faber Stapulensis, printed in Paris in 1516. This was one of the most popular editions during the 16th century. Anyway, despite his ambition to overtake the misinterpretations by Campano da Novara and Bartolomeo Zamberti, Lefèvre d'Étaples confined himself to uncritically juxtapose the two writings. An accurate study to determine whether Cardano followed Campano or rather Zamberti's interpretation is still lacking.

---

[1] See [63, page 134].

Internal references to the *Aliza* in the *Aliza*	
The Chapter	refers to
A VII	[A I.2]
A XI	A [I], III, X
A XIII	A [XI], LI
A XIV	A V
A XV	[A I]
A XVI	A XIII
XVIII	A X, XV
A XXII	[A VI]
A XXIII	A X
A XXIV	A XIII Corollary 2
A XXVIII	A XXVI
A XXXIX	A XXVI, XXVII
A XLII	A XL
A LI	A XLVI
A LIII	A II, VII, XXVIII, XXXI, XL, LVII
A LVII	A IV, XL, LIII

External references in the *Aliza*	
The Chapter	refers to
A I	DP 146 *Elements* II.9, X.20
A II	AM XXV DP *Elements* II.4, 5
A III	DP 148
A IV	OAP, book III, XIX.2 (see AMA XIX.2) *Elements* X.6, 16, 17, 33, 39, 40, 41, 54, 66, 112, 113, 114
A VII	*Elements* II.5
A VIII	*Elements* X.6, 10, 11, 33, and Definition 6
A IX	DP 143 *Elements* XI.24
A X	*Elements* II.5
A XI	*Elements* I. 43, II.4
A XII	DP 27 *Elements* VIII.9, XI.34 Apollonius' *Conics* I[.11], II[.4] Archimedes' *On the sphere and cylinder* II[.4] and Eutocius' commentary

continued on next page

continued from previous page

The Chapter	refers to
A XIII	*Elements* VI.17, XI.32
A XIV	AM XIII
A XX	*Elements* XI.34
A XXI	DP Proposition 2, 34 *Elements* XI.32
A XXII	*Elements* II.4, 7
A XXIII	*Elements* X.20, 21
A XXIV	AM XIII Corollary 1, XXV.1 *Elements* II.4
A XXV	AM XII, XXV.3 DP 209
A XXXII	*Elements* V.8, 10, 19
A XXXIII	*Elements* V.11, 19
A XXXIV	*Elements* I.44, II.1, V.19, VI.12, 16
A XXXVI	AM
A XXXIX	AM XXXIX.2 DP Definition 18, Propositions 135, 143 *Elements* II.5, 7, VI.1, 10, 11

continued on next page

continued from previous page

The Chapter	refers to
A XL	AM XXV.2
A XLI	AM [XXXIX Problem IX]
A XLIII	*Elements* XI.34
A XLVI	AM
A XLVII	*Elements* XI
A XLVIII	AM XIII
A XLIX	AM XVIII [OAP] , Book III
A LII	*Elements* VI.20
A LIII	AM XXV
A LIV	*Elements* II.1, 4
A LVI	*Elements* I.4
A LVII	PA LI AM XXV, XXIX *Elements* I.47, X.54
A LIX	AM XXV.1 *Elements* X.115

C. List of Cardano's numerical cubic equations

Here follows the list of the cubic equations in one unknown that Cardano suggests as examples in the selection of the Chapters of the *Practica Arithmeticæ*, *Ars Magna Arithmeticæ*, *Ars Magna* that we have considered. Concerning the *De Regula Aliza*, I have taken into account all the chapters. Note that sometimes Cardano does not explicitly state the equation as such, but only gives some hints from which it can be reconstructed. In these cases, I have also included the reconstructed equations.

Cubic equations in the *Practica Arithmeticæ*		
PA LI.26	$3x^3 = 24x + 21$	
	$3x^3 = 15x + 21$	
	$3x^3 = 15x + 36$	
	$3x^3 + 21 = 24x$	
	$3x^3 + 6 = 15x$	
PA LI.27	$x^3 + 7x = 4x^2 + 4$	$2x^3 + 5x^2 = 10x + 16$
	$x^3 + x + 2 = 4x^2$	$x^3 + 3 = 4x^2 + 2x$
	$x^3 + 3x^2 = 7x - 3$	$x^3 = 4x^2 + 6x + 1$
	$x^3 + 4x = 4x^2 + 1$	$x^3 + 2x^2 = 2x + 3$
	$3x^3 + 3 = 7x^2 + 7x$	$x^3 + 2 = 5x$
PA LI.32	$x^3 + 64 = 18x^2$	

Cubic equations in the *Ars Magna Arithmeticæ*		
AMA XXI	$x^3 = 3x^2 + 45$	$x^3 = 53x + 88$
	$x^3 + 15x = 75$	$x^3 + 12 = 34x$
	$x^3 = 3x^2 + 36$	$x^3 = 34x + 12$
	$x^3 + 12x = 48$	$x^3 + 216 = 27x^2$
	$x^3 = \frac{24}{5}x^2 + \frac{288}{5}$	$x^3 + 16 = 6x^2$
	$x^3 + 12x = 30$	$x^3 + 16 = 12\sqrt[3]{3}x$
	$x^3 + 3x^2 = 21$	$x^3 + 8 = 7x^2$
	$x^3 = 3x + 19$	$x^3 + \frac{49}{4}x = 7x^2 + 1$
	$x^3 + 6x^2 = 40$	$x^3 + 40 = 8x^2$
	$x^3 = 12x + 24$	$x^3 + 16x = 8x^2 + 5$
	$x^3 + 12x = 56$	$x^3 + 5 = 12x$
	$x^3 + 72 = 48x$	$x^3 + 6x^2 = 11$
	$x^3 + 4x^2 = 225$	$x^3 + 21 = 6x^2$
	$x^3 = 4x + 15$	$x^3 + 7 = 6x^2$
	$x^3 + 2x^2 = 441$	$x^3 = 12x + 9$
	$x^3 = 2x + 21$	$x^3 + 81 = 12x^2$
	$x^3 + 3x^2 = 20$	$x^3 + 12x^2 = 175$
	$x^3 + 88 = 53x$	
AMA XXII	$x^3 + 42 = 29x$	
	$x^3 + 4 = 17x$	
	$x^3 = 17x + 4$	
	$x^3 + \frac{49}{6} = \frac{45}{6}x^2$	
	$x^3 + \frac{1}{4} = \frac{17}{4}x^2$	
	$x^3 + \frac{17}{4}x^2 = \frac{1}{4}$	

continued on next page

continued from previous page

AMA XXIII	$x^3 = 3\sqrt{12}x + 8$
	$x^3 = 6x + 6$
	$x^3 + 6x = 2$
AMA XXV	$x^3 = 6x^2 + 18$
	$x^3 + 6x^2 = 50$
AMA XXVII	$x^3 + 6x = 36$
	$x^3 = x^2 + 6$
	$x^3 + 6x = 9$
	$x^3 = 6x + 9$
	$x^3 + 6x = 2$
	$x^3 = 6x + 6$
AMA XXVIII	$x^3 + 9x = 6$
	$x^3 + 9x = 26$
	$x^3 + 6x = 10$
	$x^3 + 3x = 10$
AMA XXIX	$x^3 = \frac{27}{2}x^2 + \frac{243}{2}$
	$x^3 = 6x^2 + 18$
	$x^3 = 6x^2 + 4$
AMA XXX	$x^3 = 6x + 10$

continued on next page

continued from previous page

AMA XXXI	$x^3 = 16x + 21$
	$x^3 = 20x + 32$
	$x^3 = 29x + 52$
	$x^3 = 65x + 8$
	$x^3 = 23x + 28$
	$x^3 = (24 + \sqrt{6})x + 12$
AMA XXXII	$x^3 + 6x^2 = 36$
	$x^3 = 6x + 6$
	$x^3 + 3x^2 = 21$
	$x^3 = 3x + 19$
AMA XXXIII	$x^3 + 20x^2 = 72$
	$x^3 + 11x^2 = 72$
	$x^3 + (4 + \sqrt{10})x^2 = 40$
	$x^3 + 8x^2 = 40$
	$x^3 + 7x^2 = 50$
AMA XXXIV	$x^3 + 88 = 48x$
	$x^3 = 48x + 88$
	$x^3 + 32 = 20x$
	$x^3 + 12 = 34x$

continued on next page

continued from previous page

AMA XXXV	$x^3 + 12 = 34x$	
	$x^3 + 88 = 53x$	
	$x^3 + 10 = 23x$	
	$x^3 + 4 = 17x$	
	$x^3 + 12 = 19x$	
	$x^3 + 16 = 17x$	
	$x^3 + 88 = 48x$	
AMA XXXVI	$x^3 + 12 = 5x^2$	
AMA XXXVII	$x^3 + 50 = 7x^2$	
	$x^3 + 32 = 10x^2$	
	$x^3 + 16 = 6x^2$	
	$x^3 + 40 = 8x^2$	
	$x^3 + 10 = 9x^2$	
	$x^3 + 48 = 10x^2$	
	$x^3 + 8 = 7x^2$	
AMA XXXIX	$x^3 + 6x^2 + 12x = 22$	$x^3 + x^2 + 3 = 13x$
	$x^3 + 12x = 6x^2 + 38$	$x^3 + 2x^2 + 2 = 9x$
	$x^3 + 12x = 6x^2 + 11$	$x^3 + x^2 + 11 = 47x$
	$x^3 + \frac{9}{2}x^2 + x = \frac{1}{2}$	$x^3 + x + 7 = 5x^2$
	$x^3 + 5x^2 + 3x = 1$	$x^3 + 3x + \frac{14}{3} = \frac{16}{3}x^2$
	$x^3 + 6x^2 + 4x = 16$	$x^3 + 8x = \frac{37}{6}x^2 + \frac{7}{6}$
	$x^3 + x^2 + 49 = 35x$	$x^3 + x + \frac{1}{2} = \frac{9}{2}x^2$
	$x^3 + 2x^2 + 56 = 41x$	$x^3 + 2x + \frac{3}{4} = \frac{19}{4}x^2$
	$x^3 + 3x^2 + 63 = 47x$	$x^3 + 3x + 1 = 5x^2$

Cubic equations in the *Ars Magna*		
AM I	$x^3+6x=20$	$x^3+3x+18=6x^2$
	$x^3+16=12x$	$x^3+x^2+2x=16$
	$x^3=12x+16$	$x^3+2x+16=x^2$
	$x^3+9=12x$	$x^3=2x^2+x+16$
	$x^3=12x+9$	$x^3+2x^2+16=x$
	$x^3+21=2x$	$x^3+3x^2+6=20x$
	$x^3=2x+21$	$x^3=3x^2+20x+6$
	$x^3=3x^2+16$	$x^3+72=6x^2+3x$
	$x^3+3x^2=20$	$x^3+6x^2=3x+72$
	$x^3+20=3x^2$	$x^3+4=3x^2+5x$
	$x^3+11x^2=72$	$x^3+3x^2=5x+4$
	$x^3+72=11x^2$	$x^3+10=6x^2+8x$
	$x^3+6x^2+3x=18$	$x^3+6x^2=8x+10$
AM VI	$x^3+40=12x^2$	
	$x^3+3x^2=14x+20$	
	$x^3=x^2+8$	
	$x^3+8x=64$	
AM VII	$x^3+80=18x$	$x^3+8=9x^2$
	$x^3=9x+10$	$x^3=3x^2+10$
	$x^3=6x^2+16$	$x^3+3x=\sqrt{10}$
	$x^3+12\sqrt[3]{3}x=16$	$x^3+16=14x^2$
	$x^3+18x=8$	$x^3+49x=14x^2+2$
	$x^3=9x^2+8$	$x^3+40=8x^2$
	$x^3+8=18x$	$x^3+16x=8x^2+5$

continued on next page

continued from previous page

AM VIII	$x^3 + 3 = 10x$
AM XI	$x^3 + 6x = 20$
	$x^3 + 3x = 10$
	$x^3 + 6x = 2$
	$x^3 + 12 = 34x$
AM XII	$x^3 = 6x + 40$
	$x^3 = 6x + 6$
AM XIII	$x^3 + 12 = 34x$
	$x^3 + 12 = 10x$
	$x^3 + 3 = 8x$
	$x^3 + 60 = 46x$
	$x^3 + 21 = 16x$
AM XIV	$x^3 = 6x^2 + 100$
	$x^3 = 6x^2 + 20$
AM XV	$x^3 + 6x^2 = 100$
	$x^3 + 6x^2 = 25$
	$x^3 + 6x^2 = 16$
	$x^3 + 6x^2 = 7$
	$x^3 + 6x^2 = 40$
	$x^3 + 6x^2 = 36$

continued on next page

continued from previous page

AM XVI	$x^3 + 64 = 18x^2$	
	$x^3 + 24 = 8x^2$	
AM XVII	$x^3 + 6x^2 + 20x = 100$	
	$x^3 + 6x^2 + 12x = 22$	
	$x^3 + 3x^2 + 9x = 171$	
	$x^3 + 6x^2 + x = 14$	
	$x^3 + 12x^2 + 27x = 400$	
	$x^3 + 6x^2 + 2x = 3$	
AM XVIII	$x^3 + 33x = 6x^2 + 100$	$x^3 + 12x = 6x^2 + 8$
	$x^3 + 12x = 6x^2 + 25$	$x^3 + 12x = 6x^2 + 9$
	$x^3 + 48x = 12x^2 + 48$	$x^3 + 12x = 6x^2 + 7$
	$x^3 + 15x = 6x^2 + 24$	$x^3 + 20x = 6x^2 + 24$
	$x^3 + 15x = 6x^2 + 10$	$x^3 + 20x = 6x^2 + 15$
	$x^3 + 10x = 6x^2 + 4$	$x^3 + 20x = 6x^2 + 33$
	$x^3 + 26x = 12x^2 + 12$	$x^3 + 9x = 6x^2 + 2$
	$x^3 + 100x = 6x^2 + 10$	$x^3 + 9x = 6x^2 + 4$
	$x^3 + 5x = 6x^2 + 10$	$x^3 + 21x = 9x^2 + 5$
AM XIX	$x^3 + 6x^2 = 20x + 56$	
	$x^3 + 6x^2 = 20x + 112$	
	$x^3 + 6x^2 = 20x + 21$	
AM XX	$x^3 = 6x^2 + 5x + 88$	
	$x^3 = 6x^2 + 72x + 729$	

continued on next page

continued from previous page

AM XXI	$x^3 + 100 = 6x^2 + 24x$	
	$x^3 + 64 = 6x^2 + 24$	
	$x^3 + 128 = 6x^2 + 24x$	
	$x^3 + 9 = 6x^2 + 24x$	
AM XXII	$x^3 + 4x + 16 = 6x^2$	
	$x^3 + 4x + 8 = 6x^2$	
	$x^3 + 4x + 1 = 6x^2$	
AM XXIII	$x^3 + 6x^2 + 4 = 41x$	
	$x^3 + 6x^2 + 12 = 31x$	
AM XXV	$x^3 = 20x + 32$	$x^3 + 21 = 16x$
	$x^3 = 32x + 24$	$x^3 + 18 = 19x$
	$x^3 = 10x + 24$	$x^3 + 8 = 18x$
	$x^3 = 19x + 30$	$x^3 + 18 = 15x$
	$x^3 = 7x + 90$	$x^3 + 48 = 25x$
	$x^3 = 16x + 21$	$x^3 + 20x^2 = 72$
	$x^3 = 4x + 15$	$x^3 + \frac{45}{2}x^2 = 98$
	$x^3 = 14x + 8$	$x^3 + 48 = 10x^2$
	$x^3 + 12x = 34x$	

Cubic equations in the *De Regula Aliza*	
A I	$x^3 = 29x + 140$
	$x^3 = \frac{185}{7}x + 158$
	$x^3 = \frac{158}{7}x + 185$
	$x^3 = 35x + 98$
	$x^3 = 14x + 245$
A II	$x^3 = 7x + 90$
A III	$x^3 + 8 = 7x^2$
	$x^3 + 6 = 7x^2$
	$x^3 + 48 = 7x^2$
	$x^3 + 24 = 8x^2$
	$x^3 + 40 = 8x^2$
	$x^3 + 45 = 8x^2$
	$x^3 + 75 = 8x^2$
A V	$x^3 + 24 = 32x$
	$x^3 = 32x + 24$
	$x^3 + 12 = 34x$
	$x^3 = 34x + 12$
	$x^3 + 8 = 18x$
	$x^3 + 18x = 39$
A VII	$x^3 = 18x + 30$
	$x^3 = 18x + 58$
	$x^3 = 18x + 75$
	$x^3 = 18x + 33$
	$x^3 = 18x + 42$

continued on next page

continued from previous page

A X	$x^3 = 30x + 36$
	$x^3 = 38x + 2$
A XII	$x^3 + 256 = 12x^2$
	$x^3 + 128 = 12x^2$
	$x^3 + 192 = 12x^2$
A XIV	$x^3 + 12 = 19x$
A XV	$x^3 + 1 = 3x$
	$x^3 + 2x = \frac{8}{3}x^2 + \frac{4}{3}$
	$x^3 = \frac{10}{17}x + \frac{700}{729}$
	$x^3 + x = 2x^2 + 2$
A XVIII	$x^3 = 6x + 6$
A XIX	$x^3 = 6x + 6$
	$x^3 = 9x + 9$
A XX	$x^3 + 6x^2 = 24$
	$x^3 + 8x^2 = 24$
	$x^3 + 4x^2 = 4$
A XXIII	$x^3 + \frac{7}{64} = \frac{2}{3}x$
	$x^3 = 4x + 48$
	$x^3 = 4x + 47$
	$x^3 + 12 = 34x$
	$x^3 + \frac{59172}{4913} = 34x$
	$x^3 + 52 = 30x$
	$x^3 = 4x + 50$

continued on next page

continued from previous page

A XXIV	$x^3 = 20x + 32$	$x^3 = 29x + 60$
	$x^3 = 6x + 4$	$x^3 = 29x + 52$
	$x^3 = 29x + 28$	$x^3 = 29x + 20$
	$x^3 = 29x + 50$	$x^3 = 29x + 42$
A XXV	$x^3 = 18x + 108$	
	$x^3 = 21x + 90$	
	$x^3 = 15x + 126$	
	$x^3 = 22x + 84$	
	$x^3 = 17x + 114$	
	$x^3 = 6x + 6$	
A XXVI	$x^3 = 9x^2 + 8$	$x^3 + 408 = 200x$
	$x^3 + 8 = 9x^2$	$x^3 + 927 = 300x$
	$x^3 + \frac{8569}{1000} = 9x$	$x^3 + 8829 = 900x$
	$x^3 + 200 = 100x^2$	
A XXVII	$x^3 = 6x^2 + 400$	
	$x^3 + 32 = 6x^2$	
	$x^3 + 12 = 10x$	
A XXVIII	$x^3 + 6 = 8x,$	
	$x^3 + 8x = 6$	
A XXXI	$x^3 = 13x + 60$	
	$x^3 = 6x + 1$	
A XXXV	$x^3 = 4x + 8$	
	$x^3 = \sqrt{2}x^2 + \sqrt{8}$	
	$x^3 + 2x^2 = 8$	
	$x^3 + 2x = \sqrt{8}$	
	$x^3 = 2x + 4$	

continued on next page

continued from previous page

A XXXVII	$x^3 + 576 = 25x^2$	
	$x^3 = 22x^2 + 576$	
	$x^3 + 576 = 22x^2$	
	$x^3 + 36x = 252$	
A XXXIX	$x^3 + 16 = 9x^2$	
	$x^3 = 9x^2 + 16$	
	$x^3 + 24 = 8x^2$	
	$x^3 + x = 25$	
	$x^3 + 108 = 36x$	
	$x^3 + 64 = 36x$	
	$x^3 = 27x + 46$	
A XL	$x^3 + 24 = 32x$	$x^3 + 4 = 5x$
	$x^3 + 12 = 34x$	$x^3 + 6 = 7x$
	$x^3 + 8 = 18x$	$x^3 + 4 = 12x$
	$x^3 + 153 = 64x$	$x^3 + 8 = 8x$
	$x^3 + 4 = 36x$	
A XLII	$x^3 + 4 = 6x$	
	$x^3 + 10 = 9x$	
	$x^3 + 12x = 34$	
	$x^3 + 6 = 7x$	
	$x^3 + 8 = 8x$	
	$x^3 + 12 = 34x$	
	$x^3 + 20 = 15x$	
	$x^3 + 8 = 8x$	

continued on next page

continued from previous page

A XLIV	$x^3 + 2x^2 = 8$
A XLV	$x^3 + 252 = 78x$
	$x^3 = 26x + 60$
	$x^3 + 252 = 48x$
	$x^3 + 252 = 45x$
	$x^3 + 252 = 42x$
A XLVI	$x^3 = 24x + 5$
	$x^3 = 6x + 95$
	$x^3 = 4x + 105$
	$x^3 = 5x + 100$
	$x^3 = 3x + 120$
	$x^3 = 18x + 100$
A XLVIII	$x^3 = 13x + 60$
	$x^3 + 70 = 39x$
A XLIX	$x^3 = 6x + 40$
	$x^3 = 6x + 20$
	$x^3 + \frac{81}{4}x = \frac{81}{4}$
	$x^3 + 27x = \frac{9}{2}x^2 + 54$
A L	$x^3 = 36x + 36$
A LIII	$x^3 = 12x + 20$
	$x^3 = 12x + 34$
	$x^3 = 6x + 40$
	$x^3 + 20x = 32$
A LV	$x^3 = x^2 + x + 3$

continued on next page

continued from previous page

A LVI	$x^3 = 20x + 32$	
	$x^3 = 19x + 12$	
A LVII	$x^3 = 20x + 32$	
	$x^3 = 39x + 18$	
	$x^3 = 19x + 12$	
A LVIII	$x^3 = 20x + 16$	
	$x^3 = 6x + 6$	
A LIX	$x^3 = x$	$x^3 = 4x$
	$x^3 = 2x + 1$	$x^3 = 5x + 2$
	$x^3 = 3x + 2$	$x^3 = 6x + 4$
	$x^3 = 4x + 3$	$x^3 = 7x + 6$
	$x^3 = 5x + 4$	$x^3 = 8x + 8$
	$x^3 = 6x + 5$	$x^3 = 9x + 10$
	$x^3 = 7x + 6$	$x^3 = 10x + 12$
	$x^3 = 8x + 7$	$x^3 = 11x + 14$
	$x^3 = 9x + 8$	$x^3 = 12x + 16$
	$x^3 = 10x + 9$	$x^3 = 13x + 18$
	$x^3 = 11x + 10$	$x^3 = 14x + 20$
	$x^3 = 12x + 11$	$x^3 = 15x + 22$
	$x^3 = 13x + 12$	$x^3 = 16x + 24$
	$x^3 = 14x + 13$	$x^3 = 17x + 26$
	$x^3 = 15x + 14$	$x^3 = 18x + 28$
	$x^3 = 16x + 15$	$x^3 = 19x + 30$
	$x^3 = 17x + 16$	$x^3 = 20x + 32$
	$x^3 = 18x + 17$	$x^3 = 21x + 34$

continued on next page

continued from previous page

A LIX	$x^3 = 9x$	$x^3 = 25x + 36$
	$x^3 = 10x + 3$	$x^3 = 26x + 40$
	$x^3 = 11x + 6$	$x^3 = 27x + 44$
	$x^3 = 12x + 9$	$x^3 = 28x + 48$
	$x^3 = 13x + 12$	$x^3 = 29x + 52$
	$x^3 = 14x + 15$	$x^3 = 30x + 56$
	$x^3 = 15x + 18$	$x^3 = 31x + 60$
	$x^3 = 16x + 21$	$x^3 = 32x + 64$
	$x^3 = 17x + 24$	$x^3 = 216$
	$x^3 = 18x + 27$	$x^3 = x + 210$
	$x^3 = 19x + 30$	$x^3 = 2x + 204$
	$x^3 = 20x + 33$	$x^3 = 3x + 198$
	$x^3 = 21x + 36$	$x^3 = 4x + 192$
	$x^3 = 22x + 39$	$x^3 = 5x + 186$
	$x^3 = 23x + 42$	$x^3 = 6x + 180$
	$x^3 = 24x + 45$	$x^3 = 7x + 174$
	$x^3 = 25x + 48$	$x^3 = 8x + 168$
	$x^3 = 26x + 51$	$x^3 = 9x + 162$
	$x^3 = 16x$	$x^3 = 10x + 156$
	$x^3 = 17x + 4$	$x^3 = 11x + 150$
	$x^3 = 18x + 8$	$x^3 = 12x + 144$
	$x^3 = 19x + 12$	$x^3 = 13x + 138$
	$x^3 = 20x + 16$	$x^3 = 14x + 132$
	$x^3 = 21x + 20$	$x^3 = 15x + 126$
	$x^3 = 22x + 24$	$x^3 = 16x + 120$
	$x^3 = 23x + 28$	$x^3 = 6x + 40$
	$x^3 = 24x + 32$	

continued on next page

continued from previous page

A LX	$x^3 = 25 + 20x$
	$x^3 = 36 + 30x$
	$x^3 = 18x + 8$
	$x^3 = 11x + 3$

We remark that the equations that have $\Delta_3 < 0$ are thirteen out of sixteen in the *Practica Arithmeticæ*, sixty-eight out of one hundred and five in the *Ars Magna Arithmeticæ*, sixty-eight out of one hundred and nineteen in the *Ars Magna*, and one hundred and thirty-three out of two hundred and fourteen in the *De Regula Aliza*.

Bibliography

1. Sources

[1] Rafael Bombelli. *L'algebra.* Giovanni Rossi, Bologna, 1572.

[2] Girolamo Cardano. *Hieronymi C. Cardani Medici mediolanensis, Practica arithmeticæ, et mensurandi singularis. In quaque preter alias continentur, versa pagina demonstrabit.* Antonius Castellioneus for Bernardini Calusci, Milan, 1539.

[3] Girolamo Cardano. *De sapientia libri quinque. Eiusdem De consolatione libri tres, alias æditi, sed nunc ab eodem authore recogniti. Eiusdem De libris propriis, liber unus. Omnia locupleti indice decorata*, chapter Libellus de libris propriis, cuius titulus est ephemerus, pages 419–431. Johannes Petreius, Nuremberg, 1544.

[4] Girolamo Cardano. *Hieronymi Cardani, præstantissimi mathematici, philosophi, ac medici, Artis magnae sive de regulis algebraicis, lib. unus. Qui et totius operis de arithmetica, quod Opus perfectum inscripsit, est in ordine decimus.* Iohannes Petrius, Nuremberg, 1545.

[5] Girolamo Cardano. *Hieronymi Cardani mediolanensis medici De subtilitate libri XXI nunc demum recogniti atque perfecti.* Ludovicum Licium, Basel, 1554.

[6] Girolamo Cardano. *Liber De libris propriis, eorumque ordine et usu, ac de mirabilibus operibus in arte medica per ipsum factis.* Guillaume Rouillé, Lyon, 1557.

[7] Girolamo Cardano. *Somniorum synestesiorum, omnis generis insomnia explicantes libri IIII. Quibus accedunt eiusdem hæc etiam: De libris propriis. De curationibus et prædictionibus admirandis. Neronis encomium. Geometriæ encomium. De uno. Actio in Thessalicum medicum. De secretis. De gemmis et coloribus. Dialogus de morte. [Dialogus] de humanis consiliis, tetim inscriptus. Item ad somniorum libros pertinentia: De minimis et propinquis. De summo bono*, chapter De libris propriis, eorumque usu, liber recognitus, pages 1–116. Heinrich Petri, Basel, 1562.

[8] Girolamo Cardano. *Hieronymi Cardani mediolanensis civisque bononiensis, Ars curandi parva, quae est absolutiss. medend. methodus, et alia, nunc primum aedita, opera, in duos tomos diuisa, quae versa pagina indicabit; omnia autem qualia sint autoris epistola vere praedicat.* Officina Henricpetrina, Basel, 1566.

[9] Girolamo Cardano. *Hieronymi Cardani mediolanensis, civisque bononiensis, philosophi, medici et mathematici clarissimi, Opus novum de proportionibus numerorum, motuum, ponderum, sonorum, aliarumque rerum mensurandarum, non solum geometrico more stabilitum, sed etiam varijs experimentis et observationibus rerum in natura, solerti demonstratione illustratum, ad multiplices usus accommodatum, et in V libros digestum. Prætera Artis magnæ, sive de regulis algebraicis, liber unus, abstrusissimus et inexhaustus plane totius arithmeticæ thesaurus, ab authore recens multis in locis recognitus et auctus. Item De aliza regula liber, hoc est, algebraicæ logisticæ suæ, numeros recondita numerandi subtilitate, secundum geometricas quantitates inquirentis, necessaria coronis, nunc demum in lucem edita*, chapter De aliza regula libellus, hoc est Operis perfecti sui sive algebraicæ Logisticæ, numeros recondita numerandi subtilitate, secundum geometricas quantitates inquirenti, necessaria coronis, nunc demum in lucem editæ. Oficina Henricpetrina, Basel, 1570.

[10] Girolamo Cardano. *Hieronymi Cardani mediolanensis, civisque bononiensis, philosophi, medici et mathematici clarissimi, Opus novum de proportionibus numerorum, motuum, ponderum, sonorum, aliarumque rerum mensurandarum, non solum geometrico more stabilitum, sed etiam varijs experimentis et observationibus rerum in natura, solerti demonstratione illustratum, ad multiplices usus accommodatum, et in V libros digestum. Prætera Artis magnæ, sive de regulis algebraicis, liber unus, abstrusissimus et inexhaustus plane totius arithmeticæ thesaurus, ab authore recens multis in locis recognitus et auctus. Item De aliza regula liber, hoc est, algebraicæ logisticæ suæ, numeros recondita numerandi subtilitate, secundum geometricas quantitates inquirentis, necessaria coronis, nunc demum in lucem edita*, chapter Artis magnae sive de regulis algebraicis, lib. unus. Qui et totius operis de arithmetica, quod Opus perfectum inscripsit, est in ordine decimus. Ars magna, quam volgo cossam vocant, sive regulas algebraicas, per D. Hieronymum Cardanum in quadraginta capitula redacta, et est liber decimus suæ Arithmeticæ. Oficina Henricpetrina, Basel, 1570.

[11] Girolamo Cardano. *Hieronymi Cardani mediolanensis, civisque bononiensis, philosophi, medici et mathematici clarissimi, Opus novum de proportionibus numerorum, motuum, ponderum, sonorum, aliarumque rerum mensurandarum, non solum geometrico more stabilitum, sed etiam varijs experimentis et observationibus rerum in natura, solerti demonstratione illustratum, ad multiplices usus accommodatum, et in V libros digestum. Prætera Artis magnæ, sive de regulis algebraicis, liber unus, abstrusissimus et inexhaustus plane totius arithmeticæ thesaurus, ab authore recens multis in locis recognitus et auctus. Item De aliza regula liber, hoc est, algebraicæ logisticæ suæ, numeros recondita numerandi subtilitate, secundum geometricas quantitates inquirentis, necessaria coronis, nunc demum in lucem edita*, chapter In Librum de Propotionibus Hieronymi Cardani Mediolanensis, civisqúe Bononiensis, Medici. Oficina Henricpetrina, Basel, 1570.

[12] Girolamo Cardano. *Hieronymi Cardani mediolanensis Opera omnia in decem tomos digesta*, volume 4, chapter Hieronymi Cardani, Artis magnae sive de regulis algebraicis, liber unus. Ioannis Antonii Huguetan and Marci Antonii Ravaud, Lyon, 1663. Edited by Charles Spon.

[13] Girolamo Cardano. *Hieronymi Cardani mediolanensis Opera omnia in decem tomos digesta*, volume 4, chapter Ars magna arithmeticæ, seu liber quadraginta capitulorum et quadraginta quæstionum. Ioannis Antonii Huguetan and Marci Antonii Ravaudn, Lyon, 1663. Edited by Charles Spon.

[14] Girolamo Cardano. *Hieronymi Cardani mediolanensis Opera omnia in decem tomos digesta*, volume 1, chapter Libellus De libris propriis, cui titulus est ephemerus. Ad Hieronymum Cardanum medicum, affinem suum, pages 55–59. Ioannis Antonii Huguetan and Marci Antonii Ravaud, Lyon, 1663. Edited by Charles Spon.

[15] Girolamo Cardano. *Hieronymi Cardani mediolanensis Opera omnia in decem tomos digesta*, volume 1, chapter Liber De libris propriis, eorumque ordine et usu, ac de mirabilibus operibus in arte medica per ipsum factis, pages 60–95. Ioannis Antonii Huguetan and Marci Antonii Ravaud, Lyon, 1663. Edited by Charles Spon.

[16] Girolamo Cardano. *Hieronymi Cardani mediolanensis Opera omnia in decem tomos digesta*, volume 1, chapter Liber De libris propriis, eorumque ordine et usu, ac de mirabilibus operibus in arte medica per ipsum factis, pages 96–150. Ioannis Antonii Huguetan and Marci Antonii Ravaud, Lyon, 1663. Edited by Charles Spon.

[17] Girolamo Cardano. *Hieronymi Cardani mediolanensis Opera omnia in decem tomos digesta*, volume 1, chapter "*Liber a me conscripti, quo tempore, cur, quid acciderit*", Hieronymi Cardani De propria vita, liber (1576), pages 1–54. Ioannis Antonii

Huguetan and Marci Antonii Ravaud, Lyon, 1663. Edited by Charles Spon.

[18] Girolamo Cardano. *Hieronymi Cardani mediolanensis Opera omnia in decem tomos digesta*, volume 1, chapter Vita Cardani ac de eodem iudicium. Par Gabrielem Naudæum. Ioannis Antonii Huguetan and Marci Antonii Ravaud, Lyon, 1663. (without page numbering). Edited by Charles Spon.

[19] Girolamo Cardano. *Hieronymi Cardani mediolanensis Opera omnia in decem tomos digesta*, volume 4, chapter Practica aritmeticæ generalis omnium copiosissima et utilissima. Ioannis Antonii Huguetan and Marci Antonii Ravaud, Lyon, 1663. Edited by Charles Spon.

[20] Girolamo Cardano. *Hieronymi Cardani mediolanensis Opera omnia in decem tomos digesta*, volume 4, chapter Sermo de plus et minus. Ioannis Antonii Huguetan and Marci Antonii Ravaud, Lyon, 1663. Edited by Charles Spon.

[21] Girolamo Cardano. *Jerome Cardan. The life of Girolamo Cardano, of Milan, physician.* Chapman and Hall, 1854. Edited by Henry Morley.

[22] Girolamo Cardano. *Ars magna or the rules of algebra.* Dover, 1968. Edited and translated by T. Richard Witmer.

[23] Girolamo Cardano. De libris propriis [MS 1550], *edited by Baldi, Marialuisa and Canziani, Guido. Rivista di storia della filosofia*, 4:767–98, 1998. With the foliation (1-12r) of the MS F II.38 Nr 1, Öffentliche Bibliothek der Universität Basel.

[24] Girolamo Cardano. De libris propriis. *The editions of 1544, 1550, 1557, 1562, with supplementary material.* Franco Angeli, 2004. Edited by Ian MacLean.

[25] Girolamo Cardano. *Le grand art soit des règles algébriques.* Blanchard, 2006. Edited and translated by Jean Peyroux.

[26] Federico Commandino. *Euclidis Elementorum libri XV*. Jacobus Chriegber, Pisa, 1572.

[27] Leopold Gegenbauer. Über den sogenannten *casus irreducibilis* der Cardanishen formel. *Monatshefte für Mathematik und Physik*, 4:155–158, 1893.

[28] Thomas Heath, editor. *Apollonius of Perga. Treatise on conic sections.* Cambridge University Press, 1896.

[29] Thomas Heath, editor. *The works of Archimedes*, volume I. Cambridge University Press, 2nd edition, 1897.

[30] Otto Hölder. Über den *casus irreducibilis* bei der Gleichung dritten Grades. *Matematische Annalen*, 38:307–312, 1891.

[31] Adolf Kneser. Bemerkungen über den Sogenannten *casus irreducibilis* bei cubischen Gleichungen. *Mathematische Annalen*, 41:344–348, 1892.

[32] Jacques Lefèvre d'Étaples. *Euclidis megarensis geometricorum elementorum libri XV. Campani galli transalpini in eosdem commentariorum libri XV. Theonis alexandrini Bartholamaeo Zamberto veneto interprete, in tredecim priores, commentariorum libri XIII. Hypsiclis alexandrini in duos posteriores, eodem Bartholamaeo Zamberto veneto interprete, commentariorum libri II. Utcunque noster valuit labor conciliata sunt haec omnia, ad studiosorum non paruam (quam optamus) utilitatem: id Magnifico D. Francisco Briconneto postulante. Si haec beneuole suscipiantur, et fructum adferant quem cupidimus: alia eiusdem authoris opera prodibunt in lucem, successum praestante deo, et adiutoribus (ubiubi gentium sint) ad bonarum literarum institutionem probe affectis Gallis, Italis, Germanis, Hispanis, Anglis. Quibus omnibus prospera imprecamur: et puram pro dignitate veramque cognitionis lucem.* Henrici Stephani, Paris, 1516.

[33] Vincenzo Mollame. Sul *casus irreductibilis* dell'equazione cubica. *Rendiconto dell'Accademia delle scienze fisiche e matematiche (sezione della società Reale di Napoli) II*, 4:167–171, 1890.

[34] Vincenzo Mollame. Sulle radici primitive dell'unità negativa. *Rendiconto dell' Accademia delle scienze fisiche e matematiche (sezione della società Reale di Napoli) II*, 6:179–183, 1892.

[35] Peter M. Neumann. *The mathematical writings of Évariste Galois*. European Mathematical Society, 2011.

[36] Paolo Ruffini. *Teoria generale delle equazioni, in cui si dimostra impossibile la soluzione algebraica delle equazioni generali di grado superiore al quarto*. Stamperia di S. Tommaso D'Acquino, 1799.

[37] Paolo Ruffini. *Riflessioni intorno alla soluzione delle equazioni algebraiche generali*. Società Tipografica di Modena, 1813.

[38] Matthias Schemmel and Jackie Stedall. Topics in the manuscripts of Thomas Harriot (1560-1621). Online at `http://echo.mpiwg-berlin.mpg.de/content/scientific_revolution/harriot/harriot-bl/maps/2.4.14_cardano.pt`, last checked 6th February 2015.

[39] Simon Stevin. *L'arithmétique*. Christophle Plantin, Leiden, 1585.

[40] Niccolò Tartaglia. *Quesiti et inventioni diverse*. Ateneo di Brescia, 1959. facsimile reproduction of the 1554 second edition by Arnaldo Masotti.

[41] Paul Ver Eecke. *Les œuvres complètes d'Archimède, traduites du grec en français avec une introduction et des notes*. Desclèe, De Brouwer and co., 1921.

[42] Pierre Laurent Wantzel. Classification des nombres incommensurables d'origine algébrique. *Nouvelles annales de mathématiques 1ère série*, 2:117–127, 1842.

[43] Franz Woepke, editor. *L'Algèbre d'Omar Alkhayyami.* Duprat, 1851.

[44] Sito Girolamo Cardano. Strumenti per storia del Rinascimento nell'Italia settentrionale. Online at `http://www.cardano.unimi.it/`, last checked 6th February 2015.

2. Secondary literature

[45] Maria Luisa Baldi and Guido Canziani, editors. *Girolamo Cardano. Le opere, le fonti, la vita.* Franco Angeli, 1999.

[46] Maria Luisa Baldi and Guido Canziani. La circolazione dei manoscritti cardaniani. Alcuni documenti. In Marialuisa Baldi and Guido Canziani, editors, *Girolamo Cardano. Le opere, le fonti, la vita*, pages 477–498. Franco Angeli, 1999.

[47] Maria Luisa Baldi and Guido Canziani, editors. *Cardano e la tradizione dei saperi.* Franco Angeli, 2003.

[48] Maria Luisa Baldi and Guido Canziani, editors. *La pensée scientifique de Cardan.* Les Belles Lettres, 2009.

[49] Gian Luigi Betti. Cardano a Bologna e la sua polemica con il Tartaglia nel ricordo di un contemporaneo. *Bruniana e campanelliana*, 15:159–169, 2009.

[50] Ettore Bortolotti. I contributi del Tartaglia, del Cardano, del Ferrari e della scuola matematica bolognese alla teoria algebrica delle equazioni cubiche. *Studi e memorie per la storia dell'Università di Bologna*, 10:55–108, 1926.

[51] Carl Benjamin Boyer. *A history of mathematics.* Wiley, 1991. Second edition.

[52] Moritz Cantor. *Vorlesungen über die Geschichte der Mathematik.* Teubner, 1892. Volume 2.

[53] Karine Chemla. De la synthèse comme moment dans l'histoire des mathématiques. *Diogène*, 160:97–114, 1992.

[54] Karine Chemla. Generality above abstraction. The general expressed in terms of the paradigmatic in mathematics in ancient China. *Science in context*, 16:413–458, 2003.

[55] Sara Confalonieri. *The telling of the unattainable attempt to avoid the* casus irreducibilis *for cubic equations: Cardano's* De Regula Aliza. *With a compared transcription of 1570 and 1663 editions and a partial English translation.* PhD thesis, Université Diderot Paris 7, October 2013. Online at `http://tel.archives-ouvertes.fr/tel-00875863` and `http://www2.math.uni-wuppertal.de/~confalon/`.

[56] Pietro Cossali. *Sul quesito analitico proposto all'Accademia di Padova per il premio dell'anno 1781 di una assoluta dimostrazione della irriducibilità del binomio cubico.* per gli eredi di Marco Moroni, 1782.

[57] Pietro Cossali. *Origine, trasporto in Italia, primi progressi in essa dell'Algebra.* Reale Tipografia Parmense, 1799. Two volumes.

[58] Pietro Cossali. *Particularis methodi de cubicarum æquationum solutione a Cardano luci traditæ. Generalis posteriorum analystarum usus ex cap. I De Regula Aliza ipsius Cardani vitio luculentissime evictus. Atque mysterium casus irreducibilis post duo sæcula prorsus retecta causa sublatum specimen analyticum primum.* Caroli Palesii, 1799.

[59] Pietro Cossali. Disquisizione sui varj metodi di eliminazione con il componimento di uno nuovo. *Memorie di matematica e di fisica della Società italiana delle scienze*, XVI, first part:272–330, 1813.

[60] Pietro Cossali. *Storia del caso irriducibile.* Istituto veneto di scienze, lettere ed arti, 1966. Edited by Romano Gatto.

[61] Raffaella Franci. Tra latino e volgare. La lingua della trattatistica matematica dal 1200 al 1600. *Archimede*, 4:163–70, 1996.

[62] Veronica Gavagna. Alcune osservazioni sulla *practica arithmeticæ* di Cardano e la tradizione abachistica quattrocentesca. In Marialuisa Baldi and Guido Canziani, editors, *Girolamo Cardano. Le opere, le fonti, la vita*, pages 273–312. Franco Angeli, 1999.

[63] Veronica Gavagna. Cardano legge Euclide. I commentaria in euclidis elementa. In Marialuisa Baldi and Guido Canziani, editors, *Cardano e la tradizione dei saperi. Atti del Convegno internazionale di studi. Milano (23-25 maggio 2002)*, pages 125–144. Franco Angeli, 2003.

[64] Veronica Gavagna. Medieval heritage and new perspectives in Cardano's *Practica arithmeticæ*. *Bollettino di storia delle scienze matematiche*, 1:61–80, 2010.

[65] Veronica Gavagna. Dalla Practica arithmeticæ all'Ars magna. lo sviluppo dell'algebra nel pensiero di Cardano. In *Pluralité de l'algèbre à la Renaissance*, pages 237–268. Champion, 2012. Edited by Maria-Rosa Massa-Esteve, Sabine Rommevaux, and Maryvonne Spiesser.

[66] Albrecht Heeffer. *Philosophical aspects of symbolic reasoning in Early Modern mathematics*, chapter From the second unknown to the symbolic equation, pages 57–122. College Publications, 2010. Edited by Albrecht Heeffer and Maarten Van Dyck.

[67] Albrecht Heeffer. On the nature and origin of algebraic symbolism. Online at `http://logica.ugent.be/albrecht/thesis/PMP2007Heeffer.pdf`, last checked 6th February 2015.

[68] Israel N. Herstein. *Topics in Algebra.* Wiley, 1975.

[69] Jens Høyrup. Pre-Modern "algebra". A concise survey of that which was shaped into the technique and discipline we know. *Quaderni di ricerca in didattica*, 11:84–97, 2002.

[70] Charles Hutton. *Tracts on mathematical and philosophical subjects*, chapter Tract 33. History of algebra, pages 219–224. F. C. and J. Rivington, 1812.

[71] Eckhard Kessler, editor. *Girolamo Cardano. Philosoph, Naturforscher, Arzt*. Harrassowitz, 1994.

[72] Serge Lang. *Algebra*. Springer-Verlag, third edition, 2002.

[73] Henry George Liddel and Robert Scott. *A Greek-English lexicon*. Oxford University Press, 9th edition, 1996.

[74] Gino Loria. *Storia delle matematiche. Dall'alba della civiltà al secolo XIX*. Hoepli, 1931, 1950.

[75] Ian MacLean. Girolamo Cardano. The last years of a polymath. *Renaissance Studies*, 5(21):587–607, 2007.

[76] Silvio Maracchia. *Storia dell'algebra*. Liguori, 2005.

[77] Pier Daniele Napolitani. Archimede. Alle radici della scienza moderna. *Le Scienze. I grandi della scienza*, 22:1–109, 2001.

[78] Ferdinand Nesselmann. *Versuch einer kritischen Geschichte der Algebra*, volume 1. Die Algebra der Griechen. G. Reimer, 1842.

[79] Reviel Netz. *The shaping of deduction in Greek mathematics*. Cambridge University Press, 1999.

[80] Reviel Netz. *The transformation of mathematics in the Early Mediterranean world. From problems to equations*. Cambridge University Press, 2004.

[81] Jeffrey A. Oaks. Arabic algebra as an artificial language. *Journal of Indian Philosophy*, 35:543–575, 2007.

[82] Jeffrey A. Oaks. Polynomials and equations in Arabic algebra. *Archive for history of exact sciences*, 63:169–203, 2009.

[83] Jeffrey A. Oaks. Irrational coefficients in Renaissance algebra. Draft, 2013.

[84] Oystein Ore. *Cardano. The gambling scholar.* Princeton University Press, 1953.

[85] Marco Panza. *Analysis and synthesis in mathematics. History and philosophy*, chapter Classical sources for the concepts of analysis and synthesis, pages 365–414. Kluwer, 1997.

[86] Marco Panza. *Newton et l'origine de l'analyse.* Blanchard, 2005.

[87] Marco Panza. What is new and what is old in Viète's *analysis restituta* and *algebra nova*, and where do they come from? Some reflections on the relations between algebra and analysis before Viète. *Revue d'histoire des mathématiques*, 13:85–153, 2007.

[88] Marco Panza. *Philosophical aspects of symbolic reasoning in Early Modern mathematics*, chapter What more there is in early modern algebra that its literal formalism, pages 193–228. College Publications, 2010. Edited by Albrecht Heeffer and Maarten Van Dyck.

[89] Shahid Rahman, Tony Street, and Hassam Tahiri, editors. *The unity of science in the Arabic tradition.* Springer, 2008.

[90] Roshdi Rashed and Bijan Vahabzadeh, editors. *Al-Khayyām mathématicien.* Blanchard, 1999.

[91] Daniele Sabaino. Gerolamo Cardano musicus. In *Gerolamo Cardano nel suo tempo*, pages 89–124. Edizioni Cardano, 2003.

[92] Jacqueline Stedall. *From Cardano's great art to Lagrange's reflections. Filling a gap in the history of algebra.* EMS, 2011.

[93] Massimo Tamborini. Per una storia dell'Opus Arithmeticæ Perfectum. In *Cardano e la tradizione dei saperi. Atti del Convegno*

internazionale di studi. Milano (23-25 maggio 2002), pages 157–190. Franco Angeli, 2003. Edited by Marialuisa Baldi and Guido Canziani.

[94] Rosalind C. H. Tanner. The alien realm of the minus. Deviatory mathematics in Cardano's writings. *Annals of science*, 37:159–178, 1980.

[95] Rosalind C. H. Tanner. The ordinary regiment of the minus sign. Off-beat mathematics in Harriot's manuscripts. *Annals of science*, 37:127–158, 1980.

MIX
Papier aus verantwortungsvollen Quellen
Paper from responsible sources
FSC® C105338

If you have any concerns about our products,
you can contact us on
ProductSafety@springernature.com

In case Publisher is established outside the EU,
the EU authorized representative is:
Springer Nature Customer Service Center GmbH
Europaplatz 3, 69115 Heidelberg, Germany

Printed by Libri Plureos GmbH
in Hamburg, Germany